Smart Innovation, Systems and Technologies

Volume 64

Series editors

Robert James Howlett, KES International, Shoreham-by-sea, UK
e-mail: rjhowlett@kesinternational.org

Lakhmi C. Jain, University of Canberra, Canberra, Australia;
Bournemouth University, UK;
KES International, UK
e-mails: jainlc2002@yahoo.co.uk; Lakhmi.Jain@canberra.edu.au

About this Series

The Smart Innovation, Systems and Technologies book series encompasses the topics of knowledge, intelligence, innovation and sustainability. The aim of the series is to make available a platform for the publication of books on all aspects of single and multi-disciplinary research on these themes in order to make the latest results available in a readily-accessible form. Volumes on interdisciplinary research combining two or more of these areas is particularly sought.

The series covers systems and paradigms that employ knowledge and intelligence in a broad sense. Its scope is systems having embedded knowledge and intelligence, which may be applied to the solution of world problems in industry, the environment and the community. It also focusses on the knowledge-transfer methodologies and innovation strategies employed to make this happen effectively. The combination of intelligent systems tools and a broad range of applications introduces a need for a synergy of disciplines from science, technology, business and the humanities. The series will include conference proceedings, edited collections, monographs, handbooks, reference books, and other relevant types of book in areas of science and technology where smart systems and technologies can offer innovative solutions.

High quality content is an essential feature for all book proposals accepted for the series. It is expected that editors of all accepted volumes will ensure that contributions are subjected to an appropriate level of reviewing process and adhere to KES quality principles.

More information about this series at http://www.springer.com/series/8767

Jeng-Shyang Pan · Pei-Wei Tsai
Hsiang-Cheh Huang
Editors

Advances in Intelligent Information Hiding and Multimedia Signal Processing

Proceeding of the Twelfth International Conference on Intelligent Information Hiding and Multimedia Signal Processing, November, 21–23, 2016, Kaohsiung, Taiwan, Volume 2

Springer

Editors
Jeng-Shyang Pan
Fujian Provincial Key Laboratory
 of Big Data Mining and Applications
Fujian University of Technology
Fujian
China

Pei-Wei Tsai
Fujian University of Technology
Fujian
China

Hsiang-Cheh Huang
National University of Kaohsiung
Kaohsiung
Taiwan

ISSN 2190-3018   ISSN 2190-3026   (electronic)
Smart Innovation, Systems and Technologies
ISBN 978-3-319-84347-6   ISBN 978-3-319-50212-0   (eBook)
DOI 10.1007/978-3-319-50212-0

Printed on acid-free paper

This Springer imprint is published by Springer Nature
The registered company is Springer International Publishing AG
The registered company address is: Gewerbestrasse 11, 6330 Cham, Switzerland

Preface

Welcome to the 12th International Conference on Intelligent Information Hiding and Multimedia Signal Processing (IIH-MSP 2016) held in Kaohsiung, Taiwan on November 21–23, 2016. IIH-MSP 2016 is hosted by National Kaohsiung University of Applied Sciences and technically co-sponsored by Tainan Chapter of IEEE Signal Processing Society, Fujian University of Technology, Chaoyang University of Technology, Taiwan Association for Web Intelligence Consortium, Fujian Provincial Key Laboratory of Big Data Mining and Applications (Fujian University of Technology), and Harbin Institute of Technology Shenzhen Graduate School. It aims to bring together researchers, engineers, and policymakers to discuss the related techniques, to exchange research ideas, and to make friends.

We received a total of 268 papers and finally 84 papers are accepted after the review process. Keynote speeches were kindly provided by Prof. Chin-Chen Chang (Feng Chia University, Taiwan) on "Some Steganographic Methods for Delivering Secret Messages Using Cover Media," and Prof. Shyi-Ming Chen (National Taiwan University of Science and Technology, Taiwan) on "Fuzzy Forecasting Based on High-Order Fuzzy Time Series and Genetic Algorithms." All the above speakers are leading experts in related research area.

We would like to thank the authors for their tremendous contributions. We would also express our sincere appreciation to the reviewers, Program Committee members, and the Local Committee members for making this conference successful. Finally, we would like to express special thanks for Tainan Chapter of

IEEE Signal Processing Society, Fujian University of Technology, Chaoyang University of Technology, National Kaohsiung University of Applied Sciences, and Harbin Institute of Technology Shenzhen Graduate School for their generous support in making IIH-MSP 2016 possible.

Fujian, China   Jeng-Shyang Pan
Fujian, China   Pei-Wei Tsai
Kaohsiung, Taiwan   Hsiang-Cheh Huang
November 2016

Conference Organization

Honorary Chairs

- Lakhmi C. Jain, University of Canberra, Australia and Bournemouth University, UK
- Chin-Chen Chang, Feng Chia University, Taiwan
- Xin-Hua Jiang, Fujian University of Technology, China
- Ching-Yu Yang, National Kaohsiung University of Applied Sciences, Taiwan

Advisory Committee

- Yoiti Suzuki, Tohoku University, Japan
- Bin-Yih Liao, National Kaohsiung University of Applied Sciences, Taiwan
- Kebin Jia, Beijing University of Technology, China
- Yao Zhao, Beijing Jiaotong University, China
- Ioannis Pitas, Aristotle University of Thessaloniki, Greece

General Chairs

- Jeng-Shyang Pan, Fujian University of Technology, China
- Chia-Chen Lin, Providence University, Taiwan
- Akinori Ito, Tohoku University, Japan

Program Chairs

- Mong-Fong Horng, National Kaohsiung University of Applied Sciences, Taiwan
- Isao Echizen, National Institute of Informatics, Japan
- Ivan Lee, University of South Australia, Australia
- Wu-Chih Hu, National Penghu University of Science and Technology, Taiwan

Invited Session Chairs

- Chin-Shiuh Shieh, National Kaohsiung University of Applied Sciences, Taiwan
- Shen Wang, Harbin Institute of Technology, China
- Ching-Yu Yang, National Penghu University of Science and Technology, Taiwan
- Chin-Feng Lee, Chaoyang University of Technology, Taiwan
- Ivan Lee, University of South Australia, Australia

Publication Chairs

- Hsiang-Cheh Huang, National University of Kaohsiung, Taiwan
- Chien-Ming Chen, Harbin Institute of Technology Shenzhen Graduate School, China
- Pei-Wei Tsai, Fujian University of Technology, China

Finance Chairs

- Jui-Fang Chang, National Kaohsiung University of Applied Sciences, Taiwan

Local Organization Chairs

- Jung-Fang Chen, National Kaohsiung University of Applied Sciences, Taiwan
- Shi-Huang Chen, Shu-Te University, Taiwan

Invited Session Organizers

- Feng-Cheng Chang, Tamkang University, Taiwan
- Yueh-Hong Chen, Far East University, Taiwan
- Hsiang-Cheh Huang, National University of Kaohsiung, Taiwan
- Ching-Yu Yang, National Penghu University of Science and Technology, Taiwan
- Wen-Fong Wang, National Yunlin University of Science and Technology, Taiwan
- Chiou-Yng Lee, Lunghwa University of Science and Technology, Taiwan
- Jim-Min Lin, Feng Chia University, Taiwan
- Chih-Feng Wu, Fujian University of Technology, China
- Wen-Kai Tsai, National Formosa University, Taiwan
- Tzu-Chuen Lu, Chaoyang University of Technology, Taiwan
- Yung-Chen Chou, Asia University, Taiwan
- Chin-Feng Lee, Chaoyang University of Technology, Taiwan
- Masashi Unoki, Japan Advanced Institute of Science and Technology, Japan.

- Kotaro Sonoda, Nagasaki University, Japan
- Shian-Shyong Tseng, Asia University, Taiwan
- Hui-Kai Su, National Formosa University, Taiwan
- Shuichi Sakamoto, Tohoku University, Japan
- Ryouichi Nishimura, National Institute of Information and Communications Technology, Japan
- Kazuhiro Kondo, Yamagata University, Japan
- Shu-Chuan Chu, Flinders University, Australia
- Pei-Wei Tsai, Fujian University of Technology, China
- Mao-Hsiung Hung, Fujian University of Technology, China
- Chia-Hung Wang, Fujian University of Technology, China
- Sheng-Hui Meng, Fujian University of Technology, China
- Hai-Han Lu, National Taipei University of Technology, Taiwan
- Zhi-yuan Su, Chia Nan University of Pharmacy and Science, Taiwan
- I-Hsien Liu, National Cheng Kung University, Taiwan
- Tien-Wen Sung, Fujian University of Technology, China
- Yao Zhao, Beijing Jiaotong University, China
- Rongrong Ni, Beijing Jiaotong University, China
- Ming-Yuan Cho, National Kaohsiung University of Applied Sciences, Taiwan
- Yen-Ming Tseng, Fujian University of Technology, China

Contents

Part I
Image and Video Signal Processing

The election of Spectrum bands in Hyper-spectral image classification

Yi Yu, Yi-Fan Li, Jun-Bao Li, Jeng-Shyang Pan*, and Wei-Min Zheng

Innovative Information Industry Research Center, Shenzhen Graduate School,
Harbin Institute of Technology, Shenzhen 518005, China
College of Information Science and Engineering, Fujian University of Technology,
Fuzhou 350118, China
Department of Automatic Test and Control, Harbin Institute of Technology, China
Institute of Information Engineering, University of Chinese Academy of Sciences,
China

Abstract. this paper present a framework for the pre-process of hyper-spectral image classification, it seems to be proved an important and well method in this kind of field. We construct a new calculating method in preprocess which use for reference of digital image process, such us morphology method and relations of position. Because of the data of the spectrum bands of hyper-spectral has some degree of redundancy. Some similar bands have the similar information or even same information, so it would waste many counter resource. On the other hands, redundancy means when you matching every model, you will easily addicted to Hughes phenomenon. So bands electing before we use classification model begin to classify directly is very important. It not only can decrease time-cost also improve the accuracy and degree of stability. Our experiment proved our idea practice in real classification get a gorgeous effect, it shows in different complex classification surroundings, our method also perform great.

Keywords: spectrum bands election, hyper-spectral image classification, SVM kernel classification model, pre-process in hyper-spectral, integration of image data by use morphology method and relations of position.

1 Introduction

Remote Sensing, RS is a kind of method that use every kinds of camera and Optical imaging device, which can use satellite or other rockets take a hyper-spectral camera taking a series kind of special digital image through reflect from light spectrum bands which include almost every kinds of spectrum both visible light and invisible light.

The introduction of election of spectrum bands in hyper-spectral image classification start a new important field for future developments in which spectrum

---

© Springer International Publishing AG 2017   3
J.-S. Pan et al. (eds.), *Advances in Intelligent Information Hiding
and Multimedia Signal Processing*, Smart Innovation, Systems and Technologies 64,
DOI 10.1007/978-3-319-50212-0_1

bands information should be elected before put all of them into classifier. And the neighbor information of every single pixel is also very valuable. We introduce a new framework for the election of spectrum bands to choose those who have the most typical data for most character, then use them to continue classify experiments.

1) First and foremost, our method must recognize that to compare with taking all spectrum bands in hyper-spectral image into classification, how to use its morphological character seems much more significant.
2) Use different operator into different kinds of data. So I design two kinds of operator and statistical method associated with them.
3) Finally, although use SVM model as the final classifier, but how choose a lot of accuracy parameters composite an appropriate classifier need to have a great time to design, include choose kernel function.

So this paper is organized as follows: Section 2 briefly reviews the recently popular and relevant hyper-spectral image process. Section 3 introduces the main idea about how to elect spectrum bands in hyper-spectral image classification. Section 4 depicts details and effect of Election process and calculating process. Section 5 gives the experiment results on the public domain Indian Pines hyper-spectral dataset, and compare with non-electing algorithm. Section 6 concludes the paper.

Data: the public domain Indian Pines hyper-spectral dataset which has been previously used in many different studies. This image was obtained from the AVIRIS imaging spectrometer at Northern Indiana on June 12, 1992 from a NASA ER2 flight at high altitude with ground pixel resolution of 17 meters. The dataset comprises 145*145 pixels and 220 bands of sensor radiance without atmospheric correction. It contains two thirds of agriculture (some of the crops are in early stages of growth with low coverage), and one third of forest, two highways, aril lane and some houses. Ground truth determines sixteen different classes (not mutually exclusive). Water absorption bands (104-108, 150-163 and 220) were removed (Tadjudin and Landgrebe, 1998), obtaining a 200 band spectrum at each pixel.

2   proposed approach modules

SVM algorithm is established on the statistic-learning theory, it is a new kind of machine-learning algorithm. The advantage of SVM algorithm concentrates on solving the linear inseparable problem. It solves the linear inseparable problem by calculating the inner-product operation in high dimensional space by importing one special function named kernel function. Suppose the sample set is $(x_i, y_i), x \in R^d, y \in \{1, -1\}\, i = 1, ..., n.$ is the number of classification. In a d dimensional space, when we use SVM model to deal problem, first we must use a non-linear mapping $\Phi : R^d \rightarrow H, x \rightarrow (x)$, mapping the data from original space R^d into a high dimensional space H(named kernel space), by importing kernel function $K(x_i, y_i)$, make SVM model can classify sample data through a linear classified method in this high dimensional space. So SVM model also called learning algorithm based on kernel function. We considered SVM model is a linear-classification, only when we want to deal with problem among non-linear classifications, we can map the low dimensional data into high dimensional data and continue linear classify in high dimension space instead of problem of non-linear classifications in original space. When we handle a non-linear classification problem, SVM use kernel function $K\left(x_i, y_i\right)$, replace inner-product operation in original space and then the optimal classification function in high dimensional is:

$$f(x) = \mathrm{sgn}\left[\sum_{i=1}^{n} a_i y_i k(x_i, x) + b\right] ; \qquad (1)$$

$a_i > 0$ is Lagrange factor,b is threshold value.

Choose the kernel functions which meet the requirement of Mercer conditions (whether the function is linear or nonlinear) kernel function can easily achieve inner production in high dimensional space through the vector operation, thereby avoid the directly difficult calculating in high space. Kernel function is called kernel or reproducing kernel.

At present, the most kernel function mainly has following three categories:

(1) polynomial kernel function:

$$K(x, x_i) = [(x \circ x_i) + 1]^q \qquad (2)$$

(2) Gauss radial basis kernel function:

$$K(x, x_i) = e^{\left\{-\frac{\|x - x_i\|}{2\delta^2}\right\}} \qquad (3)$$

(3) Sigmoid kernel function:

$$K(x, x_i) = \tanh(v(x \circ x_i) + c) \qquad (4)$$

$v > 0, c < 0$.

Fig.1. is the classification result of using hyper-plane by achieving polynomial kernel classifiers in high-dimensions space. Example is assume the category of the two concentric circular, just can be classified by using the second round curve. But in the practical application of classifier is more complicated than this. SVM Kernel classifier result preview:

3   process of spectrum bands election algorithm

Bands election is a very important process in Hyper-spectral image pretreatment process. If we can elect some important bands or elect some bands which can represent the result directly, and make the sample space more representative, we will save a lot of time in next classify work.

The whole procedure of bands election is as Fig.2. :

 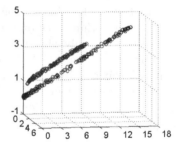

Fig. 1. Two concentric distribution data points(Left) classified easily through mapped to high-dimensional space (right)

4 detail of spectrum bands election algorithm

Hyper-spectral image bands election focus on calculating two kind of guideline: The first guideline is calculation of every direction of viscosity characteristic for every image pieces. Assume a 5*5 calculating operator, according to the order from top to bottom and from left to right slide every image piece. And obtain the max viscosity characteristic as the final direction result. (Software is Matlab2014, take the column order as the primary order). Operator in calculating is as follows:

$$\begin{pmatrix} A(3,3)-A(1,1) & A(3,3)-A(1,2) & A(3,3)-A(1,3) & A(3,3)-A(1,4) & A(3,3)-A(1,5) \\ A(3,3)-A(2,1) & A(3,3)-A(2,2) & A(3,3)-A(2,3) & A(3,3)-A(2,4) & A(3,3)-A(2,5) \\ A(3,3)-A(3,1) & A(3,3)-A(3,2) & 10000 & A(3,3)-A(3,4) & A(3,3)-A(3,5) \\ A(3,3)-A(4,1) & A(3,3)-A(4,2) & A(3,3)-A(4,3) & A(3,3)-A(4,4) & A(3,3)-A(4,5) \\ A(3,3)-A(5,1) & A(3,3)-A(5,2) & A(3,3)-A(5,3) & A(3,3)-A(5,4) & A(3,3)-A(5,5) \end{pmatrix} \quad (5)$$

In the center 3*3 matrix is the direction of viscosity characteristic, and direction is marked as 1-8.as Fig.3.

And after compute all the 25 pixels, obtain a new 5*5 matrix, and get the minimum gradient value and the minimum change of 8 directions from center 3*3 pixels as Fig.4.

Next, in order to judge the direction of position of gradient value whether it is in a normal range. And compute the extend direction of position of gradient value and whether they are all positive or all negative as Fig.5.

In the default condition, the scale of noisy data is much less than the normal image data. So assume most data is normal, and the aim of algorithm is judge whether the center pixel is a noisy point. According to the above two kind of condition, the corresponding numerical position (according to the order of the columns 2,4,6,8 and 2, 8) and the minimum gradient value are not in the same position or same negative, then conclude the position of min tag is a noisy pixel. Delete this pixel, next time when calculating gradient value, set a bigger value (like 100000), repeat computing new minimum gradient value.

When all gradient value are deleted, means after next several computing find every gradient values of position are all max value (100000). At this time, rebuild again, electing the first change position as the direction or value record.

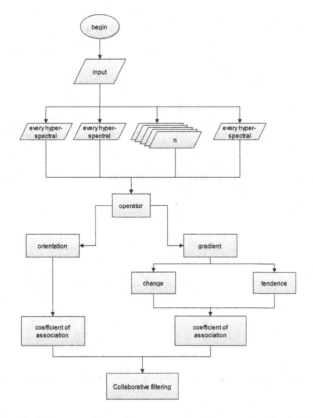

Fig. 2. The procedure of hyper-spectral images band selection

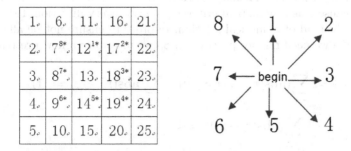

Fig. 3. Two matrixs of direction and its corresponding positions

Another guideline is calculation of every value of viscosity characteristic for every image pieces, is calculating the difference of every direction of viscosity characteristic through operator. Because if only record the direction information can not represent the most similar direction, that is most impossible trend of changing direction, most related place. When designing the direction of viscosity characteristic, the final gradient value has been record.

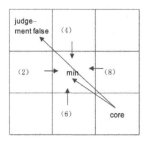 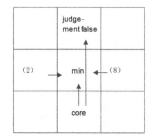

Fig. 4. two kind of operator of direction of gradient value in hyper-spectral image bands election

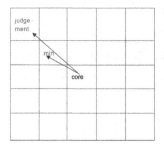 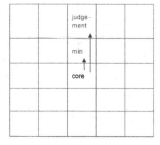

Fig. 5. Two matrixs of direction and its direction of computing gradient

All above process is counting the new assessment criteria, after obtain the final result, must continue calculating the relevancy of criteria. For the direction criteria, because it is a numerical criteria (1-8), for every direction, they are on equal terms instead of numerical relation. In order to ensure not be affected by number. Here we need further processing. So we introduce a new pretreatment:

$$\delta = 2\sum_{i,j}(\mathbf{abs}(r_i - r_j) == 0) + \sum_{i,j}(\mathbf{abs}(r_i - r_j) == 1)$$
$$+ \sum_{i,j}(\mathbf{abs}(r_i - r_j) == 7) \tag{6}$$

$i \neq j$;

This formula set the weight of same direction to 2; set the weight separated by a 45 degrees angle to 1. And set others to 0, because other direction is more different than the original direction, and its contribution degree is 0 or negative. After above calculating, must set correlation coefficient normalization range from 0 to 1. Here normalization method is put every row factor divide to factor in main principal diagonal. Because self-relation is the most similar, so the result of main principal diagonal is 1. And other factor is range from 0 to 1. And then choose upper triangular matrix, and set principal diagonal to 0, because they are all symmetrical and we dont want to choose bands oneself into judge criteria. Next get the whole matrix absolute value and sort them. Extract some bigger value from them get correspondent abscissa, ordinate, that means which two bands is most related, choose them out. The bands which most related will develop into cluster (it will appear that most related bands have been selected means some adjacent bands will also been selected, sounds like there exist a group bands will be selected, generate a bands block). And then sort them as ascend, send to next step.

As for numerical gradient value, we adopt normal normalize method standard score also named Z-score then calculate correlation coefficient. The standard score of a raw score x is:

$$Z = (x - \mu)/\delta \qquad (7)$$

Here μ is the mean of the population, δ is the standard deviation of the population.

The absolute value of Z represents the distance between the raw score and the population mean in units of the standard deviation. Z is negative when the raw score is below the mean, positive when above.

And then calculate correlation coefficient. Here we adopt Pearson product-moment correlation coefficient. It is widely used in the sciences as a measure of the degree of linear dependence between two variables. Here is its formula and its character.

$$r = \frac{\sum_{i=1}^{n}(x_i - \bar{x})(y_i - \bar{y})}{\sqrt{\sum_{i=1}^{n}(x_i - \bar{x})^2 \cdot \sum_{i=1}^{n}(y_i - \bar{y})^2}} \qquad (8)$$

After finished calculate all correlation coefficient from two parts will get two kinds of numerical matrix. As the same processing, only keep the upper triangular matrix and set the element of principal diagonal 0. Select some bigger correlation coefficient, record their bands id and gradient-value, after sorting, send to statistical calculation.

At last, design some bands cluster and their bands field. Select some part of bands from bands cluster and let them represent the whole bands cluster. So we can get a series bands both from direction of viscosity characteristic and value of viscosity characteristic. Because of getting different result from many kinds of hyper-spectral images, we must construct all images collaborative filtering. So we must get intersection band elements as the final results from many different images. And get all intersection from all final result as the final elected bands.

In order to test our idea, we adapt test directly, put an original into elect model, and get final elected bands, delete all bands cluster, but leave other bands which not construct a bands block for keep a great special sample space. Finally, we put the elected bands image into SVM-kernel classification model, test its accuracy rate and time cost.

5 result of experiment

Select 20 times bands elected and non-elected hyper-spectral as a same SVM-kernel classification algorithms input and its accuracy rate and time cost table.

Table 1. 20 times experiment non-elected hyper-spectral.

	1	2	3	4	5	6	7	8	9	10
accuracy	0.7329	0.6658	0.6662	0.7499	0.7416	0.6692	0.7168	0.7356	0.6623	0.7736
time	6.9844	6.6557	7.0614	6.6947	6.8867	7.0768	7.3948	7.1917	6.8642	6.7598
	11	12	13	14	15	16	17	18	19	20
accuracy	0.7275	0.7499	0.6850	0.7547	0.7770	0.7201	0.7314	0.7273	0.6885	0.6999
time	6.7063	7.1231	7.2725	6.8709	6.7529	6.7837	7.0426	6.8185	7.2252	6.5508

Table 2. 20 times experiment elected hyper-spectral.

	1	2	3	4	5	6	7	8	9	10
accuracy	0.7387	0.725	0.6759	0.6976	0.7447	0.7138	0.7470	0.7202	0.7361	0.7485
time	5.9734	5.2412	5.8792	5.347	5.355	5.5321	5.5753	5.3516	5.5829	5.6704
	11	12	13	14	15	16	17	18	19	20
accuracy	0.7361	0.7226	0.6468	0.7087	0.7212	0.7699	0.7145	0.6890	0.7449	0.6938
time	5.4165	5.5870	5.5217	5.4318	5.7021	5.3346	5.5144	5.3554	5.2956	5.4526

It is obviously find that in 200 bands, elected 15 bands, 136 intersection bands
(bands cluster), so only send 200-136+15=79 bands into algorithm.
At table 1. Non-elected hyper-spectral average accuracy : 0.7187; average calculating time: 6.9358; MSE of accuracy: 0.0359;MSE of calculating time: 0.2303.
At table 2. Elected hyper-spectral average accuracy : 0.7200; average calculating
time : 5.5060; MSE of accuracy : 0.0288; MSE of calculating time: 0.1914.
Because of above result, no matter in the field of accuracy or in the field of
time, even in the field of Stability of algorithm, the bands elected hyper-spectral
algorithm is better than before. From the result we also can conclude that there
exist a huge scale of redundant bands or information, and also some degree of
Hughes phenomenon of over-fitting, so electing bands during the pretreatment
is very necessary in the problem of classification of hyper-spectral.

References

1. Jun Li, Prashanth Reddy Marpu, Member, IEEE, Antonio Plaza, Senior Member, IEEE, Jos M. Bioucas-Dias, Member, IEEE, and Jn Atli Benediktsson, Fellow,
   IEEE: "Generalized Composite Kernel Framework for Hyperspectral Image Classifi-
   cation" In: IEEE TRANSACTIONS ON GEOSCIENCE AND REMOTE SENSING
   VOL.51,NO 9 SEPTEMBER 2013, 0196-2892 (2013).
2. Vineet Kumar, Jurgen Hahn, Abdelhak M.Zoubir : "BAND SELECTION FOR
   HYPERSPECTRAL IMAGES BASED ON SELF-TUNING SPECTRAL CLUS-
   TERING" In: EUSIPCO 2013 1569744031.
3. C.-C. Chang and C.-J. Lin, "LIBSVM: A library for support vector machines"
   ACM transactions on Intelligent Systems and Technology, vol.2 pp. 27:1-27:27, 2011,
   Software available at http://www.csie.ntu.edu.tw/~cjlin/libsvm.
4. This work is supported by Program for New Century Excellent Talents in University
   under Grant No. NCET-13-0168,and National Science Foundation of China under
   Grant No.61371178.

Evaluating a Virtual Collaborative Environment for Interactive Distance Teaching and Learning: A Case Study

Wen Qi

Donghua Univerity,
No.1882, Yan'an Road, Shanghai, 200051, China
design_wqi@sina.com
http://www.dhu.edu.cn

Abstract. This paper presents a case study of experimenting an interactive collaborative environment for distance teaching and learning activities. It begins with reviewing the existing multiple-users collaborative environment including the Open Wonderland platform, and then outlines the requirements of using a collaborative environment to assist distance teaching and learning, and how these requirements could be met with the Open Wonderland platform. We investigated the system in our local network settings, with a particular emphasis on embedding pedagogy content into of the system. We found that such environment could allow students to have full support throughout their studies with a tutor to guide, advise and offer direct comprehensive guidance. Finally, we summarized the results from the formative and summative evaluations, and presented the lessons learned that can guide future usage of this immersive education space within Higher Education.

Keywords: Distance teaching and learning, virtual environment, Open Wonderland, pedagogy

1 Introduction

Traditional learning and teaching activities in large part show the social characteristics. The format of distance learning is an unique, but advanced way, which allows a learner to study in his own time, at the places wherever he chooses reading, watching or listening to material supplied, doing excises and assignments with regular support from his tutors. The activities described in distance learning are quite different compared with traditional lecture-style study at normal schools. Bring social interaction into normal distance learning activities among students is desired in courses that are based on team projects because a high degree of cooperation is required. The cooperation needed is challenging to achieve among distance learners, therefore requires a particular tool or environment that can assist the teaching and learning process. In this paper, we present our initial exploration study on technical feasibility of using Open Wonderland as a collaborative environment for distance learning and teaching. The issues we aimed

© Springer International Publishing AG 2017

J.-S. Pan et al. (eds.), *Advances in Intelligent Information Hiding and Multimedia Signal Processing*, Smart Innovation, Systems and Technologies 64,
DOI 10.1007/978-3-319-50212-0_2

to address were about how we could enrich learning and teaching experiences for distributed learners and teachers who want to exchange knowledge in the academic setting of distance learning and teaching.

2 Related Work

Social interactions have been one of the key aspects studied in educational technology research. Some researchers have proposed the approaches of using asynchronous applications including email and synchronous applications like instant messaging; other suggested using more sophisticated approaches, for example, interactive collaborative game environments (ICGE) in which learners can interact with their teammates or teacher in a multiple-user set-up. The immediate benefit of a 3D collaborative environment is that it allows learners to simultaneously observe the behaviors of their peers in the groupspace. Well-designed collaborative environments allow seamless interaction between distributed users. Below is a list of several research projects and systems in the area of a collaborative environment for learning and teaching. These works have tried to find out the potential factors that affect effective collaboration and the important issues that may result from insufficient interaction and support among collaborators [1]. Prasolova-Frland investigated the mechanisms that can enhance social awareness in educational environment [2], [3] and has concluded that classic cooperative tools including ICQ and email are not sufficient technically, and the mechanism provided by collaborative environment is a promising supplement to the existing mechanisms. Bouras et al. [4], [5] described a robust environment that supports collaborative education. Nevertheless, the limitation of such environment is that editing VRML files is the only way to modify it whenever new contents need to be introduced. A typical teacher may lack of such technical expertise. Furthermore the world cannot be changed while it is in use. Okada et al. introduced a study on building an ICGE system for ecological education [6]. Although virtual areas can be added at runtime, it is complex and time-consuming. Oliveira et al. introduced collaborative learning into industrial training and e-commerce [7]. The environment allows video on demand without synchronization. Mansfield et al. describe their Orbit system, which supports a groupspace model [8]. However, it is not an ICGE and data or metadata visualization is absent. The Orbit system only allow access to data on the server.

3 The Open Wonderland

As an open-source toolkit, Open Wonderland is designed with client-server architecture and embedded with a set of technologies for creating virtual- or augmented-reality environments. The toolkit is built upon several software components or middlewares that include the Project Darkstar multiplayer game server, VoiceBridge for adding immersive audio and jMonkeyEngine (JME) to generate graphical scene. There are also other libraries around JME, such as a Collada loader, which enables users to import 3D objects or models on Google

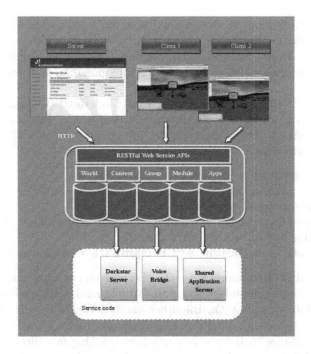

Fig. 1. The overall architecture of Open Wonderland

Warehouse. Additional objects and components (such as a camera device to record audio and video stream from a client) in Project Wonderland make use of other technologies, such as the Java Media Framework. Open Wonderland offers a rich collection of ob-jects for building up environments, and supports shared software applications, such as word processors, web browsers, and document-presentation tools. For instance, one or several users can draw on a virtual white-board and view PDF documents and presentations. Every user, represented by an avatar, can communicate to others through the avatar in the same world by means of a headset or microphone and speaker or by the use of a dedi-cated chat window for text messages. The generated scene within Wonderland can be viewed from a first-person or several third-person perspectives. Fig. 1 illustrates the overall architecture of Open Wonderland. It complies with the Representational State transfer (REST) style of architecture. With the help of Java WebStart, there is a Wonderland client running on every users PC. Java Runtime Environment (JRE) should be available while using Java WebStart. A web page becomes accessible when it is downloaded initiated by Java Web-Start. The administration tasks can also be performed via the same link, which includes selecting the initial environment, saving copy of the current environ-ment and adding extra components. Components may be software libraries that extend the core functionality of the system, for instance, a Video Camera. The administrative facilities start and terminate several core modules remotely. The properties of those modules can also be changed. There is an online games engine called Darkstar. Audio is possible in Open Wonderland by the Voice Bridge via Voice over IP with various audio qualities. The Shared Application Server makes the in-world sharing of X11 applications possible, such as Chrome and Clipboard.

The original objective of developing the Wonderland platform by Sun Microsystems was to create a tool that enables cooperative working by the companys employees [9]. As the consequence, the main advantages of this platform can be summarized as:

- Real-time application sharing;
- Tight integration with business data;
- Deployment internally or externally;
- Scalability of Darkstar server: from very large to very small implementations;
- Open source and extensible: 100% pure Java;
- Spatial Audio: a core feature with extensive telephony integration.

Wonderland is often compared to the Second Life or OpenSim platform. It is possible to customize and integrate the Wonderland platform into an organizations own infrastructure. On the other hand, Second Life and OpenSim are publicly online services that are accessible by lots of users. These users can make use them to organize their lives. In particular, teaching institutions has already used Second Life extensively to carry out online teaching (for details see [10]). Although Second Life has become first choice in terms of assisting online teaching and learning, it does have privacy and security issues around its use, for example, when a participant takes part in an online session. Furthermore, it is in doubt whether organizations have sufficient controls when they use Second Life as part of their formal education infrastructure.

4   Experimental Design

As one of best universities that is focusing on fashion and costume design education, our institute desires to offer students an unique, top-class distance learning experience. We believe that online learning enables a student to study in his own time, at home or wherever he prefers reading, watching or listening to material supplied, doing course excises and assignments with direct support from his/her tutors. The learning activities described are quite different compared with our traditional lecture-style study for regular students. Therefore, we have particular requirements on tools or environments that can assist such teaching and learning process. Research conducted in this paper was built upon the hypothesis that an interactive collaborative virtual space can support and facilitate these kinds of teaching and learning activities. Our goal is to experiment the effectiveness of a selected platform. We list major functional requirements while selecting an experimental platform:

- Emphasis on formal and informal social in-teraction with improved communication

- Strong feeling of social presence, enabling to exchange their opinions about learning contents,
- Collaboration oriented, allowing active social interactions to build mutual understanding, particularly before and after scheduled events,
- Share document easily without switching contexts,
- Extremely extensible,enabling to introduce any sort of new feathers into the platform.

Fig. 2. The virtual auditorium

Since Open Wonderland has fulfilled most of our requirements, it has been selected as the interactive collaborative game environment for this study. The virtual classroom was realized using the Open Wonderland toolkit version 0.5. The virtual world was constructed using a collection of 3D objects. Users can launch the virtual class easily through hyper-links. The initial models for the environment were created using Google SketchUp. These models were then exported into either Collada format as .dae files or Google Earth format as .kmz files, both of which Open Wonderland is able to import. They were uploaded into the main scene and positioned according to our design. Also applications planned to be shared were placed in the virtual class scene to build up the context or meet task requirements (for example, a web browser was used to access a crossword that was completed collaboratively by the students). Fig. 2 illustrates the overview of our virtual auditorium. The world itself consists of a wide open space surrounded by mountains and a representation of the auditorium building. The auditorium contains a lecture presentation board, a note board, a sketch board, standing seats, and etc. The auditorium environment also has surrounding sound and allows application sharing. This representation of a virtual learning environment offers possibilities to build up various contexts in which social interactions betweens participants are possible. There are a document reader, a note stick, and a chat box as shown in Fig. 3. Teachers and students are able to pick up a particular persona as desired by their educational tasks. Users can set their preferred view perspectives (see Fig. 3).

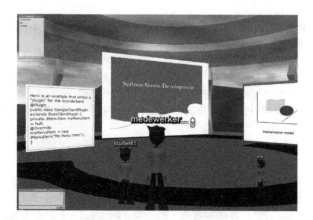

Fig. 3. The Auditorium components

One of the advantages of this environment is that a user can drag any content (documents) located in his desktop into the auditorium world. These contens are then treated as objects within the 3D Wonderland world. As long as an object is Collada compliant, it can be imported and displayed within the auditorium. Such feature is the functionality teachers needs to customize an auditorium world with new objects. It meet the requirements discussed before. The Google SketchUp 3D warehouse is a publicly available and popular repository that contains thousands of Collada content, which are freely accessible. Another advantage is that the world we experimented is extensible that allows users and developers to design their own worlds and introduce new features into it in the form of software components. The key achievement of this study was that we successfully demonstratde the feasibility of the Open Wonderland for collaborative interaction in a social learning environment. The environment we exercised could augment an existing distance education practice because it creates a sense of community amongst students who are located remotely. Such scenario is common in a non-traditional instructive higher education setting. In addition, the environment we designed allows students to reach a tutor and receive his full support throughout their studies. A tutor can supervise their learning activities and offer comprehensive opinions to their coursework. The tutor can lead group tutorials and seminars at runtime within the virtual game environment.

5   Discussion

Schlosser and Anderson pointed out that distance learning can be equally effective as learning through face-to-face [11]. In essence, there is no difference between good distance teaching pedagogy and good conventional teaching techniques. Schlosser and Anderson listed several studies that clearly proved that distance education was effective for education. However, from their points of view, they emphasized distance teachers should offer an environment that allows structured note-taking, interactive study guides, and visuals. In another survey, Threlkeld and Brzoska reviewed a number of studies and concluded that the importance of the media itself to instruction was not the same as the other factors, like the characteristics and objective of learners [12]. However, they discovered it is very critical that various supports are available for a distance learner, such as, the feedback from an instructor as well as direct access to library materials and other auxiliary resources. Indications from our studies are that online learning environments should provide means for online note taking, and online communication among students and teachers should be smooth. In addition, hypermedia-based contents should be easily incorporated. With the development of new technologies including interactive 3D collaborative environments, a pronounced shift has taken place in learning theories from behaviorism to constructivism. With the deployment of new technology, students can improve their thinking skills and develop their personal knowledge [13, 14] that can be shared immediately with anyone around the world [14]. The new technologies or features of Open Wonderland exercised in our study not only offer it's users with possibilities to break the bottleneck in conventional learning and teaching, but, more importantly, to illustrate a new way of education. The Wonderland platform as a tool urges educators to re-evaluate and rethink the existing educational models and eventually to establish innovative ones. Certainly, the platform is still evolving and poses several constraints, such as lack of rich and fluent interaction style and difficult deployment in network environment with a complex firewall. The next step could be to put the virtual Wonderland in real pedagogical environment with concrete learning activities to investigate the pitfall and usability issues in details.

6 Conclusion

This paper presents a case study of exercising the Open Wonderland platform as an interactive collaborative tool for distance teaching and learning activities. The role of Open Wonderland in this study is to provide a test bed for evaluating its teaching and learning capabilities. We described our user experience and especially focused on the social aspects of such system. We investigated how we can support good practice of learning and teaching with the use of such 3D gaming environments in education. We presented our findings on the technical feasibility and pedagogical value of using 3D collaborative environments for distance teaching and learning. The issue we aimed to address is that there is a need to improve the learning experiences and outcomes for students who are located remotely at different places.

7 Acknowledgement

This research has been supported by The Program for Professor of Special Appointment (Eastern Scholar) at Shanghai Institute of Higher Learning (No.TP2015029) from Shanghai Municipal Education Commission.

References

1. Daradoumis, T., Xhafa, F., and Manuel, Marqus J., Evaluating Collaborative Learning Practices in a Virtual Groupware Environment, Computers and Advanced Technology in Education, 2003, June 30 July 2, Rhodes, Greece.
2. Prasolova-Frland, E., Supporting Social Awareness in Education in Collaborative Virtual Environments, International Conference on Engineer-ing Education, 2002.
3. Prasolova-Frland, E., Supporting Awareness in Education: Overview and Mechanisms, ICEE, Manchester, UK, 2002, 18th-22nd, August.
4. Bouras, C., Psaltoulis, D., Psaroudis, C., and Tsiatsos, T., An Educational Community Using Collaborative Virtual Environments, Proceeding ICWL '02 Proceedings of the First International Conference on Advances in Web-Based Learning , 2002, pp180-191.
5. Bouras, C., Philopoulos, A., and Tsiatsos, T., E-Learning through Distributed Virtual Environments, Journal of Network and Computer Applica-tions, Academic Press, 2001, July.
6. Okada, M., Tarumi, H., and Yoshimura, T., Distributed Virtual Environ-ment Realizing Collaborative Environment education, Symposium on Ap-plied Computing archive, Proceedings of the 2001 ACM symposium on Applied computing, Las Vegas, Nevada, United States, 2001, pp83 88.
7. Oliveira, C., Shen, X., and Georganas, N., Collaborative Virtual Envi-ronment for Industrial Training and e-Commerce, Workshop on Application of Vir tual Reality Technologies for Future Telecommunication Systems, IEEE Globecom 2000 Conference, 2000, Nov-Dec.
8. Mansfield, T., Kaplan, S., Fitzpatrick, G., Phelps, T., Fitzpatrick, M., and Taylor, R., Evolving Orbit: A Process Report on Building Locales. In Pro-ceedings of the international ACM SIGGROUP conference on supporting group work: the integration challenge (GROUP '97). ACM, New York, NY, USA, 1997, pp241-250.
9. Yankelovich, N., Walker, W., Roberts, P., Wessler, M., Kaplan, J., and Provino, J., Meeting Central: Making Distributed Meetings More Effective. ACM Conference on Computer Supported Cooperative Work, November 610, Chicago, IL, USA, 2004
10. Coffman, T.& Klinger, M.B. Utilizing Virtual Worlds in Education: The implications for practice. International Journal of Social Sciences, 1(2), 50-54, 2007, July.
11. Schlosser, C. A., & Anderson, M., Distance Education: Review of the literature. Ames, Iowa, Iowa Distance Education Alliance, 1994.
12. Threlkeld, R., & Brzoska, K., Research in Distance Education. Dis-tance Education: Strategies and Tools. B. Willis. Englewood Cliffs, New Jersey, Educational Technology Publications, 1994, pp4166
13. Means, B., & Olson, K., Technology's Role within Constructivist Classrooms. A symposium: Teachers, Technology, and Authentic Tasks: Lessons From Within and Across Classrooms. American Educational Re-search Association. April, San Francisco, CA, 1995.
14. Kizlik, R., Connective transactions - Technology and Thinking Skills for the 21st Century. International Journal of Instructional Media, 23, 1996,pp115122.

The Linear Transformation Image Enhancement Algorithm Based on HSV Color Space

Maolin Zhang[1,3], Fumin Zou[1,3], and Jiansheng Zheng[2]

[1]School of Information Science and Engineering,
FuJian University of Technology,
No3 Xueyuan Road, University Town, Minhou,Fuzhou City, Fujian Province, China
[2]School of electronic information,WuHan University,
Luojia Hill, Wuchang District, Wuhan 430072, Hubei, China
[3]The Key Laboratory for Automotive Electronics and Electric Drive of Fujian
Province, Fujian University of Technology, Fuzhou Fujian 350108,China
{mailzml}@163.com
{fmzou}@fjut.edu.cn
{zjs}@whu.edu.cn

Abstract. For the complex and changeable recording scene and lighting
condition, which makes images uneven in light and short of contrast ratio,
a linear transform image enhancement method based on HSV color space
transform is proposed to improve the image quality. In order to maintain
the image color invariant, firstly conversing the traditional RGB color
space to HSV color space, then analyzing the relationship of S and V
components between the HSV color space model and license plate colors,
calculating linear factor and in making use of the relationship between
the components, α and γ adjusting the coefficients, finally conversing the
HSV model parameters to RGB space. The experimental results show
that, compared with the bidirectional histogram equalization method
and the Retinex method, the proposed method can be effectively used
in the image enhancement for driving recorder, not only enhanced in
contrast and bright of the license plate region, and has the strong real-
time processing capability.

Keywords: Image enhancement; HSV color space; Linear transformation; Automobile data recorder

1 Introduction

With the improving of humans life, motor vehicle ownership geometric growth,
the popularization of cars is inevitable trend. Therefore, people gradually pay
more and more attention on intelligent transportation system, only make use
of information technology, sensor technology, data communication technology
and other technology effectively in traffic management systems, could truly the
increasingly serious traffic problems solve[1]. Currently, the automobile data
recorder has been widely used in the automotive industry, reducing traffic accidents as powerful evidence, even can be basis of traffic violation penalties, so

© Springer International Publishing AG 2017
J.-S. Pan et al. (eds.), *Advances in Intelligent Information Hiding
and Multimedia Signal Processing*, Smart Innovation, Systems and Technologies 64,
DOI 10.1007/978-3-319-50212-0_3

as to achieve the purpose of correcting driving behavior. But due to light and shooting scene of automobile data recorder is complicated, the image captured exist light and dark uneven, insufficient contrast and other issues, so that unable to extract license plate extraction or extract error.

The existing image enhancement techniques can be roughly divided into two major categories of the transform domain method and the spatial domain method [3,17]. Frequency domain conversion is to transform the original image, enhancing image in the transform domain, draw widespread attention in the field of image enhancement due to excellent separation characteristics of signal and noise, usually used in enhancement of image details and specific edge, such as Retinex algorithm, but it is complicated so that processing rate slow, not suitable for the dynamic testing system of image enhancement [1, 5-7]. The spatial domain method is directly to deal with the original image, this method includes bidirectional histogram equalization, gray level transformation, unsharp masking, etc. Although its processing rate is fast, but the noise will be amplified in processing, result in visual effect weakened and hue change. Bidirectional histogram equalization is a fast and effective image enhancement method, but the grayscale merger easy to cause the loss of image detail information and image enhanced excessive, enhancement effect is poorer to low brightness contrast image [8-10]. The adaptive filter is an effective method for enhancing edge and details of image, which is usually used for some specific details or edges, generally used only for certain types of image enhancements, such as fingerprint images, poor universality[15,16]. Unsharp masking is a common enhancement method in fault diagnosis, image adaptive enhancement method based on unsharp masking use the transformation function instead of traditional sharpen enhancement coefficient, so as to realize adaptive image enhancement [11].

In order to further improve contrast and brightness of image taken by the automobile data recorder under different background and the light intensity, this paper proposes a linear transformation image enhancement based on HSV color space transform method. To maintain constant image tone, make the traditional RGB space conversion to HSV color space at first, and then analyses the relationship between the component S and V of the HSV color space model and the license plate color. Then using the relationship between the various components to calculate sum of the linear factor and adjust each component coefficients. Finally, make the HSV model parameters adjusted conversion to the RGB space. In order to verify the performance of this image enhancement method, give the experimental results of the method, the Bidirectional histogram equalization and the Retinex algorithm, compare and analyze according to image enhancement overexposed, underexposed image enhancement algorithms and processing times. We can get a conclusion after analysis, the method can be effectively used in the image captured by driving recorder enhancement, not only can enhance the contrast and brightness of the license plate area, and processing speed rate than bidirectional histogram equalization and Retinex algorithm.

2 Linear transformation algorithm base on HSV color space

2.1 HSV color space

HSV (Hue, Saturation, Value) and RGB (Red, Green, Blue) are different representations of color space. RGB color model is a kind of color space for equipment, the value of R, G, B and the three attribute of color are not directly linked. HSV, the color model for visual perception, makes use of three basic properties of color to represents color, where H represents hue, S for saturation, V represents Brightness [12, 14]; Hue, the main acceptable color of the observer , is an attribute of describing pure color; Saturation is a kind measure of the degree that pure color is diluted by white light; Value is a subjective description , it embodies the concept of a colorless intensity, and is a key parameter of describing color feeling. HSV color space can be obtained by RGB conversion, the expression as following, in which, T_{max} is the maximum among R, G, B, T_{min} is the minimum [8].

$$V = T_{max} \tag{1}$$

$$S = \begin{cases} 0, \text{if} T_{max} = 0 \\ \frac{T_{max} - T_{min}}{T_{max}} = 1 - \frac{T_{min}}{T_{max}}, \text{ otherwise} \end{cases} \tag{2}$$

$$H' = \begin{cases} 60 \times \frac{G-B}{T_{max}-T_{min}}, \text{ if } (S \neq 0) \text{ and } (T_{max} = R) \\ 60 \times \left(\frac{B-R}{T_{max}-T_{min}} + 2\right), \text{ if } (S \neq 0) \text{ and } (T_{max} = G) \\ 60 \times \left(\frac{R-G}{T_{max}-T_{min}} + 4\right), \text{ if } (S \neq 0) \text{ and } (T_{max} = B) \end{cases} \tag{3}$$

$$H = \begin{cases} H', \text{ if } H' \geqslant 0 \\ H' + 360, \text{ otherwise} \end{cases} \tag{4}$$

The HSV color model corresponds to a conical subset of the cylindrical coordinate system, as shown in figure 1.

2.2 Linear transformation algorithm based on HSV Space

According to the equation (1)-(4), RGB space of image could convert into HSV space. In the HSV color space model, when the H component is fixed, HSV color space model can be simplified into a triangle model which only containing the S and V, as shown in figure 2.

Therefore, when the H component is fixed, the average saturation and brightness component of the input image, which expressed as point $P(X_i, Y_i)$. While the saturation and the brightness component of license plate, which can be expressed as $Q(X_t, Y_t)$. According to the HSV color space model defined, images linear transformation only changes the image saturation and brightness, which doesnt affect the hue. Therefore putting forward equation.

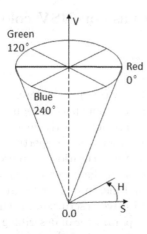

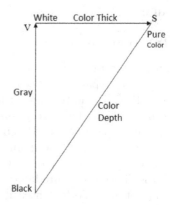

Fig. 1. HSV color space model   **Fig. 2.** S and V triangular model

$$Q\left(X_t, Y_t\right) = \alpha \cdot P\left(X_i, Y_i\right) + \gamma \tag{5}$$

Wherein α and γ are linear transformation factor, α factor which represents the original brightness and saturation direction to stretched or shortened, while γ factor represents the increase or decrease in brightness. The linear transformation model is shown in figure 3.

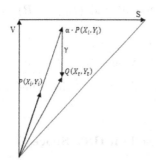

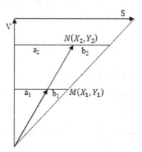

Fig. 3. Linear transformation model   **Fig. 4.** Linear transformation of α factor

Because the saturation S is a proportional value, range from 0-255, it is expressed as a ratio that between the selected color purity and the largest color purity, only gray when S=0. As shown in figure 4, multiplying the point $M\left(X_1, Y_1\right)$ determined by the saturation and brightness and α will get the point $N\left(X_2, Y_2\right)$, due to $\frac{a_1}{a_1+b_1} = \frac{a_2}{a_2+b_2}$, thus the factor α has no changement to component S. But we can observe the influence to brightness of image due to α.

Factor γ indicates the increase or decrease in brightness, the image pixels in V upward or downward shift. As shown in figure 5, add the point determined by saturation and brightness and γ the factors, get the point $N\left(X_2, Y_2\right)$. Because $\frac{a_1}{a_1+b_1} > \frac{a_2}{a_2+b_2}$, thus we can get a conclusion that improve the brightness V will sacrifice their saturation S.

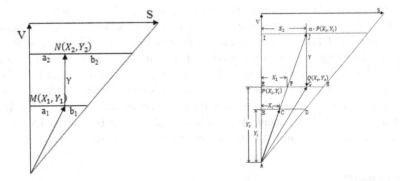

Fig. 5. Linear transformation of γ Factor **Fig. 6.** Linear transformation deduction model

According α and γ is a transformed result of linear transformation factor, we can construct the linear transformation deduction model. As shown above the figure 6, the triangle ABC and ADE are similar triangles, thus equation $\frac{Y_t}{Y_i} = \frac{X_t}{X_i}$ establish, the equation may be converted as $X_i = X_i\frac{Y_t}{Y_i}$. Multiply point $P\left(X_i, Y_i\right)$ and α can get a point $P' = \alpha \cdot P\left(X_i, Y_i\right)$, and X_2 can be represented as $X_2 = \alpha X_i = X_t$. According to HSV color space model, the V-direction represent brightness value, so brightness of G point $V_t = Y_t$, brightness of C point $V_i = Y_i$. S is the saturation ratio value, the saturation of point C is the same as the point F as $S_1 = \frac{EF}{EH}$, the saturation of G point is $S_t = \frac{EG}{EH}$, thus the saturation ratio of point F and the point G can be expressed as $\frac{EF}{EG} = \frac{S_i}{S_t}$, at the same time it can represent as $\frac{EF}{EG} = \frac{X_1}{X_t} = \frac{X_i\frac{Y_t}{Y_i}}{\alpha X_i}$. Therefore $\frac{EF}{EG} = \frac{S_i}{S_t} = \frac{X_i\frac{Y_t}{Y_i}}{\alpha X_i}$ equation is established, and derivation of the formula (6).

$$\alpha = \frac{S_t V_t}{S_i V_i} \tag{6}$$

According to the equation5can get $\gamma = Q\left(X_t, Y_t\right) - \alpha \cdot P\left(X_i, Y_i\right)$, while $\alpha X_i = X_t$, further simplify as formula (7).

$$\gamma = V_t - \alpha V_i \tag{7}$$

3   Experimental results and analysis

3.1   Image captured at evening enhancement effect comparison

As shown in figure 7which is the result of image captured by automobile data recorder enhancement in the evening light. In figure7a, due to insufficient light conditions, license plate almost drowned in the background, so that it is difficult to locate the specific location of the license plate and license plate information extraction. In figure 7b, although the image color lack fidelity after treatment, but the partial details of the license plate has been significantly restored. In figure 7c, the image has been restored well, which improved remarkably including the license plate details. Figure 7d compared with figure 7c and 7b, image enhancement effect of linear transformation algorithm based on the HSV color space is not same as histogram equalization algorithm which cause color distortion, while image enhancement effect of license plate area is same to another two algorithms, can realize adjustment of the brightness and saturation.

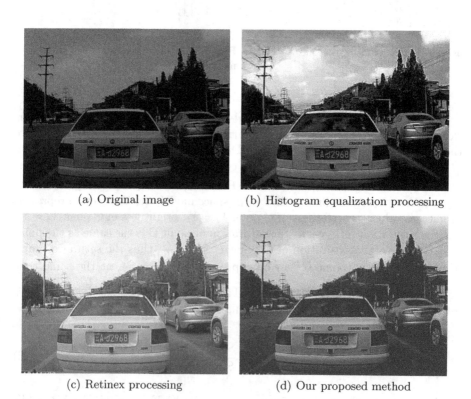

(a) Original image   (b) Histogram equalization processing

(c) Retinex processing   (d) Our proposed method

Fig. 7. Image captured at evening enhancement effect

3.2 Image captured at noon enhancement effect comparison

Such as figure 8 shows the enhancement results of image taken by automobile data recorder in the midday light. In figure 8a, since noon shooting, outdoor lit so that the captured image overexposed, in other word, the image was high brightness, low contrast and the image information is difficult to distinguish, such as license plates and so on. In figure 8b, although histogram equalization can improve the contrast of the image to a certain extent, at the same time enhancing details also introduced a lot of noise, bring new problems for subsequent processing. In figure 8b, we can see from the image, saturation and brightness to get a better adjustment, and can enrich the license plate details effectively. Compared the enhancement effect of figure 8b with figure 8d, it can be achieved with less noise introduced, the part of license plate in figure 8d after image enhancement contrast and brightness are similar to figure 8c, and the three methods are able to get a better enhancement.

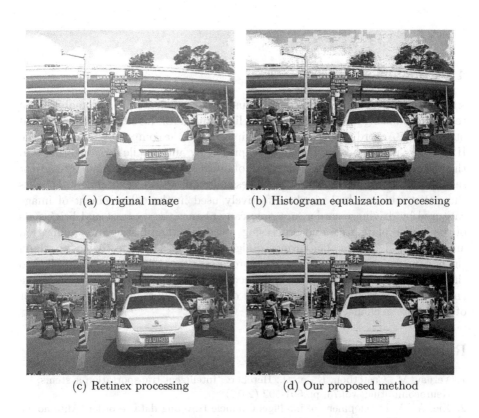

(a) Original image   (b) Histogram equalization processing

(c) Retinex processing   (d) Our proposed method

Fig. 8. Image captured at noon enhancement effect

3.3 Image enhancement processing time comparison

As shown in table 1, we need 5ms to deal with an image consist of 24 bit, 96 dpi resolution, 400   300 pixels by use of linear transformation image enhancement algorithms based on the HSV color space. However we need 6ms to deal with the same image based on the same hardware by use of histogram equalization, and needs 33ms for the Retinex algorithm. Comparing the three kinds of image enhancement algorithms, the processing time of the algorithm in this paper is shortest, so as to further application in real-time processing system of automobile data recorder.

method time	Histogram equalization processing	Retinex processing	This paper processing
Time	6ms	33ms	5ms

Table 1. Processing time

4 Conclusion

Aim to the low quality color images caused by complicated of automobile data recorders shooting scene and uneven of light and dark, insufficient contrast and other issues. According to the relationship between the component S and V of the HSV color space model in the image captured by automobile data recorder and the license plate color. Then using and of linear transformation factors to adjust saturation and brightness of image processing required. The experimental results show that, this method can be effectively used in the enhancement of image captured by driving recorder, not only can enhance the contrast and brightness of the license plate area, and processing speed fast than histogram equalization and Retinex algorithm. The proposed algorithm shows the effectiveness both on the enhancement and on the real-time processing capability, it can be applied to the image when its target color is defined, but the linear transformation image enhancement algorithm requires further discussion and study if the target color of the image is changeable.

References

1. Vernay, M., Castetbon, K., et al.: Hercberg. Intelligent transportation systems. In: Neurocomputing, vol.73, pp.591-592 (2010)
2. Xie, A.: A. Development of intelligent vehicle traveling data recorder. Automobile Parts (2010)
3. Zhao, Y., Fu, X., et al.: A new image enhancement algorithm for low illumination environment. Computer Science and Automation Engineering (CSAE), 2011 IEEE International Conference on. IEEE, pp.625-627 (2011)

4. Yoo, Y., Im, J., Paik, J.: Low-Light image enhancement using adaptive digital. Pixel Binning. Sensors vol.15, pp.14917-14931 (2015)

5. Mishra, N., Kumar, P.S., Chandrakanth, R., et al.: Image enhancement using logarithmic image processing model. Iete Journal of Research. vol.46, pp.309-313 (2015)

6. Yu, Q., Ma, S.Q., Ma, D.M.: Maintaining image brightness adaptive local contrast enhancement. Computer Engineering and Applications, vol. 7, pp. 160-164 (2015)

7. Zhang, L.B., Ge, M.L.: A kind of image enhancement algorithms combination of spatial and transform domain. Electronics Optics and Control, vol. 12, pp. 45-48, (2014)

8. Qin,X.J., et al: Retinex Structured Light Image Enhancement Algorithms in HSV Color Space. Journal of Computer-Aided Design and Computer Graphics, vol.4, pp. 488-493. (2013)

9. Chen, C.,: Improved single scale retinex algorithm in image enhancement. Computer Applications and Software, vol. 4, pp. 55-57. (2013)

10. Zhan, K., Teng, J., Shi, J., et al.: Feature-linking model for image enhancement. Neural Computation, pp. 1-29. (2016)

11. Wang, S., Li, B., Zheng, J.,: Parameter-adaptive nighttime image enhancement with multi-scale decomposition. Iet Computer Vision. (2016)

12. Centeno, J.A.S., Haertel, V.: An adaptive image enhancement algorithm. Pattern Recognition, vol. 30(7), pp. 1183-1189. (2014)

13. Qin,X.J., et al: Retinex Structured Light Image Enhancement Algorithms in HSV Color Space. Journal of Computer-Aided Design and Computer Graphics, vol.4, pp. 488-493. (2013)

14. Li, Y., Hu, J., Jia, Y.: Automatic SAR image enhancement based on nonsubsampled contourlet transform and memetic algorithm. Neurocomputing, vol. 134(9), pp. 70-78. (2014)

15. Lee, J., Pant, S.R., Lee, H.S.: An adaptive histogram equalization based local technique for contrast preserving image enhancement. International Journal of Fuzzy Logic and Intelligent Systems, vol. 15(1), pp. 35-44. (2015)

16. Prema, S., Shenbagavalli, A.: Image enhancement based on clustering. International Journal of Applied Engineering Research, vol. 10(20), pp. 17367-17371. (2015)

17. Dong, Y.B., Li, M.J., Sun, Y.: Analysis and comparison of image enhancement methods. Applied Mechanics and Materials, pp. 1593-1596. (2015)

Synthesis of Photo-Realistic Facial Animation from Text Based on HMM and DNN with Animation Unit

Kazuki Sato, Takashi Nose, and Akinori Ito

Graduate School of Engineering
Tohoku University, 6-6-05 Aramaki Aza Aoba, Aoba-ku, Sendai, Miyagi 980-8579,
Japan. kazuki.satou.p3@dc.tohoku.ac.jp, tnose@m.tohoku.ac.jp,
aito@spcom.ecei.tohoku.ac.jp

Abstract. In this paper, we propose a technique for synthesizing photo-realistic facial animation from a text based on hidden Markov model (HMM) and deep neural network (DNN) with facial features for an interactive agent implementation. In the proposed technique, we use Animation Unit (AU) as facial features that express the state of each part of face and can be obtained by Kinect. We synthesize facial features from any text using the same framework as the HMM-based speech synthesis. Facial features are generated from HMM and are converted into intensities of pixels using DNN. We investigate appropriate conditions for training of HMM and DNN. Then, we perform an objective evaluation to compare the proposed technique with a conventional technique based on the principal component analysis (PCA).

Keywords: Photo-realistic facial animation, face image synthesis, hidden Markov model, deep neural network, Animation Unit

1 Introduction

In recent years, various studies are conducted about conversational agents using speech recognition and speech synthesis technologies [9, 3, 2]. A conversational agent usually understands user's utterance and returns a reply using synthesized voice, text information, or animation of the agent's character. An agent is typically an animated character, which often looks like animals or celebrities caricatured like a character of comics; however, it is not common to use a photo-realistic human image. Therefore, in this study, aiming at the realization of a human-like conversational agent, we examine a technique to synthesize a photo-realistic facial animation image.

Sako et al. [6] applied a speech synthesis technique based on hidden Markov model (HMM) to image synthesis and proposed a technique for synthesizing a lip-animation image from a text using a pixel-based approach. It does not have to add information such as the position and form of the lips of training data and the acquisition of training data is easy by using eigen lips model using the

© Springer International Publishing AG 2017
J.-S. Pan et al. (eds.), *Advances in Intelligent Information Hiding and Multimedia Signal Processing*, Smart Innovation, Systems and Technologies 64,
DOI 10.1007/978-3-319-50212-0_4

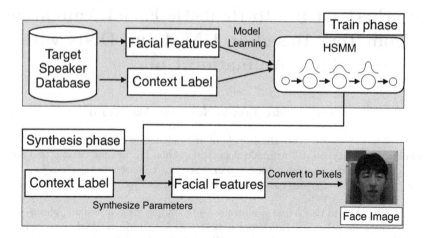

Fig. 1: The block diagram of the proposed facial animation synthesis.

principal component analysis (PCA) for a brightness of pixels. This approach has a problem that the synthesized image series become blurred because of data smoothing at the time of learning. An advantage is that the synthesize lips animation has smooth and natural motion by considering the dynamic features. Anderson et al. [1] developed a face model with good quality of face and tooth motion by expanding active appearance model (AAM), which is widely used as a two-dimensional face model. They proposed a technique for synthesizing a facial animation by using cluster adapting training (CAT) and HMM.

In this paper, we first overview the PCA-based facial animation synthesis based on HMM as a previous study. We then propose a novel synthesis technique using Animation Unit (AU) parameter which expresses the state of each part of the face and is acquired in Kinect v2 as facial features. We compare the performance of the conventional and proposed techniques.

2 Feature Representations for Photo-Realistic Face Image

Figure 1 shows the block diagram of the proposed facial animation synthesis system. This system synthesizes the facial animation from any texts. In the training phase, We model the facial features phoneme by phoneme using HMM. In the synthesis phase, we generate the features from trained HMM and any texts, and convert them into the image sequence. In this section, we show the details of the used facial features and this system.

2.1 PCA-Based Feature

As a method of expressing facial features using a low dimensional vector, eigenfaces approach is known in the field of face recognition [8]. This method regards

Table 1: Definition of Animation Unit parameter.

Number	Parameter	Number	Parameter
AU 0	Jaw Open	AU 9	Left cheek Puff
AU 1	Lip Pucker	AU 10	Right cheek Puff
AU 2	Jaw Slide Right	AU 11	Left eye Closed
AU 3	Lip Stretcher Right	AU 12	Right eye Closed
AU 4	Lip Stretcher Left	AU 13	Right eyebrow Lowerer
AU 5	Lip Corner Puller Left	AU 14	Left eyebrow Lowerer
AU 6	Lip Corner Puller Right	AU 15	Lower lip Depressor Left
AU 7	Lip Corner Depressor Left	AU 16	Lower lip Depressor Right
AU 8	Lip Corner Depressor Right		

the brightness values of a face image as a vector and represents the face image by a linear combination of the eigenface vectors obtained by PCA. Utilizing that the facial features are represented well by the limited number of eigenvalues in descending order, it is possible to reduce the number of dimensions by using the weight coefficients of eigenface vector with large eigenvalues. In the previous study [6], PCA is performed for the 176×144-pixel lips image, and the coefficients of the 32 eigenvectors are used as the features.

2.2 Animation Unit (AU) Parameter

AUs are 17 parameters that express the state of each part of the face that can be acquired by Kinect v2. The AU parameters are robust against the motion of the entire face because it focuses on each part of the face, and it can express the motion of the part of the face more clearly than PCA coefficients. Table 1 shows the states of AU parameters.

2.3 Modeling and Generation of Facial Features Using HMM

We use HMM for modeling and generation of facial features, which is the same as HMM-based speech synthesis. In the HMM-based speech synthesis, it is known that more smooth and natural speech can be synthesized by training context-dependent model [10]; thus we also train the face image HMM as a context-dependent one. In the training phase, we prepare the facial features and the corresponding context-dependent labels. In this study, the labels are the same as that for speech synthesis, which describes the information such as phoneme environment and prosody. We need to divide the training facial feature sequence phoneme by phoneme so that phoneme-based synthesis is available. The phoneme segmentation is performed using forced alignment with speech features, i.e., MFCC. The features are PCA coefficients for the conventional method and the AU parameters for the proposed method, respectively. The dynamic features are used to generate a smooth trajectory of facial features.

In the synthesis phase, the input text is converted into the context-dependent labels, and the facial feature sequence is generated from the context-dependent

HMM corresponding to the labels using a parameter generation algorithm. Here, there is a problem that the synthesized feature sequence from the HMM is over-smoothed. In the speech synthesis field, it is known that more natural speech features are synthesized by variance compensation considering the global variance [7]; so we also perform variance compensation in the conventional and proposed methods using the global affine transformation [4]. We synthesize the face image by converting the generated feature sequence into the image sequence. In the conventional method, the PCA coefficients are converted into the brightness values my multiplying the corresponding eigenvectors. In the proposed method, the generated AU parameters are converted into the image sequence using a neural network as described in Section 2.4.

2.4   Mapping from AU Parameters to Pixel Image Using DNN

In the proposed technique, an AU parameter sequence is generated using the HMM and converted into a pixel-based image using the deep neural network (DNN). In our previous study for face image conversion [5], we proposed two methods to convert the AU parameters into the brightness values of the pixels. The first one is the "two-step" method, which converts the AU parameters into the PCA coefficients of the brightness values using a Neural Network (NN), and then converts the PCA coefficients into the brightness value by multiplying the eigenvectors. The second one converts the AU parameter into the brightness value directly using a DNN. From the result of a subjective evaluation, the second method was found to give higher quality face images. Therefore, we convert the AU parameters into the brightness values using a DNN directly.

3   Experiments

3.1   Database

As a database of the color face images for the following experiments, we recorded movie samples of a male speaker using Kinect v2, who uttered 103 sentences of phonetically balanced sentences included in the ATR Japanese speech database. The format of the image was 400×400-pixel brightness value sequences, along with 17-dimensional AU parameters, their time stamps, and the audio data. We used the built-in microphone array of the Kinect for recording the voice. The speaker's head was fixed so that the position of the face did not move. Since the interval of the AU parameters is not stable, we interpolated the parameters and brightness value using the cubic spline interpolation so that the frame rate becomes 60 frame/s. Moreover, we resized the brightness value to 200×200 pixels after trimming the peripheral face by using template matching. We used 48, 25 and 30 sentences out of the database as the training data set, the development data set for determining optimum parameters of DNN, and the validation data set, respectively.

Table 2: Structure of DNN in experiments.

Input layer units	17	Activating function	tanh
Output layer units	120000	Batch size	100
Hidden layers	1,2,3	Epochs	100
Hidden layer units	512, 1024, 2048	Dropout	0.5
Optimizer	Adam		

3.2 Relation between the amount of training data and the objective quality

We first investigated the relation between the amount of data for training and the quality of the generated data for both the HMM and DNN in the proposed method. We calculated the root mean square error (RMSE) between the parameters of the generated and the original data while changing the amount of training data, and investigated how the error values changed. The AU parameters were used for evaluation of the HMM and the brightness values for the DNN. In the experiment of HMM, we changed the number of sentences for training. The training sentences were chosen from the training data set randomly. We repeated the experiment five times for each number of sentences, and the average of the RMSE values was used. The facial feature vectors consisted of 17-dimensional AU parameter, and their first and second temporal derivative dynamic features. The facial model was 3-states left-to-right HSMM. Each state had a single Gaussian distribution with a diagonal covariance matrix. The phoneme duration was separately modeled. In the experiment of the DNN, we changed the number of frames for training. The frames for training were chosen from the training data set randomly. We also repeated the experiment five times and calculated the average of RMSE. Table 2 shows the conditions of the DNN. We also chose the optimum number of hidden layers from 1, 2 or 3 and the optimum number of hidden units from the 512, 1024 or 2048. For other conditions, we used the same setting in our previous study [5].

Figure 2 shows the results. From these results, we can confirm that the error decreased with the increase of the amount of training data, and their values become minimum at 44 sentences for the HMM and 2048 frames for the DNN. The optimal number of hidden layers and hidden layer units was 3 and 512, respectively; hence we used these settings in the next experiment.

3.3 Performance comparison between the conventional and proposed techniques

We compared the synthesis performance of the facial animation images by the conventional and the proposed techniques through the objective quality measurement. The performance was measured according to the RMSE of the brightness values of the original image samples and synthesized samples. To investigate the influence of the variance compensation, we compared the following 4 conditions:

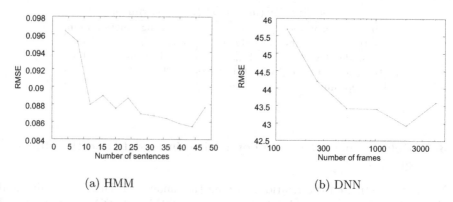

(a) HMM   (b) DNN

Fig. 2: Objective distortions between the original and generated data, i.e., the AU parameters for (a) HMM and the brightness values for (b) DNN, when changing the amount of the training data.

Table 3: Comparison of objective distortions (RMSE) between conventional and proposed techniques.

Technique	Without variance compensation	With variance compensation
Conventional	42.64	43.06
Proposed	41.69	41.80

conventional/proposed technique with/without the variance compensation. The facial feature vectors in the conventional technique consisted of 100-dimensional PCA coefficients, and its delta and delta-delta dynamic features. Other conditions were the same as the proposed technique. The cumulative contribution ratio of the 100-dimensional eigenvector was about 90%. The amount of training data was 48 sentences for HMM and 4096 frames for DNN, respectively.

Table 3 shows the result. From the result, we see that the RMSE values of the proposed techniques are smaller than those of the conventional techniques with or without the variance compensation. It is also observed that the RMSE increased when introducing the variance compensation, which might be caused by emphasizing the error such as the positional shift of the face.

Figure 3 shows examples of the captured and synthesized image by the conventional and proposed techniques with variance compensation. Fig. 3a shows the mouth-open frames. From these images, we can see that the proposed technique can reproduce the shape of the mouth better than the conventional technique. Figure 3b shows the mouth-closed frames. These examples show that the conventional technique and the proposed technique do not give much difference when synthesizing the mouth-closed face. So we consider that most of the error shown in Table 3 was derived from the mouth-open frames.

(a) Mouth-open frames

(b) Mouth-closed frames

Fig. 3: Comparison of the captured and synthesized frames. From left: captured, synthesized by the conventional technique, and the proposed technique.

4   Conclusions and future work

In this paper, we aim to realize a photo-realistic conversational agent. For this purpose, we proposed a technique for synthesizing pixel-based facial animation based on HMM and DNN with AU parameters. In the experiments, we investigated the relation between the quality of the generated parameters and the amount of the training data for both HMM and DNN. We also carried out a comparative experiment for the conventional and proposed techniques. The experimental results showed that the error of synthesized facial images by the proposed technique was smaller than that by the conventional technique. Additionally, we investigated the influence of variance compensation, and it became clear that the error increases by introducing the variance compensation. In the future, we will carry out a subjective evaluation experiment in order to examine the perceived quality of the synthesized facial animation. The other work to be carried out is to reproduce various facial expressions such as emotional expressions since the current work only focused on the reproduction of the whole face and the mouth movement.

Acknowledgment

Part of this work was supported by JSPS KAKENHI Grant Number JP15H02720.

References

1. Anderson, R., Stenger, B., Wan, V., Cipolla, R.: Expressive visual text-to-speech using active appearance models. In: Proc. Computer Vision and Pattern Recognitioan (CVPR). pp. 3382–3389 (2013)
2. Bickmore, T.W., Utami, D., Matsuyama, R., Paasche-Orlow, M.K.: Improving Access to Online Health Information With Conversational Agents: A Randomized Controlled Experiment. Journal of Medical Internet Research 18(1) (2016)
3. Horiuchi, H., Saiki, S., Matsumoto, S., Nakamura, M.: Virtual Agent as a User Interface for Home Network System. International Journal of Software Innovation (IJSI) 3(2), 13–23 (2015)
4. Nose, T., Ito, A.: Analysis of spectral enhancement using global variance in HMM-based speech synthesis. In: Proc. INTERSPEECH. pp. 2917–2921 (2014)
5. Saito, Y., Nose, T., Shinozaki, T., Ito, A.: Facial image conversion based on transformation of Animation Units using DNN. IEICE technical report (in Japanese) 115(303), 23–28 (2015)
6. Sako, S., Tokuda, K., Masuko, T., Kobayashi, T., Kitamura, T.: HMM-based text-to-audio-visual speech synthesis. In: INTERSPEECH. pp. 25–28 (2000)
7. Toda, T., Tokuda, K.: A speech parameter generation algorithm considering global variance for HMM-based speech synthesis. IEICE Trans. Information and Systems 90(5), 816–824 (2007)
8. Turk, M.A., Pentland, A.P.: Face recognition using eigenfaces. In: Proc. Computer Vision and Pattern Recognition (CVPR). pp. 586–591 (1991)
9. Yonezawa, T., Nakatani, Y., Yoshida, N., Kawamura, A.: Interactive browsing agent for the novice user with selective information in dialog. In: Proc. Soft Computing and Intelligent Systems (SCIS). pp. 731–734 (2014)
10. Yoshimura, T., Tokuda, K., Masuko, T., Kobayashi, T., Kitamura, T.: Simultaneous modeling of spectrum, pitch and duration in HMM-based speech synthesis. In: Proc. Eurospeech. pp. 2347–2350 (1999)

Silhouette Imaging for Smart Fence Applications with ZigBee Sensors

Kaide Huang, Zhengshun Zhang, Jinwen Cai

Computer Network Information Center & Chinese Academy of Science,
Guangzhou Sub-Center, Guangzhou 511458, China
{huangkaide,zhangyihang,caijinwen}@cnicg.cn
http://www.cnicg.cn

Abstract. This paper proposes a novel silhouette imaging method for smart electronic fence applications by using a ZigBee array. Particulary, the received signal strength (RSS) measurements collected by the ZigBee array is utilized to reconstruct the real-time attenuation images of a vertical profile, and then synthesize the sequential silhouette images of the target. These silhouette images provide a variety of valuable features, such as height, shape, posture, and especially attenuation density for intrusion detection, people management or other monitoring tasks. A proof-of-concept system is built to validate the proposed method.

Keywords: Silhouette imaging, smart fence, ZigBee array, received signal strength, attenuation density.

1 Introduction

Smart electronic fences can play a key role in the field of intrusion detection, intelligent monitoring and management, identity recognition, and so on. A variety of technological means have been proposed for electronic fence monitoring applications [1][2][3]. Among these means, the camera-based means are widely studies [4]. Popularly, they first extract the target features, such as human face or target shape, from high-resolution images, and then make a decision by pattern recognition algorithms. Another common means are using active near-IR sensors [1]. The state of IR sensor is change when an object is present in the sensor's field of view (FOV). It can be directly used for triggering the alarm. However, such a way cannot provide the information to discriminate the types of objects. To overcome this limitation, the novel concept of IR profile imaging is introduced [5]. Specifically, the silhouette or profile images of humans, vehicles and animals are created by collecting the state information of IR sensors. In this paper, we explore the potential ability of using a ZigBee sensor array in electronic fence monitoring applications.

Our methodology is to utilize the received signal strength (RSS) measurements between many pairs of ZigBee nodes for producing the silhouette images of objects (e.g. humans). The ZigBee nodes are installed on the fence, and transmit the beacon sequentially. When an object passes through the fence, it may

J.-S. Pan et al. (eds.), *Advances in Intelligent Information Hiding and Multimedia Signal Processing*, Smart Innovation, Systems and Technologies 64,
DOI 10.1007/978-3-319-50212-0_5

shadow, reflect, or diffract the radio frequency (RF) waves, thus resulting in the change of RSS. By developing radio tomography algorithm, the RSS attenuation image is reconstructed at each time step. Then the 3D silhouette image can be obtained by synthesizing these sequential attenuation images.

Our research on silhouette imaging with ZigBee sensors is motivated by the technique of radio tomographic imaging (RTI) [6][7]. It produces the RSS attenuation images of the monitored area caused by the presence of persons, and then locate the persons from the images. This technique can be implemented with a low-cost ZigBee network. More importantly, it relies on RF waves, and thus can work in obstructed environments, which is preferable in practical applications. Here, we extend the ability of RTI in capturing the object position to object shape. Correspondingly, the 2D RTI is extended to 3D RTI.

In contrast to camera system that uses a dense focal-plane array, our silhouette imaging system only needs a relatively sparse and low-cost ZigBee array. Additionally, our system is able to see through obstructions and work in dark. As compared to IR imaging system [5], our system has the following characteristics. First, the FOV of ZigBee sensor with omnidirectional antenna is much wider than that of IR sensor. In other words, the one-to-one transmitting/receiving mode in IR array becomes many-to-many mode in ZigBee array. Many-to-many mode can provide richer sensing information than one-to-one mode with a fixed number of sensors. Second, the change of RSS measurement can be regarded as a measure of how much an object blocks the line between transmitter and receiver. It is more useful than the two-value IR outputs (i.e. ON and OFF) that only indicate whether the obstruction is happen or not. Thus, our RSS-based imaging method has the potential to detect multiple targets/obects that pass through the fence simultaneously, which cannot be achieved by IR-based method. Third, our method can offer 3D images for more reliable monitoring classification or decision, rather than 2D images offered by IR method.

The paper is organized as follows. In Section 2, we briefly describe the principle and deployment of smart fence with ZigBee sensors. In Section 3, we present the silhouette imaging method of our fence. The experimental studies are conducted to demonstrate the feasibility of the proposed system and method in Section 4.

2   Smart Fence via ZigBee Sensors

Consider the influence on RSS of a radio link when a target moves into its propagation environment. The target may shadow, reflect or diffract the RF signal. Consequently, additional shadow fading occurs if the target is present on the line-of-sight (LOS) path, while multipath fading interferences is induced if the target is present on the non-LOS path. These fading effects make the change in RSS measurement. An illustration is offered in Fig. 1. A ZigBee transmitter sends a beacon to a receiver every 200ms, and the receiver measure the RSS. A person walks through the LOS path vertically as shown in Fig. 1. It can be seen that the RSS values fluctuate in a relatively small range when the person is located on the non-LOS path. It is caused by the multipath fading interferences. When the person stands on the LOS path, the RSS attenuates significantly due to additional shadow fading. It means that the RSS value is sensitive to the presence of a target, especially for detecting the target passing through the LOS path. That is the basis of our silhouette imaging fence.

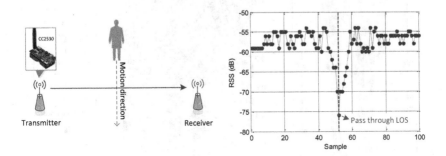

Fig. 1. The RSS measurements in the presence of a target.

The architecture of our fence is shown in Fig. 2. ZigBee nodes with omnidi-rectional antennas are deployed in a vertical plane. Pairs of nodes can build radio links for collecting RSS data. Based on the RSS sequences, a series of attenuation images are reconstructed. Then, by synthesizing these attenuation images, we can obtain 3D silhouette images, which are used for intrusion decision.

The ZigBee nodes use 2.4GHz IEEE 802.15.4 standard for communication. In addition, a base station is put besides the ZigBee array to transmit the RSS data to the data processing center. In order to achieve real-time monitoring tasks, a simple yet effective token ring protocol is adopted for RSS collection. The main rules are given as follows: 1) at each time step, only one ZigBee node (i.e. the node that gets the token) transmits the beacon; 2) the transmitter broadcasts a beacon that includes its ID number, and the RSS data array prepared before; 3) the other nodes measure the RSS value and put it in the RSS data array, and also check the token status; 4) the base station listen and then transmit the beacon to the data processing center. In practice, the time period for token transmission between two nodes is about 10ms.

3   Silhouette Imaging Method

3.1   Sensing Model

Consider a ZigBee array with K nodes and $M = K(K-1)/2$ radio links is utilized. For a link $i \in \{1, 2, \cdots, M\}$, the change in RSS at time t can be denoted as

$$y_i(t) = r_i(t) - \bar{r}_i$$
$$= \Delta S_i(t) + n_i(t) \tag{1}$$

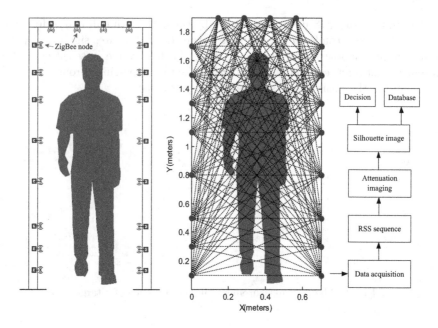

Fig. 2. The architecture of our electronic fence using ZigBee sensors.

where $r_i(t)$ is the RSS measurement at time t, $\bar{r}_i$ is the RSS reference value when the target is not present, $\Delta S_i(t)$ represents the additional shadow fading, $n_i(t)$ represents the noise component, which consists of measurement noise and multipath fading interferences due to the presence of target. Indeed, the target-induced shadowing information in RSS change is the key to monitor the target.

As shown in Fig. 3, the sensing area surrounded by the ZigBee nodes can be divided into N pixels. Let $x_j(t)$ represents the attenuation value in pixel j at time t. Then $\Delta S_i(t)$ can be regarded as a sum of $x_j(t)$ with corresponding weight w_{ij}, where $j \in \{1, 2, \cdots, N\}$. That is

$$\Delta S_i(t) = \sum_{j=1}^{N} w_{ij} x_j(t) \tag{2}$$

According to the RF propagation principle, the weight is given by

$$w_{ij} = \frac{1}{\sqrt{d_i}} \begin{cases} 1, & \text{if } d_{ij}(1) + d_{ij}(2) < d_i + \lambda \\ 0, & \text{otherwise} \end{cases} \tag{3}$$

where d_i is the length of link i, $d_{ij}(1)$ and $d_{ij}(2)$ are the distances between the centre of pixel j and the transmitting/receiving locations of link i, λ is a tunable parameter describing the related area. A non-zero weight indicates that the target in associated pixel can induce the change in shadow fading and thus RSS attenuation. As can be seen, only the pixels near the LOS path may contribute to RSS attenuation. Up to now, the RSS-based sensing model is built.

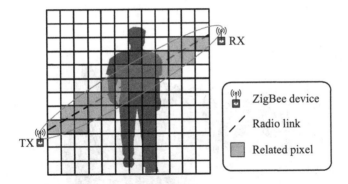

Fig. 3. An illustration of RSS-based sensing model.

3.2 Imaging Algorithm

In general, the pixel number N is greater than the link number M, meaning that the image reconstruction problem is an under-determined inverse problem. In this work, we adopt a back-projection algorithm for solving such a problem. Mathematically, the back-projection result of x_j at time t is denoted as follows:

$$\hat{x}_j'(t) = \sum_{i=1}^{M} w_{ij} y_i(t) \qquad (4)$$

Further, we find that the number of the radio links that travel through different pixels are varying. In other words, the link distribution in the sensing area is non-uniform. Fig. 4 shows an example of the link number distribution. As can be seen, the value varies in a wide range (from 3 to 80). The radio links are fairly sparse in the bottom area due to lack of ZigBee sensors. To alleviate the effect of non-uniform link distribution, it is necessary to normalize the back-projection results. Thus we have

$$\hat{x}_j(t) = \begin{cases} \hat{x}_j'(t)/\alpha, & \text{if } a_j < \alpha \\ \hat{x}_j'(t)/a_j, & \text{otherwise} \end{cases} \qquad (5)$$

where $\hat{x}_j(t)$ is the normalized result, a_j is the link number of pixel j, α is a positive parameter. At this point, the attention image at time t is obtained.

Let $\hat{\mathbf{X}}(t)$ be the matrix form of $\{\hat{x}_j(t)\}_{j=1}^{N}$. Then the synthetic 3D image at time t can be denoted as

$$\mathbf{I}(t) = \{\hat{\mathbf{X}}(t-T), \cdots, \hat{\mathbf{X}}(t)\} \qquad (6)$$

where T is the size of time window. Thus the silhouette images of the target in front view, side view, and top view are produced, which can be used for the purpose of intrusion detection, people management, and so on.

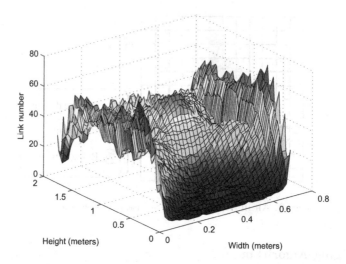

Fig. 4. An example of the link number distribution.

4 Results and Discussions

4.1 Results

To evaluate the proposed method, we build a real-life imaging system. There are 32 ZigBee nodes deployed on a fence, as shown in Fig. 5. A person (i.e., intruder) walks through the fence. We use the method above for imaging. An attenuation image is reconstructed every 100ms.

Fig. 6 shows a series of reconstructed images. We can see the whole processes that the person passes through the fence: left leg → main body → right leg. Besides the shape and behavior information, the images also provide the gray (i.e., attenuation density) information that indicates the obstruction degree by the target. It is advantageous in detecting multiple targets simultaneously.

The different views of the synthetic 3D images are shown in Fig. 7. The crude silhouette of intruder can be seen in front view, while the walking posture is presented in side view. In addition, the walking process of intruder is offered in top view. These features demonstrate the effectiveness of our prototype system in monitoring applications.

4.2 Discussions

The RSS-based silhouette images can provide various features of the target, such as height, shape, posture and attenuation intensity information. We can utilize rich features for intrusion detection, people counting, or other monitoring applications. Note that the attenuation density information is greatly valuable, which offer the potential to discriminate one target and multiple targets. Moreover, it may be helpful in identity recognition applications. The application demos will be developed in the further.

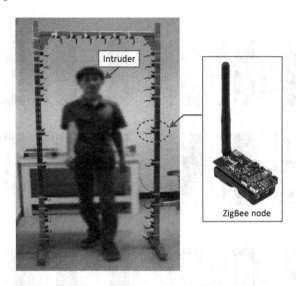

Fig. 5. A prototype silhouette imaging system using ZigBee sensors.

5   Conclusion

In this paper, a novel silhouette imaging method using ZigBee sensors is presented for electronic fence monitoring applications. The advantages of our method are
twofold. On one hand, the ZigBee nodes are cheap enough to deploy ubiquitous-
ly, and the RSS data is light-weight, which is convenient to signal transmission
and processing. On the other hand, the RSS-based silhouette images can provide
rich features of the target including height, shape, and especially the attenuation
density. Real-life experiment results demonstrate the feasibility of our method.

References

1. Sartain, R.: Profiling Sensor for ISR Applications. In Proc SPIE, pp. 69630Q–
   69630Q-11 (2008)
2. Mahapatra, E., Sathishkumar, et al.: An Optoelectronic Profiling and Ranging Sensor for Monitoring of Perimeters. IEEE Sensors Journal 15(7), 3692–3698 (2015)
3. Zha, C., Li, Y., Gui, J., et al.: Compressive Imaging of Moving Object Based on
   Linear Array Sensor. Journal of Electrical and Computer Engineering 7 (2016)
4. Foresti, G., Micheloni, C., Piciarelli, C., et al.: Visual Sensor Technology for Advanced Surveillance Systems: Historical View, Technological Aspects and Research
   Activities in Italy. Sensors 9(4), 2252C-2270 (2009)
5. Russomanno, D., Chari, S., et al.: Near-IR Sparse Detector Sensor for Intelligent
   Electronic Fence Applications. IEEE Sensors Journal 10(6), 1106–1107 (2010)
6. Wilson, J., Patwari, N.: Radio Tomographic Imaging with Wireless Networks. IEEE
   Transactions on Mobile Computing 9(5), 621-632 (2010)
7. Wang, Q., et al. Localizing Multiple Objects using Radio Tomographic Imaging
   Technology. IEEE Transactions on Vehicular Technology 65(5), 3641–3656 (2016)

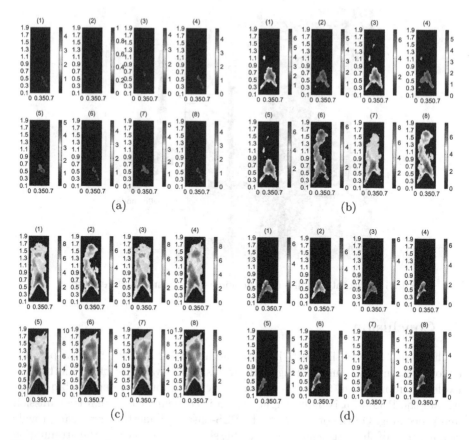

Fig. 6. A series of reconstructed attenuation images: (a) Group 1; (b) Group 2; (c) Group 3; (d) Group 4.

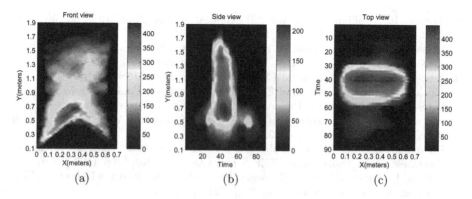

Fig. 7. Silhouette images in different views: (a) Front view; (b) Side view; (c) Top view.

Gender Recognition Using Local Block Difference Pattern

Chih-Chin Lai[1], Chih-Hung Wu[1], Shing-Tai Pan[2], Shie-Jue Lee[3], and Bor-Haur Lin[1]

[1] Department of Electrical Engineering
National University of Kaohsiung, Kaohsiung, Taiwan 81148
[2] Department of Computer Science and Information Engineering
National University of Kaohsiung, Kaohsiung, Taiwan 81148
[3] Department of Electrical Engineering
National Sun Yat-Sen University, Kaohsiung, Taiwan 80424
cclai@nuk.edu.tw

Abstract. Determining the gender of a person in a given image or video by a machine is a challenging problem. It has been attracting research attention due to its many potential real-life applications. As the human face provides important visual information for gender perception, a large number of studies have investigated gender recognition from face perception. In this paper, we present a method which uses Local Block Difference Pattern for feature extraction to identify the gender from the face images. The recognition is performed by using a support vector machine, which had been shown to be superior to traditional pattern classifiers in the gender recognition problem. Experimental results on the FERET database are provided to demonstrate the proposed approach is an effective method, compared to other similar methods.

Keywords: Gender Recognition, Local Block Difference Pattern, Support Vector Machine.

1 Introduction

Automatic gender recognition from face images has been receiving much research interest for many years due to its various practical and successful applications, such as biometric identification, security control, intelligent surveillance, human-machine interaction, and other face analysis tasks.

The framework of a gender recognition is similar to one of a facial expression recognition system. It consists of three modules: face acquisition, facial feature extraction, and classification, respectively. The feature extraction is crucial to the whole recognition process. If inadequate features are used, even the best classifier could fail to achieve accurate recognition.

A wide variety of gender recognition has been proposed. Jabid *et al.* [1] presented a new local texture descriptor based on local directional pattern (LDP) code for gender classification. The LDP codes are computed from the edge response values and hence provide robust feature to represent facial appearance.

© Springer International Publishing AG 2017
J.-S. Pan et al. (eds.), *Advances in Intelligent Information Hiding and Multimedia Signal Processing*, Smart Innovation, Systems and Technologies 64,
DOI 10.1007/978-3-319-50212-0_6

Ylioinas *et al.* [2] proposed a novel facial representation combining local binary pattern (LBP) and contrast information. The extensive experiments on the gender classification problem showed significant performance enhancement compared to popular methods such as basic LBP method or using Haar-like features with AdaBoost learning. Perez *et al.* [3] proposed a new method for gender classification from frontal face images using feature selection based on mutual information and fusion of features extracted from intensity, shape, texture, and from three different spatial scales. Li *et al.* [4] proposed a gender classification framework which utilizes not only facial information, but also hair and clothing. They extract features on facial components instead of on the whole face, which gives robustness against occlusions, illumination changes and noise. They also prove that clothing information has discriminative ability, and design feature representations for hair and clothing information that is discarded in most existing work due to high variability. Ardakany *et al.* [5] focused on improving the discriminative power of feature descriptors used in gender classification and proposed a new feature descriptor which is an extension to the local binary patterns (LBP). In order to represent the difference between different genders or ethnicities, Huang *et al.* [6] proposed a novel local descriptor called local circular patterns (LCP). Experimental results demonstrated that the LCP is more discriminative and more robust to noise than LBP-like features in the gender and ethnicity classification problems. In the current literature, most of the gender classification systems use the same face databases for obtaining the training and testing samples. Andreu *et al.* [7] presented a comprehensive experimental study on gender classification techniques using non-distorted and distorted faces. An extensive comparison of two representation approaches, three types of features and three classifiers has been provided by means of three statistical tests applied to two performance measures.

In this paper, we propose an appearance-based approach which considers a local texture descriptor called local block difference pattern (LBDP) [8] to extract more detailed facial texture information while enhancing the discriminative capability. The recognition performance is evaluated using the FERET database with a support vector machine (SVM) classifier. Experimental results are provided to illustrate that our approach yields improved recognition rate against other methods.

2   Background Review

Extracting proper features to form a good representation of the object is the most critical step in the pattern recognition problems. Due to its discriminative power and computational simplicity, the LBP [9] texture operator has become a very popular technique in various visual pattern recognition applications. However, LBP has two shortcomings: 1) it is not very robust against local changes in the texture, and 2) it may not work properly for noisy images or on flat image areas of constant gray level [9]. In order to overcome the shortcomings of the existing LBP operator, Wang *et al.* proposed a novel texture descriptor called LBDP, which decreases the influence resulting from intensity change by expanding the encoding range [8].

Although the concept of LBDP is similar to that of the multi-block local binary pattern (MB-LBP) [10], the encoding mechanism of LBDP is totally different from that of MB-LBP. The LBDP generation procedure is briefly described as follows.

Step 1. Obtain the center block and its p-th neighboring block, where $p = 1, \ldots, 8$, with the block size being set as $n \times n$ in an image.

Step 2. Convert the n^2 elements in a block to form a vector with the raster scanning technique. The converted vector of a given center block is compared with the p-th converted vector corresponding to its p-th neighboring block with radius R. Compute the difference of each pair of the n^2 elements of the converted vectors corresponding to the two selected blocks, and then form a difference vector.

Step 3. Convert the difference vector to a binary vector by checking the sign of each element in the former. If the difference value of an element is larger than 0, it is set as 1; otherwise, it is set as 0.

Step 4. Transform the binary vector obtained in step 3 to a binary value by thresholding the number of 1's elements in the binary vector. If the number of 1's elements is larger than a pre-defined threshold, the binary vector is transformed to a value 1; otherwise, it is transformed to a value 0. The threshold can be set as at least larger than half of the number of elements.

Step 5. The encoded binary value formed by the center block and its p-th neighboring block is the representative value of the p^R-th neighboring pixel of the center pixel. The LBDP of the center pixel is obtained as $(1^R, 2^R, \ldots, 8^R)$ and then the LBDP of each pixel in an image is converted into a decimal value.

The LBDP texture descriptor is defined as

$$LBDP_{P,R}(x_c, y_c) = \sum_{i=1}^{P} S(B_P^R - B_C^R) \times 2^{i-1}, \qquad (1)$$

$$S(z) = \begin{cases} 1, \left\| \sum_{i=1}^{n^2}(B_{P_i}^R - B_{C_i}^R) > 0 \right\| \geq T \\ 0, \quad \text{otherwise} \end{cases}, \qquad (2)$$

where B_C^R is a given center block with radius R centering at the center pixel, B_P^R is the p-th neighboring block with radius R centering at p^R, $\| \cdot \|$ is used to compute the number of elements with value 1 in the difference vector, and T is a threshold value. More details can be found in [8]. Figure 1 is an example illustrating the generation of LBDP.

3   The Proposed Approach

A gender recognition system consists of two components: facial feature extraction and classifier design. The feature extraction phase represents a key component of any pattern recognition system. In this study, we use LBDP descriptor as facial feature. Figure 2 illustrates the proposed feature extraction framework.

1^4				2^4				3^4
	1^3			2^3			3^3	
		1^2		2^2		3^2		
			1^1	2^1	3^1			
8^4	8^3	8^2	8^1	C	4^1	4^2	4^3	4^4
			7^1	6^1	5^1			
		7^2		6^2		5^2		
	7^3			6^3			5^3	
7^4				6^4				5^4

Fig. 1. The center block and its neighboring blocks [8]

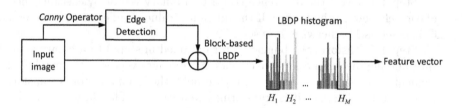

Fig. 2. The proposed facial features extraction framework

The edge is the most basic feature of image. If we can enhance the edge in the face image by sharpening it, and then we can bring out more of the detailed information in the face image. Since the Canny edge-detection operator [11] is perhaps the most popular edge-detection technique at present, we apply it in our approach. Additionally, in order to capture the local information of the micro-patterns in the image, we divide a face image into M non-overlapping regions $\{R_1, R_2, \ldots, R_M\}$ and then apply the LBDP operator on every pixel in every region. The histogram for every region is calculated, and then all H_j together are concatenated to obtain the final LBDP feature of the given image:

$$V = \{H_1, H_2, \ldots, H_M\}. \tag{3}$$

After feature extraction, the next task is to classify the different input patterns into distinct defined classes with a proper classifier. A SVM is a very popular technique for data classification in the machine learning community. The concepts of behind it are Statistical Learning Theory and Structural Minimization Principle [12]. SVM has been shown to be very effective because it has the ability to find the optimal separating hyperplane that gives the maximum margin between the positive and negative samples.

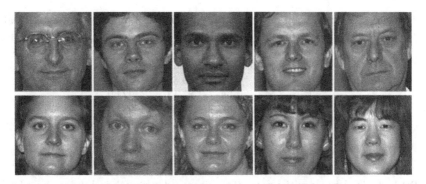

Fig. 3. Some sample images in the FERET database.

Given a training set of labeled samples $\{(x_i, y_i), i = 1, \ldots, h\}$, where $x_i \in \mathcal{R}^n$ and $y_i \in \{+1, -1\}$, a new test data x is classified by:

$$f(x) = \text{sign} \left(\sum_{i=1}^{h} \varphi_i y_i K(x_i, x) + c \right),\tag{4}$$

where φ_i are Lagrange multipliers of the dual optimization problem, c is a bias or threshold parameter, and $K(\cdot, \cdot)$ is a kernel function. In our work, we used the SVM implementation in the public available library LIBSVM [13] in all experiments.

4   Experimental Results

The proposed method is evaluated on the FERET (Face Recognition Technology) database [14]. The FERET face database is a widely used open dataset for evaluation of face recognition algorithms, and has also been used by many researchers for face gender recognition. There are 2,691 face images, from which 1,704 belong to male and 987 belong to female subjects from FERET database, selected to evaluate the effectiveness of the proposed approach in the experiments. Some examples from the FERET database are shown in Figure 3. In our experiments, classification results are estimated by a 10-fold cross validation scheme.

4.1   Influence of Block Number

In our approach, the block number is an important impact factor that influence the recognition quality. In this experiment, we want to show that the appropriate experimental setup is necessary to obtain better recognition rate. The results are presented in Table 1. It can be seen that the recognition rate is improved as the number of blocks increase. A basic principle states that the number of blocks should be large enough so that the texture features can represented reliably. On

Table 1. Gender recognition rates with different block number on FERET Database

| | Classification Accuracy (%) | | |
Block Number	Male	Female	Overall
2 × 2	91.23	89.80	90.71
3 × 3	95.77	87.13	92.60
4 × 4	97.71	91.39	95.39
5 × 5	97.07	89.46	94.82
6 × 6	97.42	91.89	95.39
8 × 8	98.30	93.21	96.43

Table 2. Gender recognition rates of the proposed approach and the compared methods

| | Classification Accuracy (%) | | |
Method	Male	Female	Overall
LBP + VAR [2]	95.38	87.69	91.54
EH [15]	97.61	91.61	95.67
AAM [16]	94.70	92.50	93.60
MSLBP + EDA [17]	94.64	96.12	95.19
Ours	98.30	93.21	96.43

the other hand, it should be small enough to accurately describe the local textures of a face image. From our observation, in order to satisfy both recognition accuracy and computation effort requirements, the appropriate number of blocks is 8 × 8 which can provide better recognition performance.

4.2 Comparison with Other Methods

In order to show the effectiveness of the proposed approach, we compare our approach with those in [2], [15], [16], and [17]. Note that the results of different methods may not be directly comparable because of differences in the number of selected images, experimental setups and pre-processing procedures, but they can still indicate the discriminative performance of every approach. The experimental results are shown in Table 2. It can be observed that our approach can obtain satisfactory performance in comparison with other existing gender classification methods.

5   Conclusions

Gender recognition using face images is widely used in real-world applications. Similar to other face analysis tasks, extracting the discriminative facial features and deriving an effective facial representation from original face images play important roles for successful gender recognition. In this paper, we use the local block difference pattern to extract facial features for gender recognition. Experimental results of the proposed approach have shown that the improvement in the recognition performance. Further work considering other local texture patterns, head pose variations, and occlusions into our approach is in progress.

Acknowledgments. This work was supported in part by the NSYSU-NUK Joint Research Project under grant 105-P004 and the Ministry of Science and Technology, Taiwan, under grant MOST 104-2221-E-390-018-MY2.

References

1. Jabid, T., Kabir, M.H., Chae, O.: Gender Classification Using Local Directional Pattern (LDP). In: 2010 International Conference on Pattern Recognition, pp. 2162–2165, IEEE Press, Istanbul, Turkey (2010)
2. Ylioinas, J., Hadid, A., Pietikäinen, M.: Combining Contrast Information And Local Binary Patterns for Gender Classification. In: Heyden, A., Kahl, F., (eds.) SCIA 2011. LNCS 6688, pp. 676–686, Springer-Verlag, Berlin Heidelberg (2011)
3. Perez, C., Tapia, J., Estevez, P., Held, C.: Gender Classification from Face Images Using Mutual Information And Feature Fusion. International Journal of Optomechatronics 6, 92–119 (2012)
4. Li, B., Lian, X.-C., Lu, B.-L.: Gender Classification by Combining Clothing, Hair And Facial Component Classifiers. Neurocomputing 76, 18–27 (2012)
5. Ardakany, A.R., Nicolescu, M., Nicolescu, M.: An Extended Local Binary Pattern for Gender Classification. In: 2013 IEEE International Symposium on Multimedia, pp. 315–320, IEEE Press, Anaheim, CA (2013)
6. Huang, D., Ding, H., Wang, C., Wang, Y., Zhang, G., Chen, L.: Local Circular Patterns for Multi-modal Facial Gender And Ethnicity Classification. Image and Vision Computing 32, 1181–1193 (2014)
7. Andreu, Y., Garcia-Sevilla, P., Mollineda, R.A.: Face Gender Classification: A Statistical Study Where Neutral And Distorted Faces Are Combined for Training And Testing Purposes. Image and Vision Computing 32, 27–36 (2014)
8. Wang, Y., Chen, Y., Huang, H., Fan, K.: Local Block-Difference Pattern for Use in Gait-based Gender Classification. Journal of Information Science and Engineering 31, 1993–2008 (2015)
9. Pietikäinen, M., Hadid, A., Zhao, G., Ahonen, T.: Computer Vision Using Local Binary Patterns. Springer-Verlag, London Limited (2011)
10. Zhang, L., Chu, R., Xiang, S., Li, S. Z.: Face Detection Based on Multi-block LBP Representation. In: Lee, S.-W., Li, S.Z., (eds.) ICB 2007. LNCS 4642, pp. 11–18, Springer-Verlag, Berlin Heidelberg (2007)
11. Canny, J.: A Computational Approach to Edge Detection. IEEE Trans. Pattern Analysis and Machine Intelligence 8, 679–698 (1986)
12. Vapnik, V.: The Nature of Statistical Learning Theory. Springer (1995)
13. Chang, C.-C., Lin, C.-J.: LIBSVM: A Libary for Support Vector Machines. Software available at http://www.csie.ntu.edu.tw/~cjlin/libsvm
14. The Facial Recognition Technology (FERET) Database, http://www.itl.nist.gov/iad/humanid/feret/feret_master.html
15. Ardakany, A.C., Joula, A.M.: Gender Recognition Based on Edge Histogram. International Journal of Computer Theory and Engineering 4, 127–130 (2012)

16. Bui, L., Tran, D., Huang, X., Chetty, G.: Face Gender Classification Based on Active Appearance Model And Fuzzy k-Nearest Neighbors. In: The 2012 International Conference on Image Processing, Computer Vision, and Pattern Recognition, pp. 617–621, CSREA Press, Las Vegas, USA (2012)

17. Azarmehr, R., Laganiere, R., Lee, W.-S., Xu, C., Laroche, D.: Real-time Embedded Age And Gender Classification in Unconstrained Video. In: 2015 IEEE Conference on Computer Vision and Pattern Recognition Workshops, pp. 56–64, IEEE Press, Boston, MA (2015)

DBN-based Classifcation of Spatial-spectral Hyperspectral Data

Lianlei Lin, Hongjian Dong, and Xinyi Song

Department of Automatic Test and Control, Harbin Institute of Technology,
Xidazhistr. 92, 150080 Harbin, China

Abstract. In this paper, present situation because of the high spectral image spectral information and spatial information leading to an increase in the increasing demand for new classification method of depth in recent years confidence in network feature extraction process large amounts of data, such as the Chinese information extraction, and other aspects of cancer determine the success of reality Combine. Introducing depth belief networks classify hyperspectral images, in excavating the hyperspectral data space information based on the study of the neighborhood mosaic spectral and spatial information used in combination, neighborhood stitching and spectral information integration and the weighted average of the empty Cape Joint strategy, through the experimental comparison with other methods to get the optimal weighted average method empty spectrum joint conclusions.

Keywords: depth belief networks, hyperspectral image classification, joint space spectrum

1 Introduction

With the continuous development of remote sensing (RS) technology, current hyperspectral images contain very rich spectral information, with each pixel having the information of hundreds of bands. However, people find that traditional methods for multi-spectral images are not suitable for hyperspectral images[1–3]. The main issue is that traditional methods cannot deal with such high data dimensionality of the spectrum. To solve this problem, in the past decade, new methods have been introduced in hyperspectral image classification, such as random forests[4], Bayesian models[5], feature extraction[6] and neural networks (NN)[7]. Among them, support vector machines (SVMs) achieve very good results in situations with small training samples[8].

On the other hand, Yi Chen[9] proposed a method using sparsity to classify hyperspectral images. The core of this algorithm is based on the fact that the pixels of a picture can be linearly and sparsely represented by a very small number of training samples in a dictionary. In recent years, multiple kernel learning (MKL) has made great progress. However, with increasing numbers of training samples and a variety of kernel functions, the requirement of the method on computer performance and computational time usually becomes unacceptable.

© Springer International Publishing AG 2017   53
J.-S. Pan et al. (eds.), *Advances in Intelligent Information Hiding
and Multimedia Signal Processing*, Smart Innovation, Systems and Technologies 64,
DOI 10.1007/978-3-319-50212-0_7

Therefore, Yanfeng Gu[10] proposed a new method called representative MKL (RMKL), which uses statistics theory to determine kernel functions, bypassing the time consuming and burdensome steps of finding the best combination of kernels.

The cost of using traditional methods to train classifiers is high, and Hughes phenomenon[11], or the so-called curse of dimensionality, usually occurs where classification accuracy deteriorates with the increase of feature dimensionality. However, hyperspectral images are nothing but high dimensional data. With the fast development of software / hardware in recent years, and the emerging of big-data era, deep learning will be the next key technology to advance the development of human beings. As a branch of deep learning, deep belief networks (DBNs) have been successfully applied in spam processing[12], singer identification[13], cancer discriminant[14] and Chinese information extraction[15]. Moreover, with the development of hyperspectral imaging technology, the spatial resolution of the acquired image data significantly increases, making it possible to use spatial information in classification. Deep learning has been gradually applied in hyperspectral image processing, such as Lins automatic encoder method[16] and Zhaos DBN method[17]. DBN is introduced in this study to solve hyperspectral image classification problem by integrating the spatial and spectrum information of hyperspectral images.

2 Spatial and Spectral Information Based Hyperspectral Image Classification

2.1 Structure of DBN

A DBN consists of several restricted Boltzmann machine (RBM)[18] stacks. It is assumed that no links exist between the visible layer nodes and the hidden layer nodes. The RBM of deep NN is pre–trained by using unsupervised layer-by-layer method, and the results are the initial values of the fine tuning model for supervised back propagation (BP) algorithm. The purpose of pre-training is to build the statistical model for the complex layer structure of RBM and massive data sets, so that the networks can have higher order abstract features and have relatively good initial weight values, and the weights are limited in a range beneficial to the global training. The learning of RBM plays an essential role in the learning of DBN. The structure of DBN is shown in Fig. 1.

2.2 Principle of the Classification Method

The principle of the hyperspectral image classification method based on DBN is shown in Fig. 2. At the training stage, the training data set is first preprocessed to extract the spatial and spectral information from hyperspectral images. Next, the parameters of DBN model are adjusted by learning. Two steps are included, namely, pre-training and BP reverse fine tuning training. In the classification stage, the learned network is used to classify the test sample set and output the

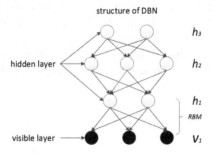

Fig. 1. The structure of DBN

classification results. When BP network is used for fine tuning and classification, softmax network is adopted for the top layer.

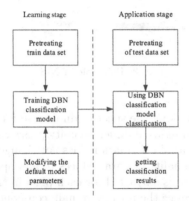

Fig. 2. Principle of hyperspectral image classification based on DBN

In both training and classification stages, preprocessing is needed for hyperspectral images. The purpose of preprocessing is to extract the spectral information and the spatial information of hyperspectral images, so that both of them can be integrated for the classification. In this study, we propose two preprocessing methods for hyperspectral images combining the spatial and spectral information. The details are as follows.

2.3 Neighborhood Information Stitching Method

The neighborhood information stitching (NIS) method mainly considers spatial information, and combines part of spectral information. Different from commonly used traditional methods which use 4–neighborhoods and 8–neighborhoods, NIS uses all pixels in the neighborhood of the pixel to be classified, and DBN autonomously learns the feature information. A 5×5 neighborhood is used in this study. However, there are hundreds of bands in hyperspectral images. If the input data are directly used, the dimensionality will be very high, thus significantly increasing the difficulty of training. Therefore, we introduce principal component analysis (PCA) to reduce the dimensionality. The principle of NIS is shown in Fig. 3.

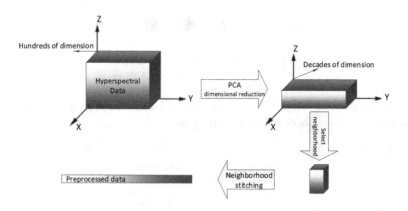

Fig. 3. Principle of NIS

First, PCA is applied along the spectral dimension for the whole hyperspectral image. The data with reduced dimensionality preserve spectral information, and the distribution of spatial information does not change. By discarding minor components, the dimensionality of data reduces to an appropriate level. Next, a small square neighborhood is extracted for each pixel in the data after PCA dimensionality reduction. Since the preserved main components of each pixel have lower dimensions (e.g., dozens of dimensions), such dimensionality is acceptable. Finally, the three-dimensional data cube extracted for each pixel is concatenated into a one-dimensional vector.

2.4   Spectral-Spatial Information Stitching Method

We further fuse the feature stitched by the spectral and spatial information of the pixel to be classified, resulting in the spectral-spatial information stitching (SSIS) method. The spectral information of a pixel is of great importance since it contains the information for recognizing different classes of geographical objects. As stated in the previous section, the stitched neighborhood information is obtained which is spatial information. Such spatial information is integrated with spectral information, forming the integrated spatial-spectral feature set. If they are directly combined, spatial information will be dominated by spectral information during normalization because the ranges of the spectral data and spatial information differ significantly. Therefore, in this method, normalization is performed separately for each of them before they are integrated. The principle of SSIS is shown in Fig. 4.

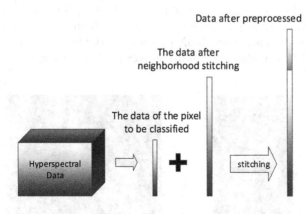

Fig. 4. Principle of SSIS

3 Experiments Results and Analysis

Two commonly used hyperspectral data sets, namely KSC and Pavia U, are adopted for the experiments of the proposed classification method. The separation method for the experimental samples is the same as that in [16], i.e., 6:2:2. The classification results are compared with SAE-LR and RBF SVM methods in [16]. The hardware environment is a computer with Intel Core i3-3110 M CPU @ 2.40GHz and 4.0G RAM, and the software environment is Windows 7 (64bit) operating system and MATLAB 2014a with Deep Learning Tool box-master tool kit.

3.1 Data Set Description

KSC data set was acquired in 1996 by National Aeronautics and Space Administration (NASA) using airborne visible / infrared imaging spectrometer (AVIRIS). The image size is 512×614 pixels. Each pixel contains 224 bands from 0.4 to 2.5 micrometers, with spatial resolution 18 m per pixel. After noise removal, 176 bands are left for analysis. There are 13 classes geographical objects.

Pavia University data set was acquired over Pavia University in 2001 by reflective optical system imaging spectrometer (ROSIS–3) sensor. The image size is 610×340 pixels. ROSIS sensor contains 115 bands within 0.43–0.86 micrometer, with spatial resolution 1.3 m per pixel. After 12 noise contaminated bands are removed, there are 103 bands left for analysis. There are 9 classes geographical objects.

3.2   Experimental and Results Analysis

The experimental results for the two data sets are shown in Tables 1 and 2. It is clear that the classification accuracy using DBN method with 5×5 sized neighborhoods (25 pixels) is comparable to that using SAE–LR method with 7×7 sized neighborhoods (49 pixels). In other words, in NIS mode and under the same classification accuracy, DBN uses less spatial information, while SAE–LR uses more. The classification results in NIS mode are shown in Fig. ??.

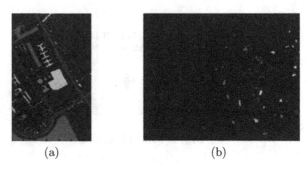

(a)   (b)

Fig. 5. Classification results in NIS mode

Table 1. Classification results for KSC data set

	NIS	SSIS	SAE–LR	RBF–SVM
OA	0.9773	0.9819	0.9782	0.9709
AA	0.9572	0.9686	0.9654	0.9571
scrub	1.0000	0.9877	0.9931	0.9862
Willow swamp	0.9783	1.0000	0.9804	0.9412
Cabbage palm hammock	0.9815	0.9683	0.9434	0.8868
Cabbage palm/oak hammock	0.8367	0.9184	0.9286	0.9762
Slash pine	0.8125	0.8276	0.8889	0.8056
Oak/broadleaf hammock	0.8809	0.9200	0.8889	0.9722
Hardwood swamp	1.0000	1.0000	1.0000	1.0000
graminoid marsh	0.9897	1.0000	0.9352	0.9908
Spartine marsh	1.0000	1.0000	1.0000	1.0000
Cattail marsh	0.9639	0.9792	1.0000	0.9730
Salt marsh	1.0000	1.0000	1.0000	0.9271
Mud flats	1.0000	0.9907	0.9919	0.9837
Water	1.0000	1.0000	1.0000	1.0000

By comparing the results in SSIS mode with those in NIS mode, it is clear that adding spectral information can improve the classification accuracy for DB-N, SAE–LR and RBF SVM methods. Among them, the overall and average accuracy of DBN method is slightly higher than that of SAE–LR method for Pavia

U data set, and lower for KSC data set. The classification results in SSIS mode
are shown in Fig. 6.

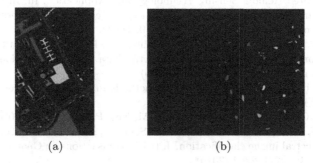

(a)   (b)

Fig. 6. Classification results in SSIS mode

Table 2. Classification results for Pavia U data set

	NIS	SSIS	SAE–LR	RBF–SVM
OA	0.9868	0.9894	0.9805	0.9787
AA	0.9862	0.9871	0.9728	0.9737
Asphalt	0.9842	0.9838	0.9194	0.9526
Meadows	0.9960	0.9981	1.0000	1.0000
Gravel	0.9620	0.9571	0.9883	0.9824
Trees	0.9800	0.9840	0.9952	0.9856
Painted metal sheets	1.0000	0.9963	0.9648	0.9569
Bare Soil	0.9815	0.9866	0.9948	0.9905
Bitumen	0.9963	0.9924	0.9830	0.9823
Self-Blocking Bricks	0.9762	0.9856	0.9535	0.9380
Shadows	1.0000	1.0000	0.9561	0.9747

4   Conclusion

In this study, DBN is introduced in hyperspectral image classification problem,
and two image preprocessing methods combining spectral and spatial information are proposed, i.e., NIS and SSIS methods. PCA is adopted to reduce the
dimensionality of the spectrums of original images. Experimental results for KSC
and Pavia U data sets demonstrate the feasibility of the DBN based joint spatial
and spectral classification method in hyperspectral image classification.

References

1. Pu, R.L., Gong, P.: Hyperspectral remote sensing and Application. Higher Education Press, Beijing (2000)

2. Tong, Q.X., Zhang, B., Zheng, F.L.: Multi disciplinary applications of hyperspectral remote sensing. Electronic Industry Press, Beijing (2006)
3. Wang, J.Y., Xue, Y.Q., Shu, R., Yang, Y.D., Liu, Y.N.: Airborne Hyperspectral and Infrared Remote Sensing Technology and Application. In: Infrared Millimeter Waves and 14th International Conference on Teraherz Electronics, Joint International Conference on IRMMW-THzpp. 9, 18–22. Sept (2006)
4. Ham, J., Chen, Y., Crawford, M.M., Ghosh, J.: Investigation of the random forest framework for classification of hyperspectral data. IEEE Transactions on Geoscience and Remote Sensing, 43, 492–501 (2005)
5. Landgrebe, David, A.: Signal theory methods in multispectral remote sensing[M]. John, Wiley., Sons (2005)
6. Burges, C.J.C.: Dimension Reduction. M. Now Publishers Inc (2010)
7. Ratle, F., Camps-Valls, G., Weston, J.: Semisupervised neural networks for efficient hyperspectral image classification. J. IEEE Transactions on Geoscience and Remote Sensing. 48, 2271–2282 (2010)
8. Fauvel, M., Chanussot, J., Benediktsson, J.A.: Evaluation of kernels for multiclass classification of hyperspectral remote sensing data. In: Acoustics, Speech and Signal Processing. pp. II–II, IEEE (2006)
9. Chen, Y., Nasrabadi, N.M., Tran, T.D.: Hyperspectral image classification using dictionary-based sparse representation . J. Geoscience and Remote Sensing. 49, 3973–3985 (2011)
10. Gu, Y., Wang, C., You, D., Zhang, Y., Wang, S., Zhang, Y.: Representative multiple kernel learning for classification in hyperspectral imagery. IEEE Transactions on Geoscience and Remote Sensing. 50, 2852–2865 (2012)
11. Zhang, X.G.: On the statistical learning theory and SVM. J. Automation Journal. 26, 32 (2000)
12. Sun, J.G.: Application of deep confidence network in spam filtering. J. Computer application. 34, 1122–1125 (2014)
13. He, Z.B.: Singer identification based on convolutional depth confidence network [D].Guangzhou: South China University of Technology (2015)
14. Furao, Shen.: Forecasting exchange rate using deep belief networks and conjugate gradient method, Neurocomputing Volume 167, 1, pp. 243–253. November (2015)
15. Chen, Y.: Chinese information extraction method based on depth confidence network[D]. Harbin: Harbin Institute of Technology (2014)
16. Lin, H.Z.: Research on feature extraction and classification of hyperspectral images based on automatic coding machine. D. Harbin: Harbin Institute of Technology (2014)
17. Zhao, X.: Classification method of hyperspectral data based on integrated depth confidence network [D]. Harbin: Harbin Institute of Technology (2015)
18. Hinton, G.E.: Training products of experts by minimizing contrastive divergence. Neural Computation. 14, 1771–1800 (2002)

Using CNN to Classify Hyperspectral Data Based on Spatial-spectral Information

Lianlei Lin and Xinyi Song

Department of Automatic Test and Control, Harbin Institute of Technology,
Xidazhistr. 92, 150080 Harbin, China

Abstract. Currently, the dimensionality of hyperspectral images is increasing, and the images have the characteristics of nonlinearity and spatial correlation, making it more and more difficult to classify these data. In this study, convolutional neural network (CNN) which has been successfully applied in image recognition and language detection is introduced. The spectral and spatial information is combined and used for hyperspectral image classification. According to the character of CNN that its input is two-dimensional image data, two methods are proposed converting the spectral and spatial information of hyperspectral images into two dimensional images. One of them converts the spatial-spectral information into gray level images and uses the varying texture features between spectral bands. The other converts the spatial-spectral information into waveforms and uses the wave characteristics of the spectral bands. Experiments on KSC and Pavia U data sets demonstrate the feasibility and efficacy of CNN in hyperspectral image classification.

Keywords: hyperspectral image classification, convolutional neural network (CNN), joint spatial and spectral feature

1 Introduction

Since the rapid development of hyperspectral remote sensing technology, the dimensionality of spectrum constantly increases, making the analysis and processing of hyperspectral data more and more difficult. Currently, the main problem of hyperspectral data analysis is classification. With the presence of the intrinsic characteristics of hyperspectral data, i.e., multi-dimensionality, correlation, nonlinearity[1] and large data amount, finding more appropriate algorithms to increase the data classification accuracy is an important task in hyperspectral image data analysis. At present, Hyperspectral image classification methods are mainly based on spectral matching[2] and statistical characteristics of data[3]. Since the hyperspectral remote sensing technology becomes increasingly mature, and the resolution is higher and higher, the spatial information can be utilized. If the spatial correlation is well adopted, the classification accuracy can be improved. Currently, hyperspectral image classification methods using joint spatial and spectral information mainly include image segmentation based methods[4–7], Markov random fields[8], and sparse representation[9–12], etc. Most methods

J.-S. Pan et al. (eds.), *Advances in Intelligent Information Hiding
and Multimedia Signal Processing*, Smart Innovation, Systems and Technologies 64,
DOI 10.1007/978-3-319-50212-0_8

integrate the idea of data dimensionality reduction. Although these methods can be used for hyperspectral images, they have drawbacks in real applications, such as high training cost, waste of high resolution information, discrepancy between high spectral image analysis / recognition accuracy and actual requirements, and inconsistency between mathematical models and the distribution laws of actual geographical objects. Therefore, it is necessary to find newer algorithms and more appropriate classification models.

In recent years, deep neural networks (DNNs)[13] have been widely applied in many fields, including biology, computer science, electronics, etc. DNN based classification has excellent performance, and is successfully applied in numerous areas[14–19]. Researchers have applied different network models of deep learning in Hyperspectral image classification tasks. In this study, the theory and model of CNN is introduced in the task of hyperspectral image classification. A CNN based hyperspectral image classification method combining spectral and spatial information is proposed, which interprets the rich information carried by hyperspectral data in an image based language.

2   CNN

CNN is inspired by biological processes[20]. It consists of neural networks that are closest to biological brains, and is the variation of multi-layer perceptron[21]. The structure of CNN includes the most essential feature extraction layer and feature mapping layer. Convolutional filtering is performed for the local receptive field of input images by filters using the idea of local receptive field. The local features of the receptive field are extracted, forming feature maps. In the feature mapping layer, the maps exist in a planar manner. The previous layer is sampled, and the idea of weight sharing is used. The weights of neurons in the same plane are the same, and then activation functions are used. In the whole process, the spatial relation between features is determined by feature extraction; the number of parameters of the whole network is reduced by weight sharing; the shift invariance of the feature map is maintained by choosing the activation function which has less influence on the kernel function; and the feature resolution is reduced by local averaging and extracting the computing layer again. The overall structure of CNN is shown in Fig. 1.

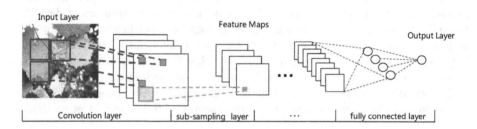

Fig. 1. The overall structure of CNN

The whole structure of lower level network is composed of alternating convolutional layers and subsampling layers. Higher level network is composed of fully connected layers, which is traditional multilayer perceptrons and logistic regression classifier. The input layer of the network is formed by the input two dimensional image data. The intermediate convolutional layers and subsampling layers are feature mapping layers. The input of the fully connected layers is the feature images extracted by convolutional layers and subsampling layers. Logistic regression, softmax regression, support vector machine (SVM) can be adopted for the final classifiers.

3   Hyperspectral Image Classification Based on CNN

3.1   Flowchart

The flowchart of hyperspectral image classification based on CNN is shown in Fig. 2. The whole process has two stages, one is learning stage,and the other is application stage. Data sets in both of two stages, train data set and test data set, need to be preprocessed. The spectral information of pixels is extracted, and processed to form the two dimensional input data suitable for CNN classification model. In learning stage, the training data set of the processed data is used to train CNN classification model, and adjust the model parameters, then get the trained classification model. In application stage, the trained CNN classification model is used to classify the test data set, generating the final classification results.

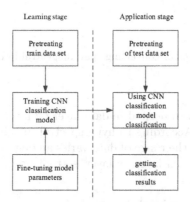

Fig. 2. Hyperspectral image classification process based on CNN

In this method, the key factor is data preprocessing, which extracts the spectral and spatial information from the original hyperspectral images and converts it into two dimensional data suitable for the input of CNN networks. Combining spatial and spectral information is to jointly use the spectral information in spectral dimension and the spatial contextual information of hyperspectral pixels. The spectral information of pixels and the spectral information of their spatial neighborhood are stitched, forming joint spatial and spectral information. In this study, the pixels in the 5×5 square area surrounding the pixel to be classified are stitched. In terms of data conversion, two methods are proposed in this study, namely, the method converting spatial spectral information into gray scale images, and the method converting spatial spectral information into waveform images, which are described in details as follows.

3.2 Method Converting Spatial Spectral Information into Gray Scale Images

This method converts the spectral vector data in a neighborhood into a gray scale image containing spectral information. The detailed data preprocessing steps are shown in Fig. 3.

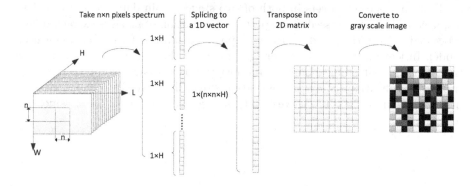

Fig. 3. The processing steps of converting spatial spectral information into gray scale images

Eq. (1) is firstly used to normalize data layer by layer for the convenience of subsequent processing. According to hyperspectral imaging theories, each layer represents a band. Since the range of data varies among different bands, normalization should be conducted layer by layer, so that no information of any band is weakened.

$$norm(x_{i,j}^k) = \frac{x_{i,j}^k - \min(x_{i,j}^k)}{\max(x_{i,j}^k) - \min(x_{i,j}^k)} (1 \leqslant i \leqslant W, 1 \leqslant j \leqslant L, 1 \leqslant k \leqslant H). \quad (1)$$

As shown in Fig. 4, taking the case with 4 bands as an example, the data values of the first band are relatively large on the whole, while the data values of the other three bands are relatively small. The spectral information of the two pixels pointed by the arrow is transposed, and the whole matrix is normalized to the range of [0, 1]. The pixel values of 6 and 16 are both close to 0 after being normalized by 65500. Therefore, the two pixels appear to be the same in the normalized gray scale image. On the contrary, the three dimensional spectral information is normalized layer by layer, and the gray difference between the pixels of converted images is large, with clear textures representing the data variation in all bands and facilitating the classification.The detailed processing steps are:

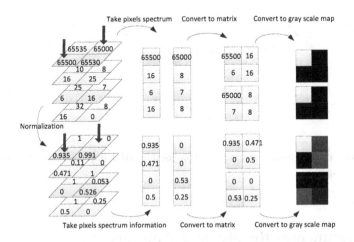

Fig. 4. Necessity analysis of data normalization within layers

Step 1. Extract the target pixel and all the pixels in the 5×5 square neighborhood, totaling 5×5 one–dimensional vectors of size $1 \times H$ each. These vectors are extended and stitched in the order of two dimensional space coordinate, forming a $1 \times (5 \times 5 \times H)$one-dimensional vector.

$$X_{i,j} = (x_{i,j}^{10}, \cdots, x_{i,j}^{1H}, x_{i,j}^{20}, \cdots, x_{i,j}^{2H}, \cdots, x_{i,j}^{n^2 0}, \cdots, x_{i,j}^{n^2 H})(1 \leqslant i \leqslant W, 1 \leqslant j \leqslant L). \quad (2)$$

Step 2. The joint spatial and spectral vector of each pixel is reshaped into a two–dimensional matrix, and this matrix is converted into a gray scale image. This process is traversed for all pixels in the whole hyperspectral image, generating a sample set with $W \times L$ gray scale images. Each gray scale image of the sample set represents a sample for classification which contains rich texture information describing the spatial and spectral data of the target pixel.

3.3 Method Converting Spatial Spectral Information into Waveform Images

The principle of the method converting spatial spectral information into waveform images is shown in Fig. 5. Similar to the other method, the spectral information of $n \times n$ pixels is stitched into a one–dimensional vector. Then, the vector is directly converted into a waveform, with the vertical coordinate representing normalized spectral values. Similar to the previous method, the hyperspectral data need to be normalized layer by layer.

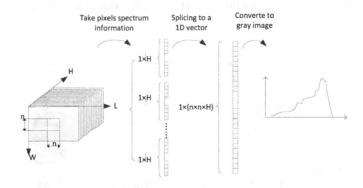

Fig. 5. Data preprocessing steps based on joint spatial and spectral classification of waveform images

4 Experimental Results and Analysis

Two commonly used hyperspectral data sets are used, i.e., KSC and Pavia U data sets, in the experiments for the proposed classification methods. The separation method of experimental samples is the same as that in [22], i.e., 6:2:2. The classification results are compared with SAE-LR and RBF SVM in [22]. The hardware environment of the experiment is a computer with Intel Core i7–6700k CPU @ 4GHz and NVIDIA GeForce GTX980Ti GPU @ 1000/1075 MHz. The software environment is Linux operating system and MATLAB 2014a with Caffe deep learning operating platform.

4.1 Data Set Description

KSC data set was acquired in 1996 by National Aeronautics and Space Administration (NASA) using airborne visible / infrared imaging spectrometer (AVIRIS). The image size is 512×614 pixels. Each pixel contains 224 bands from 0.4 to 2.5 micrometers, with spatial resolution 18 m per pixel. After noise removal, 176 bands are left for analysis. There are 13 classes geographical objects in total.

Pavia University data set was acquired over Pavia University in 2001 by reflective optical system imaging spectrometer (ROSIS-3) sensor. The image size is 610×340 pixels. ROSIS sensor contains 115 bands within 0.43–0.86 micrometer, with spatial resolution 1.3 m per pixel. After 12 noise contaminated bands are removed, there are 103 bands left for analysis. There are 9 classes geographical objects in total.

4.2 Classification Results and Analysis

Experiments on KSC and Pavia U data sets are conducted for gray scale based and waveform based methods. The results are compared with RBF-SVM and SAE–LR in [22] for verification. The results are shown in Table 5, which shows the overall accuracy (OA) and the average accuracy (AA) with 100 repeated experiments.

Table 1. Accuracy statistics of the classification method

Data set	Measurement	Greyscale image	Waveform image	RBF–SVM	SAE–LR
KSC	OA	0.9856	0.9864	0.9708	0.9776
	AA	0.9737	0.9805	0.9538	0.9625
PaviaU	OA	0.9968	0.9976	0.9725	0.9812
	AA	0.9968	0.9966	0.9666	0.9732

It can be concluded that the proposed two methods have absolute superiority in the classification ability after spatial information is introduced. Whether for KSC data set or Pavia U data set, the overall accuracy of the proposed methods is higher than the other methods, with an increase of more than 1%. Therefore, it is feasible to jointly use the spatial and spectral information of hyperspectral images, convert it into two dimensional image data, and use CNN to classify them.

5 Conclusion

In this study, CNN is introduced and the spectral and spatial information of hyperspectral images is jointly used to perform hyperspectral image classification. According to the characteristic that the input of CNN is two dimensional image data, two methods are proposed converting the spectral and spatial information of hyperspectral images into two dimensional images. Experiments on KSC and Pavia U data sets demonstrate the feasibility and efficacy of using CNN in hyperspectral image classification.

References

1. Bachmann, C.M., Ainsworth, T.L.,Fusina, R.A.: Improved manifold coordinate representations of large-scale hyperspectral scenes. J. Geo. and Remo. Sens. 44, 2786–2803 (2006).
2. Mura, M.D., Villa, A., Benediktsson, J.A., Chanussot, J.,Bruzzone, L.: Classification of hyperspectral images by using extended morphological attribute profiles and independent component analysis. J. Geos. and Remo. Sens. Lett. 8, 542–546 (2011)
3. Mianji, F.A., Zhang, Y.: Robust hyperspectral classification using relevance vector machine. J. Geos. and Remo. Sens. 49, 2100–2112 (2011)

4. Tarabalka, Y., Chanussot, J., Benediktsson, J.A.: Segmentation and classification of hyperspectral images using watershed transformation. J. Patt. Reco. 43, 2367–2379 (2010)
5. Tarabalka, Y., Benediktsson, J.A., Chanussot, J.: SpectralCspatial classification of hyperspectral imagery based on partitional clustering techniques. J. IEEE TGRS. 47, 2973–2987 (2009)
6. Valero, S., Salembier, P., Chanussot, J.: Hyperspectral image representation and processing with binary partition trees. J. IEEE TIP. 22, 1430–1443 (2013)
7. Tarabalka, Y., Benediktsson, J.A., Chanussot, J., Tilton, J.C.: Multiple spectralC-spatial classification approach for hyperspectral data. J. IEEE TGRS.48, 4122–4132 (2010)
8. Tarabalka, Y., Fauvel, M., Chanussot, J., Benediktsson, J.A.: SVM-and MRF-based method for accurate classification of hyperspectral images. J. IEEE GRSL. 7, 736–740 (2010)
9. Chen, Y., Nasrabadi, N.M., Tran, T.D.: Hyperspectral image classification using dictionary-based sparse representation. J. IEEE TGRS. 49, 3973–3985 (2011)
10. Chen, Y., Nasrabadi, N.M., Tran, T.D.: Hyperspectral image classification via kernel sparse representation. J. IEEE TGRS. 51, 217–231 (2013)
11. Fang, L., Li, S., Kang, X., Benediktsson, J.A.: SpectralCspatial hyperspectral image classification via multiscale adaptive sparse representation. J. IEEE TGRS. 52, 7738–7749 (2014)
12. Zhang, H., Li, J., Huang, Y., Zhang, L.: A nonlocal weighted joint sparse representation classification method for hyperspectral imagery. J. IEEE JSTAEORS. 7, 2056–2065 (2014)
13. Hinton, G.E., Salakhutdinov, R.R.: Reducing the dimensionality of data with neural networks. J. Science. 313, 504–507 (2006)
14. Dahl, G.E., Dong, Y., Li, D., Acero, A.: Large vocabulary continuous speech recognition with context-dependent DBN-HMMS. In: IEEE ICASSP, pp. 4688–4691. IEEE (2011)
15. Mohamed, A., Sainath, T.N., Dahl, G., Ramabhadran, B.: Deep belief networks using discriminative features for phone recognition. In: ICASSP, pp. 5060–5063. IEEE (2011)
16. Nair, V., Hinton, G.E.: 3D object recognition with deep belief nets. In: ANIPS, pp. 1339–1347 (2009)
17. Fasel, I., Berry, J., Deep belief networks for real-time extraction of tongue contours from ultrasound during speech. In: 20th ICPR, pp. 1493–1496. IEEE (2010)
18. Deselaers, T., Hasan, S., Bender, O., Ney, H.: A deep learning approach to machine transliteration. In: PFWSMT, pp. 233–241 (2009)
19. Li, D., Seltzer, M.L., Dong, Y., Acero, A., Mohamed, A.R., Hinton, G.E.: Binary coding of speech spectrograms using a deep auto-encoder In: Interspeech, pp. 1692–1695 (2010)
20. LeCun, Y., Boser, B., Denker, J.S., Henderson, D.: Backpropagation applied to handwritten zip code recognition. J. Neural Comp. 1, 541–551 (1989)
21. Rosenblatt, F.: The perceptrona probabilistic model for information storage and organization in the brain. J. Psyc. Rev. 65, 386 (1958)
22. Chen, Y., Lin, Z., Zhao, X., Wang, G., Gu, Y.: Deep learning-based classification of hyperspectral data. J. Sel. Top. in Appl. Earth Observ. and Remo. Sens. 7, 2094–2107(2014)

Multiple Kernel-Learning Based Hyperspectral Data Classification

Wei Gao and Yu Peng

Department of Automatic Test and Control, Harbin Institute of Technology,
No 92 XIDAZHI Street, 150001 Harbin, China
gaowei@hit.edu.cn

Abstract. Hyperspectral remote sensing based oil and gas exploration technology aims to extract the related information of oil and gas, to achieve the target characteristics of underground oil and gas exploration and recognition through remote sensing data processing and analysis. The rapid development of hyperspectral remote sensing technology bring the accurate detection of surface reflectance spectrum, to increase the success possibility and reduce the cost of oil and gas exploration. The increasing spectral and space resolution of hyperspectral remote sensing bring a large size of data for two problems in the practical satellite platform-based imagery processing system. The bandwidth of the communication channel limits the transmission of the full hyperspectral image data for the further processing and analysis on the ground for the oil and gas exploration. The preprocessing of hyperspectral sensing data is a feasible way through machine learning-based data analysis technology, to produce one image from the full band of hyperspectral images through classifying the spectrum curve of each pixel according to the spectrum data of oil and gas. In this paper, we present the satellite platform based kernel machine-based system for oil and gas exploration based on hyperspectral remote sensing data.

Keywords: Hyperspectral remote sensing; kernel learning; Oil and Gas Exploration

1 Introduction

The distribution of most oil and gas reservoirs is closely related to the regional geological structure information. Therefore, the extraction of structural information is of great significance to oil and gas exploration. And hyperspectral remote sensing data can be used to record the geological structure and its physical characteristics because of its high spectral information density, good continuity, and so on. In particular, it also contains special concealed information about geological structure. Therefore, hyperspectral remote sensing data is used to extract structural information, which has become an effective complement to the limitations of the traditional geological point line observation method characterized by economic and convenient. The analysis of the different distribution and combination in different geological characteristics can obtain favorable oil and gas migration and accumulation

© Springer International Publishing AG 2017   69
J.-S. Pan et al. (eds.), *Advances in Intelligent Information Hiding*
and Multimedia Signal Processing, Smart Innovation, Systems and Technologies 64,
DOI 10.1007/978-3-319-50212-0_9

conditions, to increase the success rate while reduce the cost of oil and gas exploration.

The oil and gas exploration with hyperspectral remote sensing technology mainly relies on reflection spectroscopy and oil gas micro leakage theory. Due to the internal pressure, the hydrocarbon material in the oil and gas reservoir can penetrate to the earth surface through the joint, fracture and other channels in the earth and then disperse in the near earth surface atmosphere. The dispersion of hydrocarbons to the soil, vegetation and air near the earth surface will cause spectral anomalies in the corresponding area. These spectral anomalies are usually called the hydrocarbon micro leakage halo. With remote sensing technology, the exploration of oil and gas can be achieved by detecting these halos. With the rapid development of hyperspectral remote sensing technology, it can ensure the accurate detection of surface reflectance spectrum, thus increasing the success possibility and reducing the cost of oil and gas exploration.

Hyperspectral imagery is the most popular remote sensing technology on satellite platform in recent days. It can be applied in military monitoring, energy exploration, geographic information, and so on. Hyperspectral instruments with hundreds of contiguous spectral channels can collect more information about the objects on the earth surface at the cost of a large volume of image data(known as data cube). This large volume data brings two challenges in practical applications. Firstly, the bandwidth of the satellite downlink system becomes the bottleneck to transmit the full hyperspectral image data for further analysis on the ground. Secondly, real-time processing for some applications becomes more challenging. Data compression can ease the bandwidth problem but has no benefit for real-time analysis. Onboard processing, i.e. classification, is a good option to solve the aforementioned two problems. Through hyperspectral image classification, each pixel's spectral curve is classified and labeled to a certain class based on the spectrum curves database which is collected in advance. Then the original hyperspectral image data cube can be reduced to a class label matrix, which is much smaller and can be used directly in many real-time applications.

However, there may be some inconsistency between the real collected spectral curves and those in the database. These differences are usually caused by spectral instruments errors, atmosphere variation and so on, and are usually nonlinear. So, traditional linear classification methods are not effective to hyperspectral sensing data. Machine learning methods, especially kernel learning, which are good at nonlinear classification problems, are feasible and effective methods on hyperspectral sensing data.

In recent years, many machine learning methods are proposed for data analysis and image processing. Linear Discriminant Analysis (LDA) and Principal Component Analysis (PCA) [1] are early proposed linear methods which cannot solve nonlinear problem. Kernel PCA (KPCA) and KDA [17] are developed accordingly. Many other linear learning methods are also kernelized for different applications [1][12][15][4][13]. Kernel-based manifold learning were developed [10], [7], and other improved methods [19], [6], [18], [8] [24]. To summarize, kernel methods is very popular in machine learning filed. Many researches want to solve this problem by selecting the optimal kernel parameter [14][5]. Methods focusing on the kernel functions were also proposed, i.e. data-depend kernel [16], quasiconformal kernel model based KDA [14], multiple kernel [9] and so on [21] [22] [23]. For hyperspectral image classification problem, there is still no such research.

2. Multiple kernels learning algorithm

Oil and gas exploration with hyperspectral remote sensing technology aims at extracting hydrocarbon micro leakage halo through remote sensing data processing and analysis. By finding the anomaly oil and gas exploration related information, to achieve the target characteristics of underground oil and gas exploration and recognition. The satellite based hyperspectral image system includes three main steps: image collection, image processing, and image transmission, as shown in Figure 1.

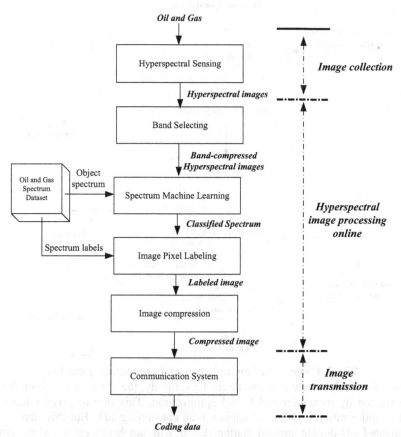

Fig. 1. Kernel Machine-Based Satellite Platform Based Hyperspectral Remote Sensing Data Processing System in Oil and Gas

In image collection step, the hyperspectral sensors get the image of the earth surface with hundreds of spectral band. Then in the image data processing step, we first select the spectral band to remove spectral feature redundancy and decrease the complexity of the computation. Then by spectrum machine learning, the image pixels are labeled according to the spectrum database collected in advance. The labeled results are compressed and send to ground in the final step.

In this work, we mainly focus on spectrum machine learning in the second step. It aims at realizing online image classification to reduce the data volume to be transferred to the ground station while increase the autonomy for many applications.

Multiple kernel, optimization via learning procedureand data-dependent kernelsare the primary approaches for optimization. Researchers attempt to solve the limitation of simple kernel methods with MKL (multiple kernel learning). It aims at learning a linear (or convex) combination of a set of predefined kernels in order to determine the best target kernel for the application. During the last decade, MKL has been widely investigated, and many algorithms and techniques were proposed in order to improve the efficiency of this new method.

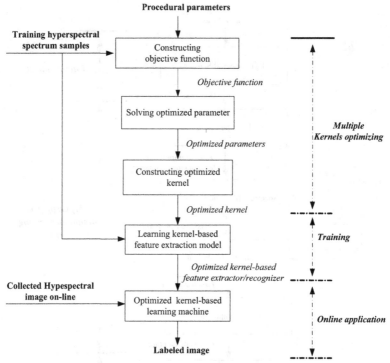

Fig. 2. Procedure of multiple kernels based learning machine

The framework mainly contains three steps. Firstly, the optimized multiple kernel is constructed by parameter and kernel optimization. This step involves solution of complex optimization problem which is a time consuming task. But this step can be implemented off-line in ground station. So it will not bring extra computational burden to the satellite. Then in the training steps, the optimal model can be got through iteration optimization which also requires lots of time and computation. Fortunately, this step can also be implemented off-line in ground station. At last, the built model is applied in online image data. And this step is less time consuming than former two steps. So it can be realized on satellites, especially considering the fact that the computation performance of satellites onboard computer is growing with the application of some new chips, i.e. FPGAs, DSPs, ARMs and so on.

The computation of the kernel matrix with the selected kernel function is the first key step in kernel learning for classification problem. In order to improve the performance of kernel learning, we use the quasiconformal kernel $k_q(\mathbf{x}, \mathbf{y})$, which is defined as follows.

$$k_q(\mathbf{x}, \mathbf{y}) = f(\mathbf{x})f(\mathbf{y})k(\mathbf{x}, \mathbf{y}) \tag{1}$$

where $\mathbf{x}$ and $\mathbf{y}$ are the sample vectors, $k(x,y)$ is basic kernel functions, i.e. polynomial kernel or Gaussian kernel, $f(\mathbf{x})$ is a positive real value function of $\mathbf{x}$ defined as

$$f(\mathbf{x}) = b_0 + \sum_{n=1}^{N_{XV}} b_n e(\mathbf{x}, \mathbf{x_n}) \tag{2}$$

where $e(\mathbf{x}, \mathbf{x_n}) = e^{-\delta\|\mathbf{x}-\mathbf{x_n}\|^2}$, and δ is a free parameter, and x_n are called the "expansion vectors (XVs)", and N_{XV} is the total number of XVs, and b_n ($n = 0,1,2,...,N_{XV_s}$) are the "expansion coefficients" associated with $\mathbf{x_n}$ ($n = 0,1,2,...,N_{XV_s}$). The expension coefficients b_n need to be computed by constraint equation.

Compared to single quasiconformal kernel, quasiconformal multi-kernels machine learning can get a better data distribution in the mapped feature space. Different from single quasiconformal kernel, the quasiconformal multi-kernels model has to get the weight parameter and expansion coefficient simultaniously. The quasiconformal multiple kernel is defined as

$$k(\mathbf{x}, \mathbf{x'}) = q(\mathbf{x}) \sum_{i=1}^{m} a_i k_{0,i}(\mathbf{x}, \mathbf{x'}) q(\mathbf{x'}) \tag{3}$$

where $\mathbf{x}, \mathbf{x'} \in P^p$, $k_{0,i}(\mathbf{x}, \mathbf{x'})$ is the i th basic kernel, such as polynomial kernel and Gaussian kernel, and m is the number of basic kernels, $a_i \geq 0$ is the weight for the i th basic kernel function, $q(\cdot)$ is the factor function defined as

$$q(\mathbf{x}) = b_0 + \sum_{i=1}^{n} b_i k_0(\mathbf{x}, \mathbf{a}_i) \tag{4}$$

where $k_0(\mathbf{x}, \mathbf{a}_i) = e^{-\gamma\|\mathbf{x}-\mathbf{a}_i\|^2}$, $\mathbf{a}_i \in P^d$, b_i is the coefficient for the combination, $\mathbf{a}_i\{i = 1,2,...,n\}$ are selected by the training samples. $k(\mathbf{x}, \mathbf{x'})$ satisfies the Mercer condition, and $k(\mathbf{x}, \mathbf{x'})$ is rewritten to $k(\mathbf{x}, \mathbf{x'}) = \sum_{i=1}^{m} d_i \left[f(\mathbf{x})k_{0,i}(\mathbf{x}, \mathbf{x'})f(\mathbf{x'}) \right]$ of quasiconformal transformation $k_{0,i}(\mathbf{x}, \mathbf{x'})$, and $q(\mathbf{x})k_{0,i}(\mathbf{x}, \mathbf{x'})q(\mathbf{x'})$ is the linear combination of kernels.

On mutiple kernel learning, the crucial step is to choose the feasible weights of multiple kernel through the comprehensive consideration on the computation efficiency and classification accuracy on hyperspectral image classification. So, some key basic kernels are chosen for the multiple quasiconformal kernel learning.

3 Experimental Results

In this section, we evaluate the proposed mothod with real hyperspectral image data. The famous Indian Pines dataset from AVRIS, as shown in Fig.3. To evaluetate the effctiveness of quasiconformal multi-kernels, we apply it in two conventiaonl classifiers which are Support Vector Classifier (SVC) and Kernel Sparse Representation Classifier (KSRC). Then we compare the classification accuracy of these two classifiers with different kernel function which are polynomial kernel, gaussian kernel, and our proposed quasiconformal multi-kernels. And according to the kernel that the classifier used, we name the six classifier as follows. For the quantitative comparison, we use the average accuracy to evaluate the performance of six classifiers. The results are shown in Table 1 and Table 2.

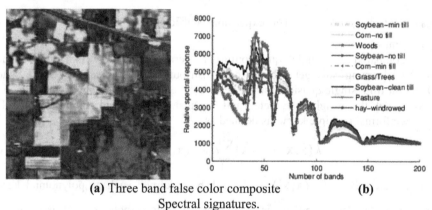

(a) Three band false color composite   **(b)**
Spectral signatures.

Fig. 3. One example of Indian Pines data

From Table 1 and Table 2, the Gaussian kernel based classifiers get better accuracy than those with polynomial kernel in most of the classes,. This results have been proved by many former researches. The classifier with quasiconformal multi-kernels get better accuray in all 12 classes than those with single kernels. Althouth the multi-kernels are the combination of basic kernels, they show better performance than single kernels. These results also demonstrate that the quasiconformal multi-kernels based classifier is feasible for hyperspectral image classification.

Table 1. Average classification accuracy of SVC (%)

Class	1	2	3	4	5	6	7	8	9	10	11	12
PK-SVC	48.3	57.7	95.4	38.2	64.8	92.6	61.9	84.3	99.5	66.8	74.3	59.4
GK-SVC	77.0	74.6	98.1	77.9	81.5	96.1	78.7	88.8	97.7	84.6	85.0	81.7
QMK-SVC	79.3	81.4	98.9	82.5	92.2	98.2	83.7	99.5	100	87.8	91.2	85.4

Table 2. Average classification accuracy of KSRC (%)

Class	1	2	3	4	5	6	7	8	9	10	11	12
PK-KSRC	52.8	58.6	97.1	48.1	79.5	94.8	64.8	85.7	100	68.5	76.2	62.7
GK-KSRC	78.8	77.4	98.1	76.5	78.0	98.4	83.7	87.7	99	84.9	87.3	83.1
QMK-KSRC	79.4	84.5	98.8	84.4	93.6	98.4	83.8	97.3	98	88.8	92.0	87.6

Notes: PK-SVC: Polynomial Kernel-SVC, GK-SVC: Gaussian Kernel-SVC, QMK-SVC: Quasiconformal Multi-kernels Based SVC, PK- KSRC: Polynomial Kernel- KSRC, GK-KSRC:Gaussian Kernel- KSRC, QMK- KSRC: Quasiconformal Multi-kernels Based KSRC.

4. Conclusions

This paper presents a hyperspectral imagery processing system on satellite for oil and gas exploration. It can ensure the accurate detection of surface reflectance spectrum, to increase the success possibility and reduce the cost of oil and gas exploration. Hyperspectral instruments with hundreds of contiguous spectral channels bring a large size of remote imagery data with the increasing spectral and space resolution. The system is to improve the performance of kernel-based system largely influenced by the function and parameter of kernel on hyperspectral sensing data. The learning framework is feasible to the hyperspectral image classification, and moreover the framework can be used to other kernel-based systems in the practical applications. The computing efficiency of training the proposed framework is the problem in the practical system, because most applications are off-line training based model learning. With the rapid development of hyperspectral remote sensing technology, it can ensure the accurate detection of surface reflectance spectrum, to increase the success possibility and reduce the cost of oil and gas exploration.

References

1. S. Amari and S. Wu, "Improving support vector machine classifiers by modifying kernel functions," Neural Networks, vol. 12, no. 6, pp. 783-789, 1999.
2. G. Baudat and F. Anouar, "Generalized Discriminant Analysis Using a Kernel Approach," Neural Computation, vol. 12, no. 10, pp. 2385-2404, 2000.
3. P. N. Belhumeur, J. P. Hespanha, and D. J. Kriegman, "Eigenfaces vs. Fisherfaces: Recognition using class specific linear projection," IEEE Trans. Pattern Analysis and Machine Intelligence, Vol. 19, No. 7, pp. 711-720, 1997.
4. Wen-Sheng Chen, Pong C. Yuen, Jian Huang, and Dao-Qing Dai, "Kernel Machine-Based One-Parameter Regularized Fisher Discriminant Method for Face Recognition", IEEE Trans. Systems, Man and Cybernetics-Part B: Cybernetics, vol. 35, no. 4, pp. 658-669, August 2005.
5. Wen-Sheng Chen, Pong C. Yuen, Jian Huang, and Dao-Qing Dai, "Kernel Machine-Based One-Parameter Regularized Fisher Discriminant Method for Face Recognition", IEEE Trans. Systems, Man and Cybernetics-Part B: Cybernetics, vol. 35, no. 4, pp. 658-669, August 2005.

6. J. Cheng, Q. Liu, H. Lua, and Y. W. Chen, "Supervised kernel locality preserving projections for face recognition," Neurocomputing, vol. 67, pp. 443-449, 2005.
7. G. Feng, D. Hu, D. Zhang, and Z. Zhou, "An alternative formulation of kernel LPP with application to image recognition," Neurocomputing, vol. 69, no. 13-15, pp. 1733-1738, 2006.
8. Jian Huang, Pong C Yuen, Wen-Sheng Chen and J H Lai. "Kernel Subspace LDA with Optimized Kernel Parameters on Face Recognition", Proceedings of the Sixth IEEE International Conference on Automatic Face and Gesture Recognition, 2004.
9. Jun-Bao Li, Jeng-Shyang Pan and Zhe-Ming Lu. "Kernel Optimization-Based Discriminant Analysis for Face Recognition", Neural Computing and Applications. vol. 18, no. 6, 2009, pp. 603-612.
10. Jun-Bao Li, Jeng-Shyang Pan, and Shu-Chuan Chu, "Kernel class-wise locality preserving projection," Information Sciences, vol. 178, no. 7, pp. 1825-1835, 2008.
11. Jun-Bao Li, Long-Jiang Yu, Sheng-He Sun, "Refined Kernel Principal Component Analysis Based Feature Extraction," Chinese Journal of Electronics. vol. 20, no.3, Page(s): 467-470, 2011.
12. Zhizheng Liang and Pengfei Shi, "Uncorrelated discriminant vectors using a kernel method", Pattern Recognition, vol. 38, pp. 307-310, 2005.
13. Juwei Lu, Konstantinos N. Plataniotis and Anastasios N. Venetsanopoulos, "Face recognition using kernel direct discriminant analysis algorithms", IEEE Transactions on Neural Networks, vol. 14, no. 1, pp.117-226, 2003.
14. J. S. Pan, J. B. Li, and Z. M. Lu, "Adaptive quasiconformal kernel discriminant analysis," Neurocomputing, vol. 71, no. 13-15, pp. 2754-2760, 2008.
15. Lei Wang, Kap Luk Chan, and Ping Xue, "A Criterion for Optimizing Kernel Parameters in KBDA for Image Retrieval", IEEE Trans. Systems, Man and Cybernetics-Part B: Cybernetics, vol. 35, no. 3, pp. 556-562, June 2005.
16. H. Xiong, M. N. Swamy, and M. O. Ahmad, "Optimizing the kernel in the empirical feature space," IEEE Transactions on Neural Networks, vol. 16, no. 2, pp.460-474, 2005.
17. M. H. Yang, "Kernel Eigenfaces vs. Kernel Fisherfaces: Face Recognition Using Kernel Methods," Proc.Fifth IEEE Int'l Conf. Automatic Face and Gesture Recognition, pp. 215-220, May 2002.
18. H. Zhao, S. Sun, Z. Jing, and J. Yang, "Local structure based supervised feature extraction," Pattern Recognition, vol. 39, no. 8, pp. 1546-1550, 2006.
19. Qi Zhua, "Reformative nonlinear feature extraction using kernel MSE," Neurocomputing, vol. 73, no. 16-18, pp. 3334-3337, 2010.
20. D. Tuia, G. Camps-Valls, G. Matasci, M. Kanevski, "Learning relevant image features with multiple-kernel classification," IEEE Transactions on Geoscience and Remote Sensing, vol. 48, no.10, pp. 3780-3791, 2010.
21. N. Subrahmanya, Y.C. Shin, "Sparse Multiple Kernel Learning for Signal Processing Applications," IEEE Transactions on Pattern Analysis and Machine Intelligence, vol. 32, no. 5, pp.788-798, 2010.
22. Sören Sonnenburg, Gunnar Rätsch, Christin Schäfer, Bernhard Schölkopf, "Large Scale Multiple Kernel Learning," Journal of Machine Learning Research, vol.7. pp. 1531-1565, 2006.
23. Marius Kloft, Ulf Brefeld, Sören Sonnenburg, Alexander Zien, "lp-Norm Multiple Kernel Learning,"Journal of Machine Learning Research, vol. 12, pp.953-997, 2011.
24. Chen Chen, Wei Li, Hongjun Su, Kui Liu, "Spectral-Spatial Classification of Hyperspectral Image Based on Kernel Extreme Learning Machine," Remote Sensing. Vol. 6, no. 6, pp.5795-5814, 2014.
25. B. Scholkopf, S. Mika, C. J. C. Burges, P. Knirsch, K.-R. Muller, G. Ratsch, and A. J. Smola, "Input space versus feature space in kernel-based methods," IEEE Transaction on Neural Network. vol. 10, no. 5, pp. 1000-1017, 1999.

Rolling Element Bearings Fault Intelligent Diagnosis Based on Convolutional Neural Networks Using Raw Sensing Signal

Wei Zhang, Gaoliang Peng, and Chuanhao Li

School of mechanics and electronics,
Harbin Institute of Technology, Harbin, China
{zw1993,pgl7782,lichuanhao}@hit.edu.cn

Abstract. Vibration signals captured by the accelerometer carry rich information for rolling element bearing fault diagnosis. Existing methods mostly rely on hand-crafted time-consuming preprocessing of data to acquire suitable features. In contrast, the proposed method automatically mines features from the RAW temporal signals without any preprocessing. Convolutional Neural Network (CNN) is used in our method to train the raw vibration data. As powerful feature exactor and classifier, CNN can learn to acquire features most suitable for the classification task by being trained. According to the results of the experiments, when fed in enough training samples, CNN outperforms the exist methods. The proposed method can also be applied to solve intelligent diagnosis problems of other machine systems..

Keywords: fault diagnosis; CNN; no preprocessing

1 Introduction

Rolling element bearings are the core components in rotating mechanism, whose health conditions, for example, the fault diameters in different places under different loads could have enormous impact on the performance, stability and life span of the mechanism. The most common way to prevent possible damages is to implement a real-time monitoring of vibration while the rotating mechanism is in operation. With the condition signals collected by the sensors, intelligent fault diagnosis methods are applied to recognize the fault types [1, 2]. The common intelligent fault diagnosis methods can be divided into two steps, namely, feature extraction and classification. Fast Fourier transform (FFT) and wavelet transform (WT) [3], which transform the raw signals from time domain into the frequency domain, are common methods to acquire representative features for the corresponding classifiers. In recent years, machine learning methods such as support vector machine (SVM) [4] and neural networks are widely used as classifiers to predict the fault types.

In recent five years, with the fast development of deep learning especially the convolutional neural networks [5] (CNNs), image classification have achieved incredible success. Some studies show that CNNs can also be used directly on

© Springer International Publishing AG 2017   77
J.-S. Pan et al. (eds.), *Advances in Intelligent Information Hiding*
and Multimedia Signal Processing, Smart Innovation, Systems and Technologies 64,
DOI 10.1007/978-3-319-50212-0_10

the raw temporal speech signals for speech recognition [6]. Inspired by those researches, we presented a novel rolling element bearings fault diagnosis algorithm based on CNNs in this paper, which doesnt involve any time-consuming time-frequency transformation, and performs all the operation on the raw temporal vibration signals. We show that not only the speech signals but also a segment of long periodic signals from any initialization can be classified with CNNs.

The remainder of this paper is organized as follows. The intelligent diagnosis method based on CNNs is introduced in Section 2. Some experiments are conducted to compare our methods with some common methods. Then discussions about the results of the experiments are presented in Section 3. We draw the conclusions and summarize the future work in Section 4.

2   Bearing Fault Intelligent Using the Proposed CNN

2.1   A Brief Introduction to CNNs

The architecture of CNNs is briefly introduced in this section, more details for CNNs can be found in [5].

The convolutional neural network is a multi-stage neural network which is composed of some filter stages and one classification stage. The filter stage is designed to extract features from the inputs, which contains two kinds of layers, the convolutional layer and the pooling layer. The classification stage is a multi-layer perceptron, which is composed of several fully-connected layers. The function of each type of layer will be described as follows.

The convolutional layer convolves the input local regions with filter kernels followed by the activation unit to generate the output features. Each filter uses the same kernel which is also known as weight sharing, to extract the local feature of the input local region. One filter corresponds to one frame in the next layer, and the number of frames is called the depth of this layer. We use K_i^l and b_i^l to denote the weights and bias of the i-th filter kernel in layer l, respectively, and use $X^l(j)$ to denote the j-th local region in layer l. The convolutional process is described as follows:

$$y_i^{(l+1)}(j) = K_i^l * X^l(j) + b_i^l \ . \tag{1}$$

where the notation $*$ computes the dot product of the kernel and the local regions, and $y_i^{(l+1)}(j)$ denotes the input of the j-th neuron in frame i of layer $l+1$.

After the convolutional operation, the Rectified Linear Unit (ReLU) which computes the function $f(x) = max\{0, x\}$ is used as the activation unit of our model to accelerate the convergence of the CNNs.

It is common to add a pooling layer after a convolutional layer in the CNN architecture. It functions as a downsampling operation which reduces the spatial size of the features and the parameters of the network. The most commonly used pooling layer is max-pooling layer, which performs the local max operation over the input features. It can reduce the parameters and obtain location-invariant features at the same time. The max-pooling transformation is described as follows:

$$p_i^{(l+1)}(j) = \max_{(j-1)W+1\leqslant t\leqslant jW}\{q_i^l(t)\} .\tag{2}$$

where $q_i^l(t)$ denotes the value of t-th neuron in the i-th frame of layer l, $t \in [(j-1)W+1, jW]$, W is the width of the pooling region, and $p_i^{(l+1)}(j)$ denotes the corresponding value of the neuron in layer $l+1$ of the pooling operation.

2.2 CNN-based Intelligent Diagnosis Method

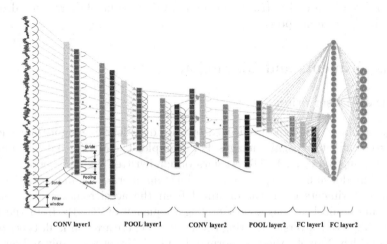

CONV layer1   POOL layer1   CONV layer2   POOL layer2   FC layer1   FC layer2

Fig. 1. Architecture of the proposed CNN model.

The architecture of the proposed CNN model is shown in Fig.1. It is composed of two filter stages and one classification stage. The input of the CNN is a segment of normalized temporal signal of bearing fault vibration. The first convolutional layer extract features from the input raw signal directly without any other transformation. The features of the input signals are extracted from two convolutional layers and two pooling layers. As shown in Fig.1, the depth of the layers becomes larger while the width of each frame becomes smaller with the increasing of the layers. The classification stage is composed of two fully-connected layers to accomplish the classification process. In the output layer, the softmax function is used to make the logits of the ten neurons accord with the probability distribution for the ten different bearing health conditions. Softmax function is:

$$f(z_j) = \frac{e^{z_j}}{\sum_k^{10} e^{z_k}} .\tag{3}$$

where z_j denotes the logits of the j-th output neuron.

Similar to MLP, CNN is also trainable. The loss function of our CNN model is cross-entropy between the estimated softmax output probability distribution and the target class probability distribution. Let $p(x)$ denotes the target distribution and $q(x)$ denotes estimated distribution, so the cross-entropy between $p(x)$ and $q(x)$ is:

$$H(p, q) = -\sum_x p(x) \log q(x) \, . \tag{4}$$

In order to minimize the loss function, the Adam Stochastic optimization algorithm is applied to train our CNN models. Adam implements straightforwardly, computes efficiently and requires little memory, which is quite suitable for models with big data or many parameters. The detail of this optimization algorithm can be found in [7]. As a common way to control overfitting, dropout is used in the training process.

3   Experiments and Discussion

3.1   Data Description

A large numbers of images are needed to train deep learning algorithms such as CNNs to be able to recognize images. For example, The MNIST dataset contains 60000 training data and 10000 test data of handwritten digits. In order to train the CNN model sufficiently, we prepared huge numbers of training samples. The original experiments data was obtained from the accelerometers on a motor driving mechanical system at a sampling frequency of 12 kHz from the Case Western Reserve University Bearing Data center. There are four fault types of the bearing health. Namely, they are normal, ball fault, inner race fault and out race fault, and sizes of the fault diameters include 0.007mm, 0.014mm and 0.021mm, so there are all together ten conditions. In this experiment, each samples contain 2400 data points. Dataset A, B and C each contain 17500 training samples and 2500 testing samples. Samples of Dataset A, B and C are collected under loads of 1, 2 and 3 hp respectively, and each of them contains ten different fault conditions. Dataset D contains 30000 training data and 7500 testing data of all three loads. In order to evaluate the necessity of large training data for the CNN dataset E which contains 1500 training is are also prepared. The details of all the datasets are described in Table 1.

3.2   Experimental Setup

Baseline System

We compare our methods with the standard ANN system with frequency features. The ANN system have one single hidden layer. The input of the networks is the normalized 2400 Fourier coefficients transformed from the raw temporal signals using fast Fourier transformation (FFT).

Hyper-parameters of the Proposed CNN

Table 1. Description of rolling element bearing datasets

Fault location		none	Ball			Inner race			Outer race			Load
Fault diameter(mm)			0.007	0.014	0.021	0.007	0.014	0.021	0.007	0.014	0.021	
Data A no.	Train	1750	1750	1750	1750	1750	1750	1750	1750	1750	1750	1
	Test	250	250	250	250	250	250	250	250	250	250	
Data B no.	Train	1750	1750	1750	1750	1750	1750	1750	1750	1750	1750	2
	test	250	250	250	250	250	250	250	250	250	250	
Data C no.	train	1750	1750	1750	1750	1750	1750	1750	1750	1750	1750	3
	test	250	250	250	250	250	250	250	250	250	250	
Data D no.	Train	3000	3000	3000	3000	3000	3000	3000	3000	3000	3000	1,2,3
	Test	750	750	750	750	750	750	750	750	750	750	
Data E no.	Train	150	150	150	150	150	150	150	150	150	150	1,2,3
	Test	750	750	750	750	750	750	750	750	750	750	

The architecture of the proposed CNN is composed of 2 convolutional layers, 2 pooling layers, and also one fully-connected hidden layer. The parameters of the convolutional and pooling layers are shown in Table 2. The number of the neurons in the fully-connected hidden layer is 1024. The experiments were implemented using tensorflow toolbox of Google.

Table 2. Parameters of the convolutional and pooling layers

Layer	Kernel width	Kernel height	Kernel depth	Stride	Frames
CONV1	20	1	1	5	32
POOL1	2	1	1	2	32
CONV2	10	1	32	2	64
POOL1	2	1	1	2	64

3.3 Results

Taken into account the sample quantity of each dataset, 2500 samples, 3000 samples and 100 samples are randomly selected as the training validation datasets A-C, D and E, respectively. Twenty trials were implemented for the diagnosis of each dataset. The fault recognition results of datasets A-E using CNN are shown in Fig. 2. The diagnosis accuracies of all trials on dataset A-D are over 99.6%. Actually, for most of the trials on dataset A-C, the diagnosis accuracies are 100%. This means that by directly feeding raw vibration signals into CNN, a very good result has been achieved when recognizing the fault conditions of the rolling element bearings. The accuracy of the CNN declined when trained with dataset E, which indicates that more training data will result in higher accuracy. The reason for this would be discussed later.

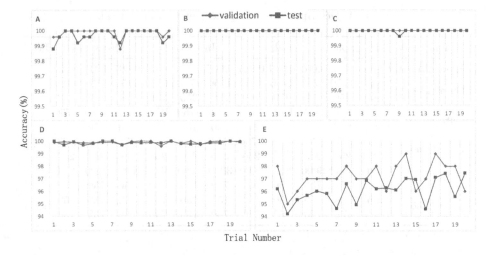

Fig. 2. Fault recognition results of datasets A-E.

For comparison, datasets D and E are also used to train the baseline system. The training samples are transformed to the frequency domain using FFT before being fed into the ANN. The comparison results between our method and the baseline system are shown in Fig. 3 and Table 3. The accuracies of these two methods trained with dataset D are high, actually both higher than 99%, and our method slightly outperforms the baseline system. The standard deviations of 20 trials on the validation samples of dataset D are very close for both methods. However, the standard deviations of the baseline system on the test samples of dataset D is almost twice that of the proposed method, which suggests that the proposed method is more stable than the baseline system. The results on dataset E are quite different, the test accuracy of the proposed method declines near 4%, while the standard deviation on test samples increases by eight times. However, the results of the baseline system remain almost unchanged in terms of test accuracy and standard deviation.

Table 3. Comparison results between our method and the baseline system

Dataset	Proposed method		FFT+ANN	
	Validation accuracy	Test accuracy	Validation accuracy	Test accuracy
D	99.9% ± 0.12%	99.86% ± 0.11%	99.90% ± 0.11%	99.47% ± 0.20%
E	97.2% ± 1.06%	96.04% ± 0.96%	99.65% ± 0.49%	99.47% ± 0.21%

3.4 Discussion

Some interesting points acquired from the experiments above are as follows:

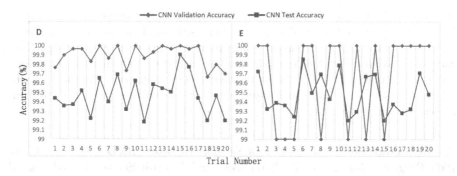

Fig. 3. Fault recognition results of the baseline system.

The results above show that CNN performs excellently when dealing with the bearing vibration signals. Although the performance of the common method is also quite good, it could be quite tricky to choose between FFT or WT as the preprocessing method of the data, because the most suitable frequency features may vary from different dataset. Whats worse, the transformation to frequency domain itself is very time-consuming. However, any dataset could be fed into a CNN without any preprocessing, which generates filter banks by learning from the training data, and then automatically acquires features suitable for the dataset without any hand crafting.

When the size of the dataset is large enough to train the CNN sufficiently, CNN has better and more stable performance than common methods. The major drawback of CNN is the need for huge amount of data, as can be seen from Table 3. This may result in overfitting and it could also make it hard to be applied to industries where the labeled data is hard to acquire. However, the labeled data of the rolling element bearing vibration is easy to acquire, so data augmentation is unnecessary. For example, when the sample frequency is 12 kHz, the length of the data is 2400 and the stride is 1, it takes only 1 second to acquire 9600 different labeled samples. So CNN is quite suitable for the bearing fault diagnosis since huge amount of data are available.

What's more, it is shown that it is possible to recognize the rolling element bearing fault type with raw periodic vibration signal as input to the CNNs. What we want to emphasize here is that the bearing vibration signal is just a representative of the periodic signals, and CNNs may be good at recognizing many other kinds of raw periodic signals, which may provide us a new approach to deal with these kinds of signals.

4   Conclusions

An intelligent diagnosis method for rolling element fault based on convolutional neural network with raw vibration data is proposed in this paper. As powerful feature exactor and classifier, CNN can learn to acquire features most suitable for the classification task by being trained. According to the results of the experiments, the proposed method is able to mine suitable fault features adaptively from the raw bearing vibration data and classify the fault classes with high accuracy and stability. Different from the deep neural networks trained by frequency spectra, the proposed CNN can be trained directly using raw temporal signals, which means this method can be applied to solve intelligent diagnosis problems of other machine systems.

Acknowledgments. The support of National High-tech R&D Program of China (863 Program, No. 2015AA042201), National Natural Science Foundation of China (No. 51275119) in carrying out this research is gratefully acknowledged.

References

1. Jayaswal, P., Verma, S.N. and Wadhwani, A.K.: Development of EBP-Artificial neural network expert system for rolling element bearing fault diagnosis. J. VIB. CONTROL. 17, 1131–1148 (2011)
2. Li, Y., Xu, M., Wei, Y. and Huang, W.: A new rolling bearing fault diagnosis method based on multiscale permutation entropy and improved support vector machine based binary tree. MEASUREMENT. 77, 80–94 (2016)
3. Inoue, T., Sueoka, A., Kanemoto, H., Odahara, S. and Murakami, Y.: Detection of minute signs of a small fault in a periodic or a quasi-periodic signal by the harmonic wavelet transform. MECH. SYST. SIGNAL. PR. 21, 2041–2055 (2007)
4. Widodo, A. and Yang, B.S.: Support vector machine in machine condition monitoring and fault diagnosis. Mechanical systems and signal processing. MECH. SYST. SIGNAL. PR. 21, 2560–2574 (2007)
5. Krizhevsky, A., Sutskever, I. and Hinton, G.E.: Imagenet classification with deep convolutional neural networks. In: 26th Annual Conference on Neural Information Processing Systems, pp. 1097–1105. (2012)
6. Czajkowski, K., Fitzgerald, S., Foster, I., Kesselman.: Convolutional Neural Networks-based continuous speech recognition using raw speech signal. In: 40th IEEE International Conference on Acoustics, Speech and Signal, pp. 4295–4299. IEEE Press, New York (2015)
7. Kingma, D. and Ba, J.:Adam: A method for stochastic optimization. arXiv:1412.6980, arXiv preprint. (2014)

Forensics of Operation History Including Image Blurring and Noise Addition based on Joint Features

Yahui Liu, Rongrong Ni, Yao Zhao

Institute of Information Science, & Beijing Key Laboratory of Advanced Information
Science and Network Technology, Beijing Jiaotong University, China
{rrni, yzhao}@bjtu.edu.cn

Abstract. Multi-manipulation is becoming the new normal in image tampering while the forensic researches still focus on the detection of single specific operation. With uncertain influences laid by pre-processing and post-processing operations, the multi-manipulation cases are hardly identified by existing single-manipulated detection methods. Here come the studies on image processing history. In this article, a novel algorithm for detecting image manipulation history of blurring and noise addition is proposed. The algorithm is based on the change of attributes correlation between adjacent pixels due to blur-ring or noise addition. Two sets of features are extracted from spatial domain and non-subsampled contourlet transform (NSCT) domain respectively. Spatial features describe the statistical distribution of differences among pixels in neighbourhood, while NSCT features capture the consistency between directional components of adjacent pixels in NSCT domain. With the proposed features, we are able to detect the particular processing history through supporting vector machine (SVM). Experiment results show that the performance of proposed algorithm is satisfying.

Keywords: NSCT, operation chain, blur, noise addition.

1 Introduction

The development of digital image processing technology promotes the popularization of image retouching and tampering. Against this background, complicate combinations of basic operations are widely used in practical applications. Under the influences of pre-processing or post-processing operations, the accuracy of detection algorithms for specific manipulation would seriously decrease. Accordingly, the concept of operation history is proposed to keep pace with the tampering situation. Operation history is defined as the combination of manipulation operations that an image has been through. Forensics on operation history refers to the detection of operation type, parameters setting and processing order, etc. Since the research is in the beginning stage, the operation history we investigate in this article contains two sorts of operations at most with operation types limited to blurring and noise addition.

Blurring is usually used to cover visual traces left by other forms of tampering like copy-move, while noise addition is applied to disturb the statistical feature left by tampering operations as an anti-forensics method. Blurring is also performed after noise addition in some occasion to diminish visual granular sensation. The operation history based on these two types of operations is of pretty realistic value.

Quite a few existing detection algorithms for operation forensics are targeting on blurring because of the wide application. These studies pursed in different domains are instructive to our research. In spatial domain, weighted local entropy is proposed

© Springer International Publishing AG 2017   85
J.-S. Pan et al. (eds.), *Advances in Intelligent Information Hiding
and Multimedia Signal Processing*, Smart Innovation, Systems and Technologies 64,
DOI 10.1007/978-3-319-50212-0_11

to detect feathering fuzzy of gray image [1], yet proportion of tonal distortion is for color image forensics [2]. In wavelet domain, evaluation of image blurriness through the feature of regularity of coefficients is devoted to an altered image [3], while homomorphic filtering is used to enhance the artificial blurring area in the detection method of mathematical morphology [4]. Apart from that, features of other transform domains are also applied to detect traces of blur operation. In contourlet domain, algorithms based on analysis of sharp edge points are proposed in different studies [5-6]. Contourlet transform decomposes an image into subbands of different scales and multiple directions. Compared to the directional limitation of wavelet transform, Contourlet transform has advantages in expressing detail information of images. However, the forensics on noise addition is paid less attention. Differential characteristic of image blocks is calculated to determine the trace of noise addition in DCT domain [7].

In this article, an algorithm is proposed to detect an operation history over a digital image, which is a simple combination of blurring and noise addition. In the algorithm, two sets of features are extracted from spatial domain and NSCT domain, and put into Support Vector Machine (SVM) for classification. Spatial features describe the differences between adjacent pixels from the statistical point of view. Both blurring and noise addition will draw influences on the differences between adjacent pixels. The difference values exhibit different statistical distributions under different combination of operations. NSCT features capture the directional consistency between a pixel and its surrounding neighbors, because the directional components in decomposed subbands are fairly sensitive to operations that change the correlation of pixels like blurring and noise addition. SVM is adopted for features training and classifying. As a result, the types and processing order of operations performed to an image are investigated and determined. From the point of processing history, each combined manipulation with particular processing order is regard as a distinct operation.

The organization of the paper is as follows: Section 2 describes the features extraction procedure in both spatial domain and NSCT domain. In section 3, parameters setting of operations and experimental results are given in details. Section 4 gives the conclusion of our research.

2   FEATURES ANALYSIS AND EXTRACTION

2.1   Spatial features

As for noise addition and blurring, identical manipulation traces are left in spatial domain after either operation is applied to an image. Traces of both operations are related to the differences between pixels in neighborhood. Since the operations we discuss in this article are both applied in a global way, it is proper to describe the tampering trace with statistical features.

As the most commonly used statistical feature of an image, histograms of the absolute difference between every pixel and its eight neighbors are formulated on image sample under blurring or noise addition. With the isolated pixels caused by noise addition and detail information loss caused by blurring, the histogram shows different distributions. Because most of the difference values remain zero after alteration, it is better to catch the manipulation traces by means of the difference distributions with the zero part eliminated.

The histograms of nonzero absolute differences between every pixel and its eight neighbors under different operation cases are shown in Fig. 2~Fig. 3. Apparently, when noise addition or blurring is applied to an image, the distribution curves for nonzero difference histogram will change in an obvious way. Compared to the original image, the upper bound of difference values under noise addition case is larger while the upper limit of distribution proportion is much smaller. However, the change caused by blur operation behaves almost the other way around. With both operations applied, both operation traces are left in histogram of the image under noise addition-blur processed, while blur-noise addition case presents much more characters caused by noise addition than blurring. Since the distribution curve of noise addition-blur case keeps partial character of curves under each single manipulated case, it is easy to identify this operation chain from other related tampering cases. Nevertheless, to tell the difference between blur-noise addition chain and single noise addition operation, pertinence features for blur operation are needed.

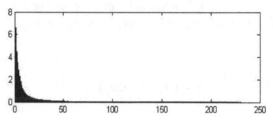

Fig. 2. Nonzero Differences Histogram for adjacent pixels of original image

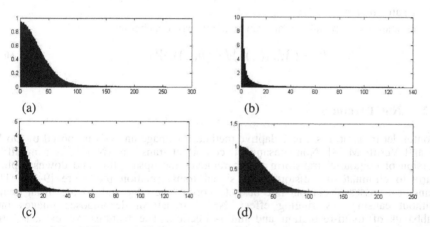

Fig. 3. Nonzero Differences Histogram for adjacent pixels of (a) image after noise addition, (b) image after blurring, (c) image after noise addition-blur processing chain, (d) image after blur-noise addition processing chain.

To capture the distribution characteristics into features with fixed dimensions, curve fitting is used in the detection algorithm to portray the curves. There are multiple function models for curve fitting, where commonly used ones like exponential function and logarithmic function can achieve satisfying fitting optimization in this case. For the purpose of achieving abundant fitting parameters, polynomial function is chosen as the fitting function for the algorithm. The function model is

$$y(a_0, a_1, \ldots, a_k) = a_0 + a_1 x + a_2 x^2 + \cdots + a_k x^k \ . \tag{4}$$

The fitting parameters along with the maximal values of abscissa and ordinate are extracted as the spatial features to describe the distribution of pixels differences. The extracting procedures are as follows:

Step 1: Histogram formulation. The histogram $H(I)$ of nonzero absolute differences between every pixel and its eight neighbors is formulated for the testing image I. If x denotes the difference value between the pixel and its neighbors, $h(x)$ denotes the corresponding proportion percentage of the difference value x.

Step 2: Curve fitting. The polynomial curve fitting calculation is performed on $H(I)$, with $h(x)$ substituting y in Eq.(4). Then the fitting parameters are obtained as Ψ:

$$\Psi = (a_0, a_1, \cdots, a_k) \ . \tag{5}$$

Step 3: Upper bounds extraction. The maximum values of x and $h(x)$ in $H(I)$ are obtained.

The features in spatial domain can be summarized as below:

$$F_s = (Max(x), Max(h(x)), \Psi) \ . \tag{6}$$

2.2   NSCT features

Contourlet transform is a nonadaptive method for image analysis proposed by Do M N and Verttli M [8]. Non-subsampled contourlet transform (NSCT) is a modified version of contourlet transform which removes the upsampling and downsampling parts to eliminate the distortion in signal transformation procedure [9-10]. The transform possesses contour-like basis to portray the edge and texture in an image without causing any ringing effect. NSCT transform decomposes images into subbands of multi-resolution and multi-direction. The features we extracted for tampering detection are based on the directional details decomposed by NSCT.

NSCT transform comprises two steps of decomposition: dimensional decomposition by Nonsubsampled Pyramid Filter (NSP) and directional decomposition by Nonsubsampled Directional Filter Bank (NSDFB). The overview of NSCT transform procedure is shown in Fig.4. NSP is a two-channel non-subsampled 2-D filter bank constructed by removing the subsample part from Laplacian Pyramid filter in contourlet transform. An image is decomposed into subbands of different scales through the NSP. NSDFB is a tree-structured filter bank constructed by non-subsampled fan filter bank. The high frequency subbands emerged by NSP in different levels are filtered by NSDFB into multiple directional subbands. For each level of dimension decomposition, numbers of directional subbands are set respectively. Therefore, NSCT is able to extract many directional subbands from an

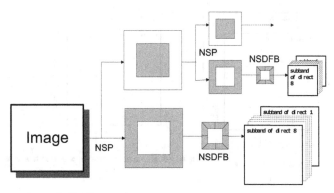

Fig. 4. Construction of NSCT transform

image compared to the wavelet transformation. With multi-resolution and multi-direction subbands obtained by NSCT transform, an image is conducted into directional details for operation trace investigation.

Suppose NSCT is performed on an image I, the number of transform decomposed level is set to be l. For decompose level t $(1 \le t \le l)$ in NSCT, directional decomposition parameter is set to be j_t, then 2^{j_t} directional subbands can be obtained from the corresponding level.

In NSCT domain, a pixel in an image corresponds to dozens of directional components in subbands of multi-resolusion and multi-direction. These components describe the directional attributes of the pixel against contour-like basis in a specific scale. These components are proper variables in analyzing operation traces. In our research, we define a direction vector to describe the directional attributes of a pixel in a specific NSCT decomposition level (Fig. 5(a)).

Consider a random pixel p in I, the directional component of p in r^{th} $(1 \le r \le 2^{j_t})$ directional subband at scale t is marked by c_{dr}. The directional vector of p at scale t is denoted by V_{dp}.

$$V_{dp} = (c_{d1}, c_{d2}, \cdots, c_{dr}, \cdots, c_{d2^{j_t}}) \ . \qquad (7)$$

The directional attribute of pixels in neighborhood has high consistency especially in the flat areas. When high frequency information of an image is waken by artificial modification, the directional attributes in neighborhood turn to be more similar due to the difference detail loss. Therefore, the algorithm is based on the consistency measure between directional vectors of adjacent pixels. We choose the consistency calculation of vectors as the features in NSCT domain to capture the tampering traces. The algorithm of local directional consistency vector extraction can be summarized as below:

Step1: Extract the directional vector for assigned pixel p_0 according to Eq.(7), and mark it as V_{d0}.

Step2: Extract the directional vectors for 8 neighbor pixels of p_0 based on Eq. (7), and mark them as $V_{d1} \sim V_{d8}$, respectively.

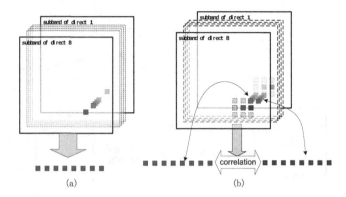

Fig. 5. The extraction of local directional similarity vector, (a) directional vector of a pixel, (b) local directional similarity vector of a pixel

Step 3: Calculate the correlation coefficient cr_i between V_{d0} and V_{di}.

Step 4: The local directional similarity vector for a selected pixel p_0 is obtained as V_{ds}:

$$V_{ds} = (cr_1, cr_2, \cdots, cr_8)$$

Since the consistency of direction vectors varies rapidly from flat to texture of the image contents, the features of NSCT domain will be extracted in both flat area and texture area respectively. In this article, we use the variance of 3×3 pixel block as the measurement of neighborhood smoothness. 10 blocks with smallest and largest block variance of an image are picked for local directional consistency vector extraction.

3   EXPERIMENTS

In the proposed detection method, two sets of features are extracted from spatial domain and NSCT domain, respectively. SVM is employed for feature training and classification, and the accuracy is adopted to evaluate our detection algorithm.

In our experiments, niche targeting image database is built to simulate tampering cases. Manipulate operations are performed by filter functions in Matlab7.1 and filter tool in Photoshop CS3. NSCT transform in feature extraction is implemented by NSCT toolbox for Matlab. Multiple classification of features is done by Libsvm toolbox.

Tampered images database contains five sets of images with different manipulation situations consisting of blurring and noise addition. In particular, they are non-processed, blur processed only, noise addition processed only, blur-noise addition processing chain and noise addition-blur processing chain. Images of the database are originally picked up from the UCID (Uncompressed Color Image Database) [11], and conducted by manipulations with multiple settings of either software. 500 TIF format origin images with different definition are screened from UCID as source samples.

Five blur operations and three noise addition operations are selected for manipulations processing over sample copies in each corresponding set, giving a total of 15 compositions for each kind of processing chain. The function selection and radius settings of both operations in particular are shown in Table 1, making tampering categories reach the count of 38.With the image sample capacity of 500, the total count of image database is 19500.

Table 1. Function and Radius settings

	PhotoShop CS3	Matlab7.1
Blur	Gaussian function 1.0 3.0	Gaussian filter 0.849 0.919 0.979
Noise addition	Noise function 5% 10% 15%	

In the NSCT decomposition procedure, NSP decompose scale is set to be 4, and the numbers of directional subbands decomposed by NSDFB in each layer are 2,4,8,8 sorted by layer scale from coarse to fine. The NSCT features are extracted from level 3 and level 4, with 8 dimensional directional vectors extracted for each pixel in a decomposed level. And the dimension of fitting parameters is set to be 20 in our case.

In the classification part, classification accuracy results of 2-fold, 3-fold and 5-fold cross-validation are obtained respectively. We are hoping to figure out the influence on classification accuracy drawn by different proportion between sizes of training set and testing set. The accuracy results are shown in Table 2.

Table 2. Classification accuracy results of n-fold cross validation

n	Total (%)	Origin	Blur	Noise addition	Blur-Noise addition	Noise addition-Blur
2	85.02	80.34	83.26	89.90	86.6	84.91
3	87.33	83.68	86.14	90.72	89.03	86.35
5	89.14	84.60	86.37	92.73	89.25	88.29

From the data in table II, the classification accuracy of our detection algorithm is satisfying in principle. With appropriate expansion of the training set, the accuracy of detection method turns to rise correspondingly. The overall accuracy of the algorithm is around 85% to 89%, while the accuracy of each processed set is floating around this number. The blur-noise addition set is more likely to be missorted as the noise addition only case due to the cover of the noise addition operation to blurring traces. So the NSCT features in our method are mostly adopted to identify the trace of blurring that may be covered by noise addition. Apparently, it's not hard to tell from the detection result, later operation in the manipulated history of blurring and noise addition lays disturbance on the pre-processed one, while it is unable to cover the former operation trace completely.

4   CONCLUSION

In this paper, a detection algorithm with two sets of features from different domains is proposed to detect the manipulation history of blur and noise addition. The spatial features describe the distribution situation of adjacent pixels differences. The NSCT features capture the directional attributes consistency between adjacent pixels.

Support vector machine is adopted for features classification. The accuracy of proposed algorithm in experiments indicates our detection method has a satisfying performance in discriminating different operation traces.

The proposed features can also be used in the detection of single operation tampering cases. In addition, we are still looking forward to extra features to improve the classification accuracy of specific processed sets in future studies.

ACKNOWLEDGMENTS

This work was supported in part by National NSF of China (61332012, 61272355, 61672090), Beijing Nova Programme Interdisciplinary Cooperation Project, Fundamental Research Funds for the Central Universities (2015JBZ002), the PAPD, the CICAEET, CCF-Tencent Open Research Fund.

REFERENCES

1. Zhang, C., Zhang, H.: Detecting Digital Image Forgeries Through Weighted Local Entropy. 2007 IEEE International Symposium on Signal Processing and Information Technology. 62--67 (2007)
2. Wang, B., Kong, L., Kong, X.: Forensic Technology Of Tonal Distortion For Blur Operation In Image Forgery. Journal of Electronics. 34(12), 2451--2454 (2006)
3. Sutcu, Y., Coskun, B., Sencar, H. T.: Tamper Detection Based On Regularity Of Wavelet Transform Coefficients.2007 IEEE International Conference on Image Processing, vol.1, pp. 397--400 (2007)
4. Zheng, J., Liu, M.: A Digital Forgery Image Detection Algorithm Based On Wavelet Homomorphic Filtering. International Workshop on Digital Watermarking. pp. 152--160 , Springer Berlin Heidelberg (2008)
5. Liu, G., Wang, J., Lian, S.: Detect Image Splicing With Artificial Blurred Boundary. Mathematical and Computer Modeling. 57, 2647--2659 (2013)
6. Wei, L. X., Zhu, J. J., Yang, X. Y.: An Image Forensics Algorithm For Blur Detection Based On Properties Of Sharp Edge Points. Advanced Materials Research. Trans Tech Publications，vol.341, pp. 743--747 (2012).
7. Cao, G., Zhao, Y., Ni, R.: Forensic Detection Of Noise Addition In Digital Images. Journal of Electronic Imaging, 23, 023004--023004 (2014)
8. Do, M. N., Vetterli, M.: The Contourlet Transform: An Efficient Directional Multiresolution Image Representation. IEEE Transactions on Image Processing, 14, 2091--2106 (2005)
9. Da Cunha, A. L., Zhou, J., Do, M.N.: Nonsubsampled Contourlet Transform: Filter Design And Applications In Denoising. IEEE International Conference on Image Processing 2005，vol.1, pp. 749--752 (2005)
10. Li, H., Zhao, Z., Chen, Y.: Research On Image Denoising Via Different Filters In Contourlet Domain. Infrared Technology, vol.30, no.8, pp.450--453 (2008)
11. UCID   -   Uncompressed   Color   Image   Database, http://vision.cs.aston.ac.uk/datasets/UCID/ucid.html

An Improvement Image Subjective Quality Evaluation Model Based on Just Noticeable Difference

Bohan Niu

School of Information and Communication Engineering,
Beijing University of Posts and Telecommunications University,
Beijing, 100876, China
bohanniu@vip.163.com

Abstract. The research of information hiding technology is a significant aspect in information security. The limitation of the development in this research is the judgment of the image quality after applying the technology. The traditional image quality evaluation standard of information hiding algorithm such as Peak Signal to Noise Ratio (PSNR) is not meeting with the human subjective assessment. Watson Just Noticeable Difference (JND) model can be used in the perceptual adjustment of the information hiding algorithm. However, JND model gives the objective quality evaluation, so setting up the image subjective quality evaluation model by just noticeable difference with subjective uniformity for information hiding algorithm is very necessary. In this paper, an improvement image subjective quality evaluation model based on just noticeable difference with Human Visual System (HVS) is proposed, the corresponding relation between ITU-R quality and impairment scales model and objective assessment by JND model is built. The experimental results show that the improvement image quality evaluation model has better agreement with the human's visual judgment.

Keywords: image quality evaluation; subjective assessment; JND objective criteria; human visual system.

1   Introduction

A good information hiding algorithm means that information is added to the images without being noticed by human's visual sense. It is a balance between high embedding rate and low distortion factor. The designer of the algorithm seeks an algorithm with a high embedding rate as well as a low distortion factor.

The research on evaluation criteria of information hiding technology is a significant aspect in information security. The limitation of the development in this research is how to build the objective index which can basically reflect the human visual system on image quality changes, also can distinguish the visual observation and well capture the difference between images.

© Springer International Publishing AG 2017
J.-S. Pan et al. (eds.), *Advances in Intelligent Information Hiding
and Multimedia Signal Processing*, Smart Innovation, Systems and Technologies 64,
DOI 10.1007/978-3-319-50212-0_12

A traditional way to judge the quality of an image uses peak signal to noise ratio (PSNR) and etc. PSNR calculates the statistical error between the carrier image and the hidden information image. Under the same embedding rate, a larger PSNR implies better quality, lower distortion factor and better algorithm capability. It is important that an objective scale can reflect the perceived image quality. For instance, simple distortion scales. PSNR or even the weighted mean-square error (WMSE) are good distortion indicators for random errors but not for structured or correlated errors. But such structured errors are prevalent in image coders, and they degrade local features and perceived quality much more than random errors do. Hence, PSNR and WMSE alone are not suitable objective scales to evaluate compressed images. There have been many studies of the construction of objective scales which represent properties of the human observer [1, 2].

To match the visual feeling, some researchers began to use the structural similarity (SSIM) method for image quality evaluation [3-5]. SSIM is to study the image object from the structure; this method can be better reflected by the similar structure of visual experience, but it cannot express the frequency changes from visual characteristics.

Just-noticeable difference (JND) refers to the minimum visibility threshold when visual contents are altered, and results from physiological and psychophysical phenomena in the human visual system (HVS) [6-8]. The use of JND facilitates effective quality evaluation of watermarked image [9], etc.

The benefits of objective evaluation are the process is easy to be calculated, and can be carried out at any time. However, human beings are the ultimate users, subjective assessment should be the best way for the image quality evaluation. But subjective evaluation is expensive, time consuming, and difficult to organize. This paper tries to build the corresponding relation between subjective evaluation criteria and objective assessment by JND model. In this paper, the objective assessment of image quality is calculated by JND, and the subjective assessment adopts ITU-R BT.500-11 RECOMMENDATION Methodology.

The rest of the paper is organized as follows. Section 2 introduces the foundation of related works, an image quality evaluation model with JND objective index and subjective assessment uniformity is then proposed in Section 3. The further improvement and experimental performance are demonstrated in Section 4, and Section 5 concludes the paper.

2   Related Work Foundation

The same modifications of an image are located at human sensitive and non -sensitive parts of the image respectively such that the difference of PSNR is very small; however, the difference of evaluations given by human viewers will be relatively large.

It is easy to know that the result of PSNR can be very far from the supervisor's evaluation by simple calculation. Fig. 1 shows that the PSNR values of Fig.1 (a) and (b) are very close. Therefore, the PSNR cannot distinguish image Fig. 1, (a) and (b). However, the Fig. 1 (a) is better than (b) according in the view of human vision.

(a)   (b)

Fig. 1. The two LENA images with almost the same PSNR values. PSNR value of (a) is obtained by calculating the relative to the original image, it has 28.3816dB; PSNR value of (b) is equal to 28.3815dB.

For many sensory modalities, over a wide range of stimulus magnitudes sufficiently far from the upper and lower limits of perception. JND is a fixed proportion of the reference sensory level, and so the ratio of the JND/reference is roughly constant. Measured in physical units, it means:

$$K = \frac{\Delta I}{I} \tag{1}$$

Where I is the original intensity of the particular stimulation, ΔI is the addition to it required for the change to be perceived, and K is a constant. This rule was first discovered by Ernst Heinrich Weber (1795–1878), an anatomist and physiologist, in experiments on the thresholds of perception of lifted weights. A theoretical rationale was subsequently provided by Gustav Fechner, so the rule is therefore known either as the Weber Law or as the Weber–Fechner law; the constant K is called the Weber constant.

The perceptual adjustment based on Watson Just Noticeable Difference model can refer to the relevant literature [6]. Following this approach, an improved estimation of the JND in terms of luminance and contrast masking can be obtained literatures [7-12].

3   Consistency Model for Image Quality Evaluation

ITU-R BT.500 is the methodology for the subjective assessment of the quality of television pictures. ITU-R BT.500 adopts a five-grade quality and impairment scales made during routine or special operations by certain supervisory engineers, it can also make some use of certain aspects of the methods recommended for laboratory assessments. The five-grade quality and impairment scale follows as Table 1.

Table 1.   ITU-R quality and impairment scales.

Impairment	Quality	Grade
Imperceptible	Excellent	5
Perceptible, but not annoying	Good	4
Slightly annoying	Fair	3
Annoying	Poor	2
Very annoying	Bad	1

In order to set up the consistency model between the subjective and JND image quality evaluations, the subjective assessment experiments are obtained as follows.

Before the formal evaluation, the reviewers should study the evaluation criteria with better understanding of the test rules and methods, and the participants will be to carry out a simulation test program.

The same monitor displays two images at the same time, one is the original image and the other is the evaluating image, when the reviewers begin to evaluate images. Each image will remain about 30 seconds, such that observers have time to evaluate the image according to predetermined criteria with the original image as a reference, and judge quality damage situation of the evaluating image and evaluate image quality. The observers should give the five-grade quality and impairment scale of the evaluating image independently.

Every observer will be evaluated no less than 5 times with memoryless, the so-called memoryless refers to the same image that will be marked a score in different time, such as the observer gives a score at 8:00, and then gives another score at 8 o'clock the next morning again on the same evaluating image without remembering the former evaluation results, repeat testing until the evaluation is finished. In order to improve accuracy of the comprehensive subjective assessment, an image will be evaluated by 10 persons with different times.

The subjective assessment of color images is similar to gray images, since the index calculation formulas for gray image and color image are different; the consistency model is established for gray and color images respectively

Before establishment of the consistency evaluation model between ITU-R BT.500 and JNG model, more than 500 cases of information hiding are evaluated by the human subjective assessment according to ITU-R quality and impairment scales evaluation criteria. Through the average score of the subjective evaluation and the objective measure, empirical mapping is established.

In the process of establishing mapping, piecewise curve fitting function is obtained between the fraction of the subjective assessment and objective measure. By adjusting the sectional size of the JND index, it can reflect the overall trend of the subjective level.

In order to accurately establish the function relationship between the subjective assessment and JND index, we obtained the mapping relationship functions from piecewise objective measure.

For the case of gray image information hiding, the following empirical mapping is the function between the JND objective index and the evaluation grade.

$$g(jnd) = \begin{cases} 0.001659019 \times jnd^2 & 0.0 \le jnd < 30.33 \\ -0.002152782 \times jnd^2 + 0.337039562 \times jnd - 8.191728494 & 30.33 \le jnd < 53.79 \\ -0.000760448 \times jnd^2 + 0.144485047 \times jnd - 1.863039754 & 53.79 \le jnd < 95.00 \\ 5.0 & 95.00 \le jnd \end{cases} \quad (2)$$

The empirical mapping for color image quality evaluation is established as follows:

$$g(jnd) = \begin{cases} 0.002158895 \times jnd^2 & 0.0 \le jnd < 23.16 \\ -0.001721206 \times jnd^2 + 0.264646129 \times jnd - 5.155974445 & 23.16 \le jnd < 62.89 \\ -0.000743879 \times jnd^2 + 0.124995901 \times jnd - 0.238847784 & 62.89 \le jnd < 80.00 \\ 5.0 & 80.00 \le jnd \end{cases} \quad (3)$$

where jnd is the index value calculated by JND model, $g(jnd)$ is the corresponding score of subjective assessment grade given by the above empirical mapping function.

In order to verify the performance of our model, simulation experiments are carried out. We first use MATLAB programming to draw the empirical mapping of our model, and then the evaluating images are judged by above subjective evaluation method. The obtained average grades by the subjective evaluation are scattered point diagram.

Fig. 2 shows the test results of the simulation experiments for gray and color images respectively.

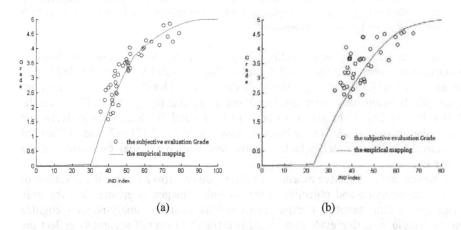

(a)   (b)

Fig. 2. Experimental performance of the consistency model for image quality evaluation. (a) is the test results of the simulation experiments for gray image based on JND index; (b) is the experimental results for color image, where the horizontal axis represents JND index of the image, and the vertical coordinate indicates the corresponding subjective evaluation level.

From the above figure, we know that subjective evaluation for the gray image have produced results essentially in agreement with the empirical mapping curve of JND, which illustrates that objective index of the JND index can well reflect the human eye to gray image perceptibility. That is to say, the reality of the JND index can be very good to reflect the human being's visual system.

The subjective evaluation results for the color image are generally around empirical mapping curve of JND from the above figure, although it is not better than the one of gray image.

4   Experimental performance and Improvement

In order to verify the performance of the model, an example will be calculated by our model. The original image and stego-images of Lena are showed in Fig. 3.

(a)   (b)   (c)

Fig. 3. The original image and stego-images of Lena. (a) is the original image of Lena, (b) and (c) are the obtained stego-images by applying two kinds of information hiding algorithms LSB and DCT to Lena's original image respectively.

It is obvious that the image quality of Fig. 3 (b) is much better than Fig. 3 (c) by attentive observation. But the PSNR values of Fig. 3 (b) and 3 (c) are 11184824.7529 dB and 11184823.591 dB respectively. It can be considered that PSNR values are equal, which means we cannot distinguish the image quality by PSNR. However, the JND values of Fig. 3 (b) and 3 (c) are 54.071040 and 50.688200 respectively and hence the quality grade for subjective assessment is 4.121467 and 3.8361789 respectively by our model, which is consistent with the visual evaluation of the human eye.

However, in the objective measures of JND, you can also note that the deviation of the subjective value and objective value for color images is greater than the gray images. By a large number of experiments and the results of analysis, we recognize that the single objective evaluation index is difficult to overall accurately reflect the image quality. Therefore, a variety of objective indicators will be considered in the evaluation model of the image quality.

The structural similarity (SSIM) [3-5] assumes that natural image signals are highly "structured", and HVS is adapted well for extracting structural information from a scene. Hence, SSIM provides a better method to evaluate the quality of the image, its objective assessment of image quality is calculated by Structure and Hue Similarity [5]. Similarly, the image subjective quality evaluation model based on Structural Similarity can be obtained from the above.

The improvement image subjective quality evaluation model is to consider the two objective grade evaluations. The comprehensive value of grade quality evaluation is the weighted sum of JND and SSIM grade. The weight value of JND evaluation is 0.6, left 0.4 to SSIM assessment.

The comprehensive grade is obtained by computing the weighted sum of JND grade and SSIM level, the subjective quality evaluation is the average value given by 6 persons with memoryless assessment according to the methodology of ITU-R BT.500. 50 pairs of the subjective and objective grade quality evaluation of gray image are showed in Fig. 4(a). Similarly, the comprehensive contrast of color images is shown in Fig. 4(b).

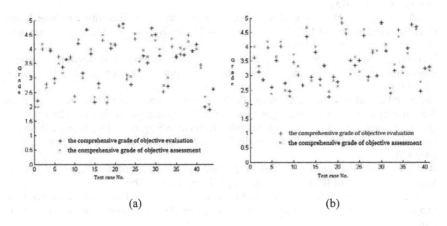

(a)   (b)

Fig. 4. Comprehensive comparison between the subjective and objective quality evaluation. (a) and (b) are the assessment contrast diagram of gray and color image respectively,

From Fig. 4, we note that the improved model has obvious effect on quality evaluation, especially for color image.

5   Conclusions

The research of information hiding technology needs to judge image quality after applying the technology. The traditional image quality evaluation standard such as PSNR is not meeting with the human subjective assessment. We built the corresponding relation between ITU-R quality model and JND objective assessment model. The experimental results show that image quality evaluation model based on JND has better agreement with the human's visual judgment.

By above experimental performance, we recognize that objective index of JND can basically reflect the human visual system on image quality changes; also can distinguish the visual observation and well capture the difference between images. Furthermore, from above experimental results, the improved comprehensive evaluation model is very consistent with the subjective evaluation.

References

1. Lukas, F.X.J., Budrikis, Z. L.: Picture Quality Prediction Based on a Visual Model. IEEE Trans. Commun., vol. COM-30, pp. 1679--1692 (1982)
2. Limb, J.O.: Distortion Criteria of the Human Viewer. IEEE Trans. Syst., Man, Cybern., vol. SMC-9, pp. 778--793 (1979)
3. Qin, C., Chang, C.C., Chiu, Y.P.: A Novel Joint Data-Hiding and Compression Scheme Based on SMVQ and Image Inpainting, IEEE Trans. on Image Processing, vol. 23, no. 3, pp.49--55 (2014)
4. Hong, W., Chen, T.S., Wu, M.C.: An Improved Human Visual System Based on Reversible Data Hiding Method Using Adaptive Histogram Modification. Optics Communications, vol. 291, pp. 87--97 (2013)
5. Shi, Y.Y., Ding, Y.D., Zhang, R.R., Li, J.: Structure and Hue Similarity for Color Image Quality Assessment. IEEE 2009 International Conference on Electronic Computer Technology, pp. 329--333 (2009)
6. Watson, A. B., Borthwick, R., Taylor, M.: Image Quality and Entropy Masking. Proceedings of SPIE - The International Society for Optical Engineering, vol. 3016, pp. 2--12 (1997)
7. Callet, P.L., Autrusseau, F., Campisi, P.: Visibility Control and Quality Assessment of Watermarking and Data Hiding Algorithms. Multimedia Forensics & Security (2008)
8. Qin, C., Chang, C.C., Lin, C.C.: An Adaptive Reversible Steganographic Scheme Based on the Just Noticeable Distortion, Multimedia Tools and Application, vol. 74, no. 6, pp. 1983--1995 (2015)
9. Wolfgang, R.B., Podilchuk, C.I., Delp, E.J.: Perceptual Watermarks for Digital Images and Video. Proceedings of IEEE, vol. 87, no. 7, pp. 1108--1126 (1999)
10. Zhang, X.H., Lin, W.S., Xue, P.: Just-noticeable Difference Estimation with Pixels in Images. Journal of Visual Communication & Image Representation, vol. 19, pp. 30--41 (2008)
11. Liu, K.C.: Just Noticeable Distortion Model and its application in color Image Watermarking. SITIS 2008 - Proceedings of the 4th International Conference on Signal Image Technology and Internet Based Systems, pp. 260--267 (2008)
12. Zhang, X.H., Lin, W.S., Xue, P.: Improved Estimation for Just-noticeable Visual Distortions. Signal Processing, vol. 85, no. 4, pp. 795--808 (2005)

More Efficient Algorithm to Mine High Average-Utility Patterns

Jerry Chun-Wei Lin[1], Shifeng Ren[1], Philippe Fournier-Viger[2], Ja-Hwung Su[3,4], and Bay Vo[5]

School of Computer Science and Technology
[1]School of Natural Sciences and Humanities
[2]Harbin Institute of Technology (Shenzhen), Shenzhen, China
`jerrylin@ieee.org,renshifeng@stmail.hitsz.edu.cn,philfv@hitsz.edu.cn`
[3]Department of Information Management
Cheng Shiu University, Kaohsiung, Taiwan
[4]Department of Information Management
Kainan University, Taoyuan, Taiwan
`bb0820@ms22.hinet.net`
[5]Faculty of Information Technology
Ho Chi Minh City University of Technology, Ho Chi Minh City, Vietnam
`vd.bay@hutech.edu.vn`

Abstract. In this paper, an efficient algorithm with three pruning strategies are presented to provide tighter upper-bound average-utility of the itemsets, thus reducing the search space for mining the set of high average-utility itemsets (HAUIs). The first strategy finds the relationships of the 2-itemsets, thus reducing the search space of k-itemsets ($k \geq 3$). The second and the third pruning strategies set lower upper-bounds of the itemsets to early reduce the unpromising candidates. Substantial experiments show that the proposed algorithm can efficiently and effectively reduce the search space compared to the state-of-the-art algorithms in terms of runtime and number of candidates.

Keywords: high average-utility pattern mining; data mining; pruning strategies; lower bound.

1 Introduction

The main purpose of data mining techniques is to reveal the important, potential or useful information from various databases and frequent itemset mining (FIM) [1] plays an essential role in data mining. In traditional FIM, only the occurrence frequency of the item/sets is considered, but the other important factors such as quantity, profit, interestingness or weight. To reveal more useful and meaningful information, the high-utility itemset mining (HUIM) was proposed by Yao et al. [15, 16], which considers both quantity and unit profit of the items. The HUIM can be considered as the extension of FIM, which has a wide range of applications [2, 11, 13, 17].

© Springer International Publishing AG 2017
J.-S. Pan et al. (eds.), *Advances in Intelligent Information Hiding and Multimedia Signal Processing*, Smart Innovation, Systems and Technologies 64,
DOI 10.1007/978-3-319-50212-0_13

Although the HUIM can reveal more information than that of the FIM, it suffers, however, a serious problem since the length of the itemset is not concerned in HUIM. To handle this situation, the high average-utility itemset mining (HAUIM) was proposed to consider the size of the itemset to reveal the high average-utility itemsets (HAUIs). Hong et al. first designed a two-phase average-utility (TPAU) algorithm to mine HAUIs. A projection-based PAI algorithm [7], a tree-based high average-utility pattern (HAUP)-tree algorithm [9], and a HAUI-tree [14] were respectively designed to efficiently mine the HAUIs based on the TPAU algorithm. The HAUI-Miner algorithm [10] was then developed to enhance the mining performance based on the similar utility-list structure [13] to directly mine the HAUIs. The HAUI-Miner algorithm still suffers, however, the costly join operation for mining the HAUIs at each level. In this paper, we present an efficient algorithm with three pruning strategies for mining the HAUIs. The designed algorithm is then compared with the state-of-the-art approaches for mining the HAUIs and the results showed that the designed algorithm has better results in terms of runtime and number of candidates.

2   Related Work

Association-rule mining (ARM) is the fundamental topic of data mining, which has been widely developed and applied to real-life situations. The Apriori algorithm [1] was first developed to mine the association rules (ARs), which belongs to the level-wise approach. To improve mining performance, a compact tree structure called frequent pattern (FP)-tree was developed and its mining algorithm called the FP-growth was designed to derive the FIs from the FP-tree structure [5]. Since the FIM or ARM considers only the occurrence frequency of the item/sets, but the other factors. Thus, high-utility itemset mining (HUIM) was developed to consider both the quality and the unit profit of the item/sets to derive the set of high-utility itemsets (HUIs). The transaction-weighted utilization (TWU) model [12] was designed to maintain the transaction-weighted downward closure (TWDC) property of the high transaction-weighted utilization itemsets (HTWUIs), thus speeding up the mining process. Li et al. [8] then designed the isolated items discarding strategy (IIDS) to further reduce the candidate size for mining the HUIs based on the TWU model. The novel HUI-Miner algorithm [13] was first developed to construct the utility-list structures for mining the HUIs without candidate generation in the level-wise manner. To further improve the mining performance, the FHM algorithm [3] was then designed to keep the relationship of 2-itemsets, thus reducing the search space for mining the HUIs.

The problem of traditional HUIM is that the utility of the item/set is increased by the size of it. Thus, the utility of an itemset increases along with the number of items within it. To reveal better utility by considering the length of the item/set, the high average-utility itemset mining (HAUIM) [6] was thus designed to provide another utility measure. The first two-phase TPAU algorithm [6] was designed and the average-utility upper bound ($auub$) property was developed to estimate the upper-bound utility of the item/set to ensure that the completeness and the correctness of HAUIM. The high average-utility pattern (HAUP)-tree structure and its mining algorithm called HAUP-growth were then designed to overcome the multiple database scans [9]. A novel HAUI-Miner algorithm [10] was then developed to mine the HAUIs based on the compact list structure. In this paper, we present a more efficient algorithm to enhance the mining performance of HAUIM.

3 Preliminaries and Problem Statement

Let $I = \{i_1, i_2, ..., i_m\}$ be a finite set of m distinct items. A quantitative database is a set of transactions $D = \{T_1, T_2, ..., T_n\}$, where each transaction $T_q \in D$ $(1 \leq q \leq m)$ is a subset of I and has a unique identifier q, called its *TID*. Moreover, each item i_j in a transaction T_q, has a purchase quantity (a positive integer) denoted as $q(i_j, T_q)$. A profit table ptable = $\{pr(i_1), pr(i_2), ..., pr(i_m)\}$ indicates the profit value of each item i_j. A set of k distinct items $X = \{i_1, i_2, ..., i_k\}$ such that $X \subseteq I$ is said to be a k-itemset, where k is the length of the itemset. An itemset X is said to be contained in a transaction T_q if $X \subseteq T_q$.

An example database is shown in Table 1. A profit table of the running example is stated as: $ptable = \{A : 5, B : 1, C : 2, D : 3, E : 4, F : 1\}$.

Table 1: A transactional database.

TID	Items with their quantities
1	A:1, B:6, C:3, D:3, F:6
2	B:2, C:3, E:2
3	A:2, C:1, D:2, E:1
4	A:1, B:9, C:3, D:2, F:2
5	A:3, B:9, C:3, D:1, E:1
6	C:4, D:1, E:1
7	A:1, E:1

Definition 1. The utility of an item i_j in a transaction T_q is denoted as $u(i_j, T_q)$, and is defined as:

$$u(i_j, T_q) = q(i_j, T_q) \times pr(i_j). \tag{1}$$

Definition 2. The utility of an itemset X in transaction T_q is denoted as $u(X, T_q)$, and defined as:

$$u(X, T_q) = \sum_{i_j \in X \wedge X \subseteq T_q} u(i_j, T_q). \tag{2}$$

Definition 3. The utility of an itemset X in a database D is denoted as $u(X)$, and defined as:

$$u(X) = \sum_{X \subseteq T_q \wedge T_q \in D} u(X, T_q). \tag{3}$$

Definition 4. The transaction utility of a transaction T_q is denoted as $tu(T_q)$, and defined as:

$$tu(T_q) = \sum_{i_j \in X} u(i_j, T_q), \tag{4}$$

in which j is the number of items in T_q.

Definition 5. The total utility of all transactions in database D is denoted as TU, and defined as:

$$TU = \sum_{T_q \in D} tu(T_q). \tag{5}$$

The above definitions are used in traditional HUIM. Since the utility of itemsets revealed in the HUIM increases by the length of it, a better framework called high average-utility itemset mining (HAUIM) [10] was designed to provide a better measure by taking the number of items within the itemset.

Definition 6. The average-utility of an item (i_j) in a transaction T_q is denoted as $au(i_j)$, and defined as:

$$au(i_j, T_q) = \frac{q(i_j, T_q) \times pr(i_j)}{1} = \frac{u(i_j, T_q)}{1}. \tag{6}$$

Definition 7. The average-utility of a k-itemset X in a transaction T_q is denoted as $au(X, T_q)$, and defined as:

$$au(X, T_q) = \frac{\displaystyle\sum_{i_j \in X \wedge X \subseteq T_q} u(i_j, T_q)}{|X| = k}. \tag{7}$$

Definition 8. The average-utility of an itemset X in the database is denoted as $au(X)$, and defined as:

$$au(X) = \sum_{X \subseteq T_q \wedge T_q \in D} au(X, T_q). \tag{8}$$

Problem Statement: An itemset X is concerned as a HAUI iff its average-utility is no less than the minimum average-utility count as:

$$HAUI \leftarrow \{X | au(X) \geq TU \times \delta\}, \tag{9}$$

where TU is the total utility in the database D and δ is the minimum average-utility threshold, which can be defined by user's preference.

4 Proposed Algorithm and Pruning Strategies

In this paper, we present an algorithm with three new pruning strategies to more efficiently and effectively mine the HAUIs. The proposed algorithm adopts the average-utility (AU)-list structure [10] to maintain the related information of the itemsets. Based on the property of AU-list structure, the simple join operation is then used to find the k-itemsets ($k \geq 2$).

In the AU-list structure, the items are then sorted in their utility-descending order and their AU-lists are respectively built. Based on this sorting property, it can easily append the item after the processed item/set as a new itemset and hold the correctness and completeness of HAUIM. For the running example, the built AU-lists are given in Fig. 1.

A			C			D			B			E		
tid	*utility*	*mtu*	*tid*	*utility*	*mtu*	*tid*	*utility*	*mtu*	*tid*	*utility*	*mtu*	*tid*	*utility*	*mtu*
1	5	9	1	6	9	1	9	9	1	6	9	2	8	8
3	10	10	2	6	8	3	6	10	2	2	8	3	4	10
4	5	9	3	2	10	4	6	9	4	9	9	5	4	15
5	15	15	4	6	9	5	3	15	5	9	15	6	4	8
7	5	5	5	6	15	6	3	8						
			6	8	8									

$$u(A) \succ u(C) \succ u(D) \succ u(B) \succ u(E)$$

Fig. 1: AU-lists of 1-*HAUUBIs*.

In traditional HUIM, the relationships of 2-itemsets [3] are kept to reduce the search space for mining the HUIs. For the first pruning strategy, we adopt the similar idea and design a matrix to keep the average-utility of 2-itemsets. This structure keeps the *auub* values of 2-itemsets, which can be used to easily prune the unpromising candidates for k-itemsets ($k \geq 3$).

Pruning Strategy 1: If the average-utility of a 2-itemset in the built matrix is less than the minimum average-utility count, any superset of X would not be the HAUI and the search space of k-itemset ($k \geq 3$) in the enumeration tree can be greatly reduced.

Definition 9 (Lower-upper-bound average-utility, lubau). The lower-upper-bound average-utility of an itemset X is denoted as $lubau(X)$, and defined as:

$$lubau(X) = \frac{u(i_1) + u(i_2) + \ldots + u(i_m)}{m}, \tag{10}$$

where $X(= i_1 \cup i_2 \cup \ldots \cup i_m)$ and $u(i_1) \succ u(i_2) \succ \ldots \succ u(i_m)$.

Pruning Strategy 2: If the $lubau(X)$ is less than the minimum average-utility count, any superset of X cannot be the HAUI and can be directly ignored in the enumeration tree, thus reducing the search space and the cost of join operations of the AU-lists.

Definition 10 (tighter-upper-bound average-utility, tubau). The tighter-upper-bound average-utility of an itemset X is denoted as $tubau(X)$, and defined as:

$$tubau(X) = \frac{u(Y) + u(i_{m+1})}{2},$$ (11)

where $X(= Y \cup i_{m+1})(= i_1 \cup i_2 \cup \ldots \cup i_m \cup i_{m+1})$.

Pruning Strategy 3: If the $tubau(X)$ is less than the minimum average-utility count, any superset of X cannot be the HAUI and can be directly ignored in the enumeration tree, thus reducing the search space and the cost of join operations of the AU-lists.

Here, we present an efficient algorithm adopted with three pruning strategies. Details of the designed algorithm are stated in Algorithm 1 and the sub-search procedure is then described in Algorithm 2.

Algorithm 1: Proposed algorithm

    Input: D, $ptable$, δ.
    Output: The set of high-average utility itemsets (HAUIs).
1   scan D to find TU;
2   **for** *each item i_j in D* **do**
3   **if** $auub(i_j) \geq TU \times \delta$ **then**
4   1-$HAUUBIs \leftarrow$ 1-$HAUUBIs \cup i_j$;

5   sort items in 1-$HAUUBIs$ in their $u(i_j)$-descending order;
6   construct $AU.matrix$;
7   **for** *each $i_j \in$1-$HAUUBIs$* **do**
8   construct $i_j.AULs$;
9   **for** *each $i_j \in$ 1-$HAUUBIs$* **do**
10   **HAUIs-Search**$(i_j, i_j.AULs, AU.matrix, \delta, TU,$ 1-$HAUUBIs)$;
11   **return** HAUIs;

Algorithm 2: HAUIs-Search $(x, x.AULs, AU.matrix, \delta, TU)$

    Input: x, $x.AULs$, $AU.matrix$, δ, TU, 1-$HAUUBIs$.
    Output: The set of high-average utility itemsets (HAUIs).
1   set $X \leftarrow \emptyset$;
2   **for** *each $i_j \subseteq x \wedge i_j \in$ 1-$HAUUBIs$* **do**
3   **if** $i_k \in$ 1-$HAUUBIs \wedge u(i_j) \succ u(i_k), i_j \neq i_k$ **then**
4   **if** $AU.matrix(i_j, i_k) \geq TU \times \delta$ **then**
5   **if** $lubau(x) \geq TU \times \delta$ **then**
6   **if** $tubau(x) \geq TU \times \delta$ **then**
7                  $X \leftarrow x \cup i_k$;
8                  $X.AULs \leftarrow$ **Construct**$(x.AULs, i_k.AULs)$;

9   **if** $\frac{u(X)}{|X|} \geq TU \times \delta$ **then**
10      $HAUIs \leftarrow HAUIs \cup X$;
11   **if** $auub(X) \geq TU \times \delta$ **then**
12   **HAUIs-Search**$(X, X.AULs, AU.matrix, \delta, TU,$ 1-$HAUUBIs)$;

The Construct algorithm is similar as the HAUI-Miner algorithm shown in [10]. We uses the breath-first search in the designed algorithm to mine the HAUIs. Due to the page limit, details are skipped in this paper.

5   Experimental Evaluation

In this section, a series of experiments are conducted to evaluate the performance of the proposed algorithms compared to the state-of-the-art algorithms of HAUIM, such as PAI [7] and HAUI-Miner [10] in terms of runtime and number of candidates. Four real-life datasets such as accidents [4], chess [4], kosarak [4], and pumsb [4] are used in the experiments. Experiments are then evaluated under different minimum average-utility thresholds.

5.1   Runtime

In this subsection, runtime of the compared algorithms are then evaluated and the results are given in Fig. 2.

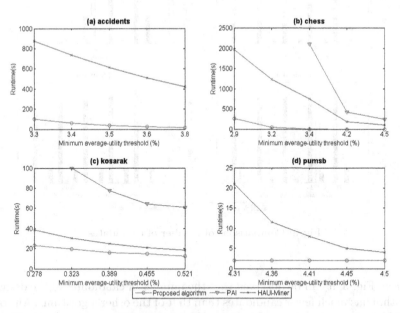

Fig. 2: Comparisons of runtime.

In Fig. 2, it is obvious to see that the runtime decreases along with the increasing of minimum average-utility threshold. This is reasonable since when the minimum average-utility threshold is set higher, fewer HAUIs are discovered; less computation is required to find the set of HAUIs. It also can be seen that the

designed algorithm perform well compared to the state-of-the-art algorithms under different minimum average-utility thresholds. Thus, the designed algorithm
with pruning strategies is efficient to speed up the mining performance.

5.2 Number of Candidates

In this section, the number of candidates is then evaluated to show the search
space for mining the HAUIs. This criteria make a significant impact on the
computational time. Note that the join operation is the most costly operation
under list-framework as said in FHM [3]. The results of the compared algorithms
are then shown in Fig. 3.

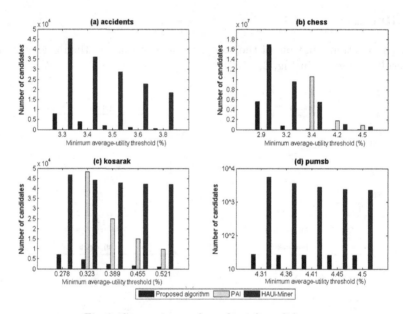

Fig. 3: Comparisons of number of candidates.

From Fig. 3, it can be observed that the number of candidates of the designed
algorithm has much fewer candidates than that of the other algorithms. Although
the HAUI-Miner algorithm adopts the AU-list structure to mine the HAUIs, the
amounts of unpromising candidates are generated without any pruning strategy
based on the *auub* property used in traditional HAUIM. The PAI algorithm
requires the most candidates for mining the HAUIs. This is reasonable since the
two-phase model was adopted in the PAI algorithm. Thanks to the developed
pruning strategies, the unpromising candidates can be greatly pruned, as well
as the search space in the enumeration can be greatly reduced.

6   Conclusion

In this paper, we present an algorithm three pruning strategies to efficiently prune the unpromising candidates and reduce the search space for mining the HAUIs. From the conducted experiments, it can be observed that the designed algorithm and the pruning strategies have great performance compared to the state-of-the-art approaches for mining the HAUIs.

Acknowledgment

This research was partially supported by the National Natural Science Foundation of China (NSFC) under grant No.61503092, by Ministry of Science and Technology, Taiwan, R.O.C. under grant No. MOST 104-2632-S-424-001 and MOST 104-2221-E-230-019.

References

1. Agrawal, R., Srikant, R. Srikant: Fast algorithms for mining association rules. International Conference on Very Large Data Bases, pp. 487-499 (1994)
2. Erwin, A., Gopalan, R. P., Achuthan, N. R.: Efficient mining of high utility itemsets from large datasets. The Pacific-Asia Conference on Advances in Knowledge Discovery and Data Mining, pp. 554-561 (2008)
3. Fournier-Viger, P., Wu, C. W., Zida, S., Tseng, V. S.: FHM: faster high-utility itemset mining using estimated utility co-occurrence pruning. International Symposium on Methodologies for Intelligent Systems, pp. 83-92 (2014)
4. Frequent itemset mining implementations repository, `http://fimi.ua.ac.be/data/` (2016)
5. Han, J., Pei, J., Yin, Y., Mao, R.: Mining frequent patterns without candidate generation: a frequent-pattern tree approach. Data Mining and Knowledge Discovery, 8, pp. 5387 (2004)
6. Hong, T. P., Lee, C. H., Wang, S. L.: Effective utility mining with the measure of average utility. Expert Systems with Applications, 38(7), pp. 8259-8265 (2011)
7. Lan, G. C., Hong, T. P., Tseng, V. S.: Efficiently mining high average-utility itemsets with an improved upper-bound. International Journal of Information Technology and Decision Making, 11(5), pp. 1009-1030 (2012)
8. Li, Y. C., Yeh, J. S., Chang, C. C.: Isolated items discarding strategy for discovering high utility itemsets. Data and Knowledge Engineering, 64(1), pp. 198-217 (2008)
9. Lin, C. W., Hong, T. P., Lu, W. H.: Efficiently mining high average utility itemsets with a tree structure. The Asian Conference on Intelligent Information and Database Systems, pp. 131-139 (2010)
10. Lin, J. C. W., Li, T., Fournier-Viger, P., Hong, T. P., Zhan, J., Voznak, M.: An efficient algorithm to mine high average-utility itemsets. Advanced Engineering Informatics, 30(2), pp. 233243 (2016)
11. Liu, Y., Liao, W. K., Choudhary, A.: A fast high utility itemsets mining algorithm. The International Workshop on Utility-based Data Mining, pp. 90-99 (2005)
12. Liu, Y., Liao, W. K., Choudhary, A.: A two-phase algorithm for fast discovery of high utility itemsets. Lecture Notes in Computer Science, pp. 689-695 (2005)

13. Liu, M., Qu, J.: Mining high utility itemsets without candidate generation. ACM International Conference on Information and Knowledge Management, pp. 55-64 (2012)

14. Lu, T., Vo, B., Nguyen, H. T., Hong, T. P.: A new method for mining high average utility itemsets. Lecture Notes in Computer Science, pp. 33-42 (2014)

15. Yao, H., Hamilton, H. J., Butz, C. J.: A foundational approach to mining itemset utilities from databases. SIAM International Conference on Data Mining, pp. 482-486 (2004)

16. Yao, H., Hamilton, H. J., Geng, L.: A unified framework for utility based measures for mining itemsets. The International Workshop on Utility-Based Data Mining, pp. 27-28 (2006)

17. Yen, S. J., Lee, Y. S.: Mining high utility quantitative association rules. International Conference on Big Data Analytics and Knowledge Discovery, pp. 283-292 (2007)

Infrared Video based Sleep Comfort Analysis using Part-based Features

Lumei Su, Min Xu, and Xiangsong Kong

Xiamen University of Technology,
No.600 Ligong Road, Xiamen, China

Abstract. This work investigated a new challenging problem: how to analyze human sleep comfort which is an urgent problem in intelligent home and medical supervision, especially in intelligent temperature control of air conditioners. To overcome this problem, a robust sleep posture feature descriptor named part-based feature descriptor is firstly proposed to analyze human sleep comfort not matter human body is covered by a sheet or not. Experiments on a custom-made database established by a remote infrared camera demonstrated that the proposed method has promising performance for on-line human sleep comfort analysis.

Keywords: Human sleep comfort estimation; Part-based features; Infrared camera

1 Introduction

Sleep is a complex behavioral state that occupies one-third of the human life span. Sleep quality is of fundamental importance to human health. Sleep gives your body a rest and allows it to prepare for the next day. It's like giving your body a mini-vacation. Sleep also gives your brain a chance to sort things out. It is important for a wide variety of applications to be able to analyze sleep comfort and sequentially improve sleep quality, especially in intelligent home and medical supervision.

Ambient temperature is a crucial factor for sleep quality. People may wake up when the ambient temperature is too high or too low. Air conditioner is widely used to control the ambient temperature in our daily life. However, different sleep stages require different ambient temperatures[1], and current automatic intelligent air conditioner can only achieve thermostatic control which cannot meet the demands of different sleep stages and would inevitable affect sleep quality. Therefore, estimating real-time comfortable levels of sleep is very useful for intelligent temperature control technology of air conditioners.

There has been several works on sleep quality analysis or sleep comfort analysis[2–6]. Some sleep analysis techniques using inertial sensors which including polysomnographic measurements, accelerometers, gyroscopes, and magnetometers have been developed for doing this [7–9].However, these sensors need to be attached to human body, which causes inconvenience during sleep and

© Springer International Publishing AG 2017

J.-S. Pan et al. (eds.), *Advances in Intelligent Information Hiding*
and Multimedia Signal Processing, Smart Innovation, Systems and Technologies 64,
DOI 10.1007/978-3-319-50212-0_14

Table 1. Sleep comfort levels

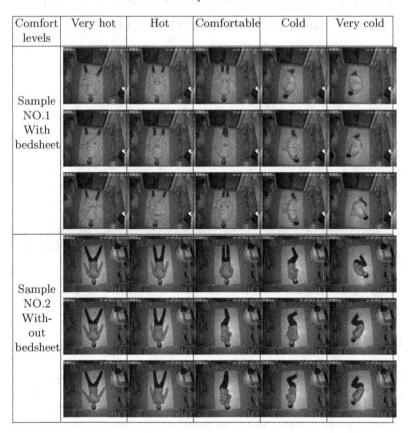

Comfort levels	Very hot	Hot	Comfortable	Cold	Very cold
Sample NO.1 With bedsheet					
Sample NO.2 Without bedsheet					

consequently decrease sleep quality. Recently, the approach of dispersed sensors embedded in the mattress has become promising. The sensors embedded mattress can record the pressure distribution of human bodies, which forms a pressure image.However, these methods for sleep posture recognition are based on local features and individualized pre-training, which requires considerable efforts to apply in a large population.

Compared with above mentioned wearable or contact-based sensors, video cameras [10] and infrared cameras [11] are more suitable for sleep analysis because these sensing modalities are unobtrusive to users and minimize the privacy concerns. They can remotely detect global human sleep posture without attaching to the human body.Heinrich et al.[10] proposed a camera-based system combining video motion detection, motion estimation, and texture analysis with machine learning for sleep analysis. The system is robust to time-varying illumination conditions while using standard camera and infrared illumination hardware. Fan et al.[12] developed a home sleep screening technique with the aim of assisting the evaluation of quality of sleep in smart TV environment using day-and-night video cameras.However, existing cameras-based approaches suffers from privacy concerns and image noise due to low visibility at night. Few effective image segmentation methods are used to clearly segment the human body from low visible background. Whats more, there has not been a suitable sleep posture descriptor or model used to estimate human sleep comfort.

In our literature review, we found that the challenges of sleep comfort analysis are to develop an effective way to well describe or extract sleep comfort features even human body is fully or partially covered, while existing sleep quality or comfort studies mostly focus on using powerful classification methods to analyze noisy sleep features and rarely investigate covered body problem. To overcome this problem, a part-based feature descriptor is firstly proposed to estimate sleep comfort not matter human body is covered by a sheet or not. The novel sleep posture feature descriptor is given based on sufficient sleep data analysis and experimental results demonstrated that the proposed method has promising performance for sleep thermal comfort estimation.

2   Proposed method

2.1   Sleep thermal comfort categories

To evaluate sleep comfort levels, the categories of sleep thermal comfort should be defined first. Actually, sleep thermal comfort analysis is a new challenge problem and there has not been an available work on the definition of the categories of sleep thermal comfort so far. In order to analyze the sleep thermal comfort categories, 120 sequences of sleep video are collected by a remote infrared camera in real situation. Some frames of sleep video are shown in Table 1.

As shown in Table 1, when the object feels hot, he automatically stretches his body with supine posture. The hotter, the outer body is stretched. In contrast, when the object fells cold, he automatically huddles with holding his arms and legs close to his body. The colder, the closer he huddles his body. When the object feels the ambient temperature is comfortable, he always lies supine on the bed with straightening his legs, putting his hand close to his body and slightly opening his feet. Sometimes the object lies on his side with slightly huddling his body when he feels comfortable. Based on this observation, five categories of sleep postures are defined corresponding to five levels of sleep thermal comfort that are very hot, hot, comfortable, cold and very cold. In Table 1, three typical frames are given for each sleep comfort level in case of sleeping with bedsheet or not.

2.2   Sleep comfort features analysis based on contours of human body

By analyzing our collected sequences, it is found that the sleep posture varies from different sleep comfort levels. When the subject feels very hot, he stretches his body up to his limit with largely opening his arms and feet. It is found that the distance from head to feet is nearly the same as his height and the area of bounding polygon is maximal. When the subject feels hot, he also stretches his body with opening his arms and feet, while the stretching range is relatively smaller than the very hot case. The same as the case of very hot, it is also found that the distance from head to feet is nearly the same as his height and the area of bounding polygon is maximal. When the subject feels comfortable, the observation could be classified into two cases. In the case of lying supine on the bed with straightening his legs, putting his hand close to his body and slightly opening his feet, it is also found that the distance from head to feet is nearly the same as his height and the area of bounding polygon is maximal. In the other case of lying on his side with slightly huddling his body, it is found that the distance from head to feet is shorter than his height and the area of bounding polygon is also smaller than other cases because of overlapping some parts of body. When the object fells cold, he lies on his side with huddling his body, putting his feet together and holding his arms close to his body. The degree of huddling is larger than other cases. Because more parts of body overlap together, the area of bounding polygon is much smaller than other cases. And the distance from head to feet is much shorter than his height because of huddling his body.

Based on the above observation, it is found that there are regular contour features not matter sleeping with or without bedsheep. The contour features of sleep postures vary from different thermal feels. First of all, the distance between feet varies from different comfort levels. The distance between feet becomes longer corresponding to hotter feel, while the distance between feet is almost invariable and shorter when objects feel cold with putting feet together. Secondly, the distance between head and feet also varies from different comfort levels. The variation of the distance between head and feet is slight when objects feel hot while the distance becomes shorter when objects feel colder. Because the bedsheet and hands always overlap together and it is difficult to clearly segment them in infrared images, the shape features of hands are unavailable as sleep postures in sleep comfort analysis.

2.3 A part-based sleep posture descriptor for sleep thermal comfort analysis

According to above posture features analysis, a novel sleep posture descriptor is proposed to describe sleep thermal comfort features. This feature descriptor mainly use the proportion of the distance between feet W to body height h and the proportion of the distance between head and feet H to body height h, which are denoted as Wh and Hh respectively. Three crucial contour points in sleep posture binary images are defined to calculate the sleep posture features W and H. They are the extreme top left corner point of sleep contour $C1(a1, b1)$, the extreme top right corner point of sleep contour $C2(a2, b2)$ and the extreme bottom left corner point of sleep contour $C3(a3, b3)$. The distance between feet W is given by $W = a3 - a1$ and the distance between head and feet H is given by $H = b2 - b1$.

Table 2. Some sleep posture features extraction results

Comfort levels	Raw sleep images	Processed images	h (/pixel)	W (/pixel)	H (/pixel)	Wh =W/h	Hh =H/h
Very hot			179	97	183	0.5419	1.0223
			179	98	185	0.5475	1.0335
Hot			179	71	182	0.4022	1.0168
			179	78	184	0.4358	1.0279
Comfortable			179	26	179	0.1453	1
			179	31	181	0.2961	1.0112
Cold			179	43	157	0.2402	0.8771
			179	51	158	0.2849	0.8826
Very cold			179	32	135	0.1788	0.7542
			179	50	129	0.2793	0.7207

Table 3. Sleep comfort level recognition rate based on BP neural network

Comfort levels		Very hot	Hot	Comfortable	Cold	Very cold
Comfort level value		1	2	3	4	5
First test	Test samples	9	11	8	13	9
	Correct sample	6	8	6	10	8
	Recognition rate (%)	66.67	72.72	75.00	76.92	88.89
Second test	Test samples	10	9	9	11	11
	Correct sample	7	7	8	8	7
	Recognition rate (%)	70.00	77.78	88.89	72.73	63.63

The learning procedure of sleep posture features W and H is detailed as following. First of all, search the first point with pixel value 1 from left to right and from top to bottom in each binary images x. Then, move down this first point with empirical value 5 pixels to offset the effect of convex point and denote it as contour point $C1(a1, b1)$. Thirdly, search the extreme right point in the same row of contour point C1 and denote it as contour point $C2(a2, b2)$. Fourthly, search the first point with pixel value 1 from left to right and from bottom to top in each binary images x and denote it as contour point $C3(a3, b3)$. Finally, for $i = 1, , 20$, spin binary image x anticlockwise around its center with -20+2i degree. Obtain contour point $C1$, $C2$ and $C3$ by former steps. Calculate Wc and Hc by $Wc = a3 - a1$ and $Hc = b2 - b1$. Calculate $W = max(W, Wc)$ and $H = max(H, Hc)$. Final sleep posture features W and H are obtained.

3   Experimental results and discussion

To investigate the performance of the proposed sleep comfort evaluation method, a sleep posture database was established by using a 120 fps infrared camera with a resolution of 960 × 582. We asked the subjects sleep under our sleep posture capture system every night for 30 days. The ambient temperature varies from $24°C$ to $29°C$. For each collected sequence, we manually labeled the sleep comfort level (very hot, hot, comfortable, cold, and very cold) for each frame.

Before extracting sleep posture features, four image processings including denoise, background segmentation, contour detection and redundant data removing are applied to raw sleep posture image. Level set algorithm is decided to detect contour of sleep posture based on the comparison results with other methods.

A novel sleep posture features extraction is proposed to analyze sleep comfort features, as illustrated in Section 2. In order to offset the effect of convex point, contour point $C1(a1, b1)$ is moved down with empirical value 5 pixels and the posture feature W is refined. The empirical value is decided by doing a lot of experiments. Some sleep posture features extraction results given in Table 2.

To evaluate sleep thermal comfort in real time, a sleep posture model based on part-based features is sequently established. Two classification methods are used to estimate sleep comfort level. One is BP neural network which has ex-

Table 4. Sleep comfort level recognition rate based on FKM algorithm

Comfort levels		Very hot	Hot	Comfortable	Cold	Very cold
Comfort level value		1	2	3	4	5
First test	Test samples	10	10	10	10	10
	Correct sample	10	10	10	10	10
	Recognition rate (%)	100	100	100	100	100
Second test	Test samples	20	20	20	20	20
	Correct sample	20	20	20	20	20
	Recognition rate (%)	100	100	100	100	100

tensive application in many fields with the advantage of simple structure and mature technique. We evaluated the results on the basis of the leave-one-out cross-validation methodology. Table 3 shows two sleep comfort level recognition test results by using BP neural network. The recognition performance of BP neural network was not satisfactory. By analyzing the misrecognition samples, it is found that these samples which are difficult to give a suitable label in between two cases are wrongly labeled before training BP neural network. For example, the value of Wh in hot case should be smaller than that in very hot case. However, the value of Wh in hot case should be bigger than that in very hot case for these wrongly labeled samples which will consequently affect recognition results.

Because of the poor recognition performance of BP network, another recognition method Fuzzy K-Means (FKM) algorithm is proposed to estimate sleep comfort levels. The value of posture features in training samples are used to obtain centers of each clustering. The recognition results by using FKM algorithm illustrated that FKM algorithm outperformed BP network in sleep comfort level estimation as shown in Table 4. The recognition rate reaches 100% for every sleep comfort level.

4   Conclusion

This work proposes a novel sleep comfort analysis method that is used to analyze sleep posture and measure the current comfortable levels of sleep based on real-time inputs from infrared video. Our proposed novel sleep comfort feature descriptor is on the basis of experimental and theoretical analysis. The results of experiments using a custom-made sleep thermal comfort dataset indicated that the proposed sleep thermal comfort estimation method can successfully classify detected sleep video into five sleep thermal comfort levels. Our research is very useful for current intelligent air conditioner technology which is limited to thermostatic control. The intelligent air conditioner assisted by our method can adaptively control the ambient temperature based on real-time sleep comfort measurement result and consequently improve sleep quality.

Future work includes applying our proposed method in thermostatic control of intelligent air conditioner technology. In real situation, multi sleep thermal comfort estimation is essential to decide a suitable thermostatic control of intelligent air conditioner. We can optimize the trade-off between multi sleep comfort feels by considering multi sleep thermal comfort estimation as a constrained optimization problem.

Acknowledgments. This research is supported by the Scientific Research Foundation of Xiamen University of Technology (Grant No. YKJ13013R)and the Scientific Research Foundation for Young and middle-aged teachers in Fujian Province (Grant No. JA15386).

References

1. Kryger and Meir, H.: Principles and practice of sleep medicine. Elsevier/Saunders (2005)
2. Merilahti, J., Saarinen, A., Parkka, J., Antila, K., Mattila, E., Korhonen, I.: Long-Term Subjective and Objective Sleep Analysis of Total Sleep Time and Sleep Quality in Real Life Settings. In: 29th IEEE Conference on Engineering in Medicine and Biology Society, pp.5202–5205. IEEE Press (2007)
3. Dafna, E., Tarasiuk, A., Zigel, Y.: Sleep-quality assessment from full night audio recordings of sleep apnea patients. In: IEEE Conference on Engineering in Medicine and Biology Society, pp.3660–3663. IEEE Press (2012)
4. Zhu, X., Chen, W., Kitamura, K. I, Nemoto, T.: Estimation of Sleep Quality of Residents in Nursing Homes Using an Internet-Based Automatic Monitoring System. In: 11th IEEE Conference on Pervasive Ubiquitous Intelligence and Computing, pp.659–665. IEEE Press (2014)
5. Butt, M., Moturu, S.T., Pentland, A., Khayal, I.: Automatically captured sociability and sleep quality in healthy adults. In: 35th IEEE Conference on Engineering in Medicine and Biology Society, pp.4662–4665. IEEE Press (2013)
6. Pino, E.J., Dorner De la Paz, A., Aqueveque, P.: Noninvasive Monitoring Device to Evaluate Sleep Quality at Mining Facilities. Industry Applications. 51, 101–108 (2014)
7. Kishimoto, Y., Kutsuna, Y., Oguri, K.: Detecting Motion Artifact ECG Noise During Sleeping by Means of a Tri-axis Accelerometer. In: 29th IEEE Conference on Engineering in Medicine and Biology Society, pp.2669–2672. IEEE Press (2007)
8. Gautam, A., Naik, V.S., Gupta, A., Sharma, S.K., Sriram, K.: An smart phone-based algorithm to measure and model quantity of sleep. In: 7th IEEE Conference on Communication Systems and Networks, pp.1–6. IEEE Press (2015)
9. Han, H., Jo, J., Son, Y., Park, J.: Smart sleep care system for quality sleep. In: IEEE Conference on Information and Communication Technology Convergence, pp.393–398. IEEE Press (2015)
10. Heinrich, A., Di Geng, Znamenskiy, D., Vink, J.P., de Haan, G.: Robust and Sensitive Video Motion Detection for Sleep Analysis. Biomedical and Health Informatics. 18, 790–798 (2014)
11. Kurylyak, Y., Lamonaca, F., Mirabelli, G., Boumbarov, O., Panev, S.: The infrared camera-based system to evaluate the human sleepiness. In: IEEE International Workshop on Medical Measurements and Applications, pp.253–256.IEEE Press (2007)
12. Fan, C. T., Wang, Yuan-Kai, Chen, Jian-Ru: Home sleep care with video analysis and its application in smart TV. In: 3rd IEEE Global Conference on Consumer Electronics, pp.42–43. IEEE Press (2014)

Fall Detection Algorithm Based on Human Posture Recognition

Kewei Zhao[1,2,3*], Kebin Jia[1,2,3*] and Pengyu Liu[1,2,3*]

1 Beijing Laboratory of Advanced Information Networks, Beijing, China
2 College of Electronic Information and Control Engineering, Beijing University of Technology,
Beijing, China
kebinj@bjut.edu.cn

Abstract. In the background of global aging, more attention should be paid to the elders' health and the equality of their life. Nowadays falls became one of greatest danger for old people. Almost 62% of injury-related hospitalizations for the old are the result of it. In this paper, we propose a new method to detect fall based on judging human's moving posture from the video. It consists of three main parts, detecting the moving object, extracting the feature and recognizing the pattern of behavior. To improve the precision and increase the speed of the detection, we adopt two layers codebook background modeling and codebook fragmentation training. Two level SVM method to recognize the behavior: In the first level of the SVM classifier, we distinguish the standing posture and other posture by the feature of moving object, such as the ratio of the major and minor axis of the ellipse. In the second level of the SVM classifier, angle of the ellipse and head moving trajectory to judge the falls and squat. The experimental results indicate that our system can detect fall effectively.

Key words: fall detection, two layers codebook background modeling, codebook fragmentation training, two level SVM classifier

1 Introduction

Based on the statistic census from the National Bureau of statistic in 2015 shows that the population of people over 60 in china is 221.82 million, 16.15 of the total population, and the population over 65 years old is 143.74 million, 10.47% of that. Compared with the sixth national census, the proportion have increased 2.89% and 1.60%, china is stepping into the population aging society. Facing the serious tendency of the population aging, government should establish a lot of now information system to ensure the safety of the old people. According to the statistic, more than 35% old people who is older than 65 years old have fall down before, 63% of the old people died by falls, it even reached 70% among the old people over 75 years old[1]. In conclusion, it is especially important to detect the falls for the protecting the safety of old people.

© Springer International Publishing AG 2017   119
J.-S. Pan et al. (eds.), *Advances in Intelligent Information Hiding
and Multimedia Signal Processing*, Smart Innovation, Systems and Technologies 64,
DOI 10.1007/978-3-319-50212-0_15

There are lots of techniques to detect fall. They can divided roughly into three categories by different kinds channel of signal[2]: one of them is based on some wearable devices, the problem of such detectors is that old people often forget to wear or charge devices; another one is based on the environment devices, setting up the equipment is so complex and expensive; the third one is based on the video, it can not only perform a 24-hour monitoring, but also can avoid the danger cause by forgetting to wear. The paper proposed a method which is based on judge the posture of the old people, it is totally implemented by the image processing techniques, it can detect the posture of fall effectively, and also satisfy the require of real-time processing on the hardware with low computing capability. The basic principle show below: it recognize the moving object by the modified codebook, Then, human characteristic matrices are constructed based on the information of human body posture extracted from human silhouette and are used as features to train SVM classifier for fall detection.

2   Related Work

Some method to detect fall have been purposed in recent research. Accelerate sensors can collect values and direction information, it is widely use in the fall detection systems based on wearable sensor. Dai et al. get the accelerator values and latitude to detect fall with the accelerator sensor in the mobile phone. After fall, it can directly alarm the family, it is easy to carry and alarm rapidly. The false-negative is 2.67% and false-positive is 8.7%[3]. Bourke et al. designed a fall detection system based on two-link gyroscope sensor. It can be bind on the chest and detects fall by the angular acceleration and velocity. The system has 100% sensitivity、 specificity and accuracy[4].

Scott et al. designed a matters fall detection system, it consists of lots of airmattress and pressure sensors, by the changes of pressure on the airmattress guesses the pressure changes of human. In this way, we can get the posture changes of the human. It is used for detection while people is sleeping or scribing, it is comfortable and convenient[5]. Litvak et al. designed a detection system based on the analyzing the shake of the ground and the sound. It combines the accelerator sensor and microphone, it doesn't need to wear anything and affects people move less. It's sensitivity is 97.5% and specificity is 98.6%[6].

In the video frame fall detection system, Jean Menunier et al. proposed an advance detection system based on MHI[7]. They extracted the human in the foreground by cutting the background, it computes the ratio of the major and minor axis、 obliquity and MHI statistic of the ellipse to detect the fall. The camera always be set at the roof to have a wide view avoiding occlusion.

3   Technical Details

In this paper, we propose a method to detect the fall of the old people. It can divided into three part: moving object detection, feature extraction and behavior recognization(Fig.1). We use a modified codebook foreground detection algorithm in making object detection, compared with the old algorithm, it improves the precision of the foreground extraction and makes the extraction more real-time. We use two level SVM classifier to recognize the fall and squat in object behavior recognization.

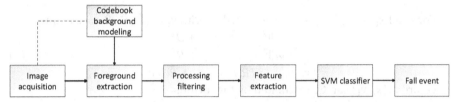

Fig.1. Flow diagram of the overall fall detection procedure.

3.1   Foreground extraction

Codebook algorithm is to establish the background model in the long-term observation sequence with quantization technique. The algorithm will make every pixel a codebook model. During the initial algorithm training, $c = \{x_1, x_2, \cdots, x_n\}$ represent the training sequence of one pixel, which is made up of RGB vectors, $\xi=\{c_1, c_2,...,c_L\}$ represents a pixel codebook which consists of L codeword, It has different numbers of code word depending on the variation of the pixel. Each codeword $c_i(i = 1 .... l)$ consist of a RGB vector $v_i = (\overline{R}_i, \overline{G}_i, \overline{B}_i)$ and $aux_i = \langle \check{I}_i, \hat{I}_i, f_i, \lambda_i, p_i, q_i \rangle$.

We match the currently codeword c_m with each instance x_t, then we use the matching codeword as the estimation of the instance. We generally using colordist and brightness to evaluate whether a codeword is in a best matching status.

Foreground-background segmentation

（1）Let $L \leftarrow 0$; $Z \leftarrow \emptyset$

（2）When t = 1....N

1. $x_t = (R, G, B), I \leftarrow \sqrt{R^2 + G^2 + B^2}$
2. If a. $colordist(x_t, v_m) \leq \varepsilon_1$

    b. $brightness\big(I, \langle \check{I}_m, \hat{I}_m \rangle\big) = true$

    Searching the codeword c_i in $\xi=\{c_1, c_2,...,c_L\}$to match x_t.

3. If $Z \leftarrow \emptyset$ or the no matching codeword having been found, then $L \leftarrow L + 1$, and

   a new codeword c_L will be greated：

    a. $v_L \leftarrow (R, G, B)$

    b. $aux_L \leftarrow \langle I, I, 1, t - 1, t, t \rangle$

4. otherwise refresh the matching codeword c_i, including $v_m = (\bar{R}_m, \bar{G}_m, \bar{B}_m)$
and $aux_m = \langle \check{I}_m, \hat{I}_m, f_m, \lambda_m, p_m, q_m \rangle$:

    a. $v_m \leftarrow \left(\frac{f_m R_m + R}{f_m + 1}, \frac{f_m G_m + G}{f_m + 1}, \frac{f_m B_m + B}{f_m + 1} \right)$

    b. $aux_m \leftarrow \langle \min\{I, \check{I}_m\}, \max\{I, \hat{I}_m\}, f_m + 1, \max\{\lambda_m, t - q_m\}, p_m, t \rangle$

（3）We compute the maximum time interval of each codeword $c_i (i = 1 .... l)$
between with every pixel having been matched again

    $\lambda_i \leftarrow \max\{\lambda_i, (N - q_i + p_i - 1)\}$

For pixel $x_t = (R, G, B)$ and a codeword c_i, with $v_i = (\bar{R}_i, \bar{G}_i, \bar{B}_i)$, existing:

$$\|x_t\|^2 = R^2 + G^2 + B^2$$
$$\|v_i\|^2 = \bar{R}_i^2 + \bar{G}_i^2 + \bar{B}_i^2 \tag{1}$$
$$(x_t, v_t)^2 = (\bar{R}_i R + \bar{G}_i G + \bar{B}_i B)^2$$

$colordist(x_t, v_i)$ can be calculated by the formula.

$$p^2 = \|x_t\|^2 \cos^2 \theta = \frac{(x_t, v_i)^2}{\|v_i\|^2} \tag{2}$$

$$colordist(x_t, v_i) = \delta = \sqrt{\|x_t\|^2 - p^2} \tag{3}$$

This paper modified the model built by codebook in following parts: we build
a module consist 30*30 pixels, each module one or several codewords will be in
the codebook. The modified codebook method can prevent failing of foreground
extraction when camera shake and it can reduce memory pressure, foreground
extraction show below Fig.2.

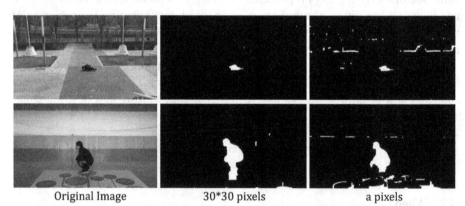

Original Image   30*30 pixels   a pixels

Fig.2. foreground extraction result

The twice codebook matching mentioned in this paper, background
subtraction library show below Fig.3. The codeword in codebook B come from
codebook A When $\lambda_i < \lambda_0$ (λ_0 is threshold).

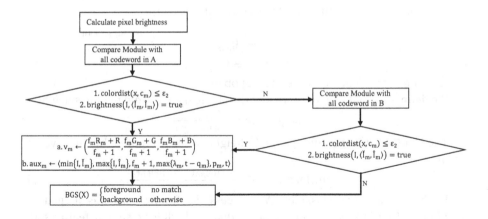

Fig.3. background subtraction library procedure

3.2   Feature Extraction And Posture Classification

Behavior recognition is a complex problem and have been researched for a long time. In paper[8], humans behavior has been modeled by two layer HMM(hidden markov model), while in paper[9],it use RBF(radial basis neural network) to model the behavior. We method concentrate on fall, so we use SVM, it is a good binary classifier. We designed a two level SVM to judge the old people's posture. The first level classifier judge whether the old man is standing, when the result shows the old man is not standing, we put it into the second level classifier. In this level, we can judge whether the old man fall, when we find the old man fall, the system will alarm, the two level SVM use RBF kernel and confirm the optimal parameters g=0.03125, penalty factor c=4 by the method of gridding search.

The first level classifier need C_{motion}, which represent the intense of the MHI(motion history image), show below Fig.4, the ratio of the major and minor axis of the ellipse, the height of the old man to train and judge.

Fig.4. motion history image

1. the ratio of the major and minor axis of the ellipse

$$I_{min} = \frac{u_{20}+u_{02}-\sqrt{(u_{20}-u_{02})^2+4u_{11}^2}}{2} \quad I_{max} = \frac{u_{20}+u_{02}+\sqrt{(u_{20}-u_{02})^2+4u_{11}^2}}{2} \quad (4)$$

$$RateY2X = I_{max}/I_{min} \quad (5)$$

2. the height feature of the moving object

We defined HeightRate to represent the changing of the old people's height.

$$HeightRate = \frac{(Y_{max}-Y_{min})}{ManHeight} \quad (6)$$

The ManHeight is the old people's height. It is computed by the average formula

$$ManHeight = \frac{Height_1+Height_2+\cdots+Height_n}{n} \quad (7)$$

In the second level classifier. We process the frames that was judged not standing by first level classifier. Because squat and fall have similar features, the second level classifier mainly distinguish squat and fall. The second level classifier uses four features: the ratio of the major and minor axis of the ellipse, the project area of the moving object, the inclination angle of ellipse, the variation tendency of the major and minor axis.

1. the project area of the moving object

The projection area represent the sum of non-zero moving pixels in the moving object after binaryzation. When squatting, the old man will carried up his body and the project area will be small, when fall, the body will stretch, the area will be larger. So we can use it to distinguish the squat and fall.

2. the inclination angle of ellipse

The changing of the inclination angle can distinguish squat and fall well, when the old people squats, the angle will not change a lot, but when the old man fall, the body change will vertical to horizontal, the angle will change a lot. Formula is as follow:

$$\theta = \begin{cases} \tan^{-1}\left\{\frac{e_{21}}{e_{11}}\right\} & \text{if } \lambda_M = \lambda_1 \\ \tan^{-1}\left\{\frac{e_{22}}{e_{12}}\right\} & \text{if } \lambda_N = \lambda_2 \end{cases} \quad (8)$$

4   Experimental Results

To evaluate the accuracy of the algorithm, we invite 10 college students to do experiment, we choose three outside scene and three inside scene. The postures include: falling but not lying, falling and lying, squat and fall to the left/right(Fig.5). we designed some postures in table1. Experimenters will act some postures with fall to accomplish the experiments. Video data will be collect by HD camera. Each experimenter will do 20 experiments, all the experimenters will accomplish 200 experiments.

The statistic results shows in table1. A_L refers to whether alarm. N_A is number of actions. N_C is number of correctly detected events. N_F is number of falsely detected events. N_T is number of no alarm. R is the recognition rate. Falling down1 is falling down forward but not lying. Falling down2 is falling down backward but not lying. Falling down3 is falling down forward and lying. Falling down4 is falling down backward and lying. Falling down5 is falling on the left/right side. From the results, we know that our method have high validity, it can get most of falls, but it has misdescription when the experimenter fall but not lie and keep balance after fall.

Table 1 RECOGNITION RATE FOR VARIOUS EVENTS

Events	A_L	N_A	N_C	N_F	N_T	R
Stand	N	50	0	0	50	100
Walk	N	50	0	2	48	96.0
Run	N	30	0	1	29	96.7
Falling down1	Y	60	55	4	1	91.7
Falling down2	Y	60	57	1	2	95.0
Falling down3	Y	60	54	4	2	90.0
Falling down4	Y	60	54	3	3	90.0
Falling down5	Y	60	53	5	2	88.3
squat	Y	60	51	4	5	85.0
ALL		490	451	24	15	92.1

Inside1 squat   Inside2 before fall   Inside3 fall after

Outside1 before fall   Outside2 squat   Outside3 before fall

Fig.5. posture video data

5 Conclusions and Future Work

This paper introduce a fall detection method based on video, the method consists three parts: moving object detection, feature extraction, behavior recognization. Combining modified codebook algorithm and denoising method in OpenCV, we implement the moving object detection. We use two level SVM classifier to recognize the behavior, it is easy to implement. Through the designed well experiments, we get 90.27% accuracy. It shows that the method is practical in some degree and lag the foundation of applying in our daily life. But the algorithm is easily influenced by the complex background, for example, the system will have misdescription when detect lots of people; codebook algorithm can't distinguish the similar color well, this makes foreground extraction has more misdescription. So the method should combined with other method to raise the accuracy.

Acknowledgment

This paper is supported by the Project for the Key Project of Beijing Municipal Education Commission under Grant No. KZ201610005007, Beijing Postdoctoral Research Foundation under Grant No.2015ZZ-23, China Postdoctoral Research Foundation under Grant No.2015M580029, 2016T90022, the National Natural Science Foundation of China under Grant No.61672064 and Computational Intelligence and Intelligent System of Beijing Key Laboratory Research Foundation under Grant No.002000546615004.

References

1. Zhu L, Zhou P, Pan A, et al. A Survey of Fall Detection Algorithm for Elderly Health Monitoring[C]// IEEE Fifth International Conference on Big Data and Cloud Computing. IEEE, 2015:270-274.
2. Mubashir M, Shao L, Seed L. A survey on fall detection: Principles and approaches[J]. Neurocomputing, 2013, 100(2):144-152.
3. Dai J, Bai X, Yang Z, et al. PerFallD: A Pervasive Fall Detection System Using Mobile Phones[C]// Eigth IEEE International Conference on Pervasive Computing and Communications, PERCOM 2010, March 29 - April 2, 2010, Mannheim, Germany, Workshop Proceedings. 2010:292-297.
4. Bourke A K, Lyons G M. A threshold-based fall-detection algorithm using a bi -axial gyroscope sensor [J]. Medical Engineering and Physics, 2008, 30(1): 84-90.
5. Schuman Sr. R J, Collins W F. Indicator apparatus for healthcare communication system: EP, US8384526[P]. 2013.
6. Zigel Y, Litvak D, Gannot I. A method for automatic fall detection of elderly people using floor vibrations and sound--proof of concept on human mimicking doll falls.[J]. IEEE Transactions on Biomedical Engineering, 2010, 56(12):2858-2867.
7. Mubashir M, Shao L, Seed L. A survey on fall detection: Principles and approaches[J]. Neurocomputing, 2013, 100(2):144-152.
8. Awaida S M, Mahmoud S A. Automatic Check Digits Recognition for Arabic Using Multi-Scale Features, HMM and SVM Classifiers[J]. British Journal of Mathematics & Computer Science, 2014, 4(17):2521-2535.
9. Zhuo K Y. Recognition of automobile types based on improved RBF neural network[J]. Journal of Computer Applications, 2011.

Vehicle Detection Algorithm Based on Modified Gradient Oriented Histogram Feature

Wen-Kai Tsai[1], Sheng-Kai Lo[1], Ching-De Su[2], Ming-Hwa Sheu[2]

[1]Department of Electrical Engineering, National Formosa University, Yunlin, Taiwan
twk@nfu.edu.tw
[2]Department of Electronic Engineering, National Yunlin University of Science & Technology,
Yunlin, Taiwan
sheumh@yuntech.edu.tw

Abstract. In this study, algorithm development was conducted in two steps. Step 1 focused on training the algorithm using positive and negative samples. To increase execution speed, principal direction was adopted as the first feature to be identified. Subsequently, vehicle regions were converted into modified histogram of oriented gradients format and entered as inputs to a support vector machine (SVM) to identify the second feature. In the vehicle detection process, the first feature was adopted to eliminate nonvehicle regions first, and SVM training results were used to identify actual vehicle regions. Experimental results indicated that the proposed algorithm can effectively detect vehicles with an accuracy rate of up to 98%; moreover, the proposed method was approximately 40% faster than an SVM-based detection algorithm using the features of histograms of oriented gradients.

Keywords: vehicle detection,

1 Introduction

Contemporary vehicle detection algorithms can be categorized into two types: those that require training samples and those that do not [1, 2]. Algorithms that do not need training samples mainly detect vehicles by features such as edges, colors, and contours. For instance, a study proposed to detect vehicles using contours and skeleton features [3]. However, this type of algorithm can only be applied to static scenes. Some algorithms first identify specific features from training samples, and then use neural networks or classifiers to train a database suitable for vehicle detection [4–10]. A study used histograms of oriented gradients (HOGs) as an image feature [4]. HOGs serve as a widely adopted image feature in complex scenes because they are not likely to be affected by local deformation. Another study used Haar features for classification; however, random variations of the sizes and positions of Haar features can result in excessively long execution times, thus image integrals may be required to reduce computational complexity. Some scholars have proposed to use principal component analysis; however, this method requires considerable memory space and has a low

J.-S. Pan et al. (eds.), *Advances in Intelligent Information Hiding
and Multimedia Signal Processing*, Smart Innovation, Systems and Technologies 64,
DOI 10.1007/978-3-319-50212-0_16

execution speed. Another study proposed to use three features, namely, gray level, HOG, and linear back projection, to identify the images, and then to adopt support vector machines (SVMs) to classify and analyze the results [10]. Although this method is relatively accurate, using multiple classifiers results in excessively high computational complexity.

2   Proposed Method

The proposed systematic algorithm includes training and detection phases. Fig. 1 presents the flowchart. For each positive sample, modified histograms of oriented gradients (MHOGs) were used to select the principal direction of each cell; then, a principal direction of each sample was determined on the basis of the principal directions of its component cells (Fig. 1 [a]). Experimental results showed that the principal directions of all positive samples were identical; consequently, this direction was selected as the first feature in vehicle detection. Next, other features were extracted using MHOGs; the results were provided to an SVM to train the second feature.

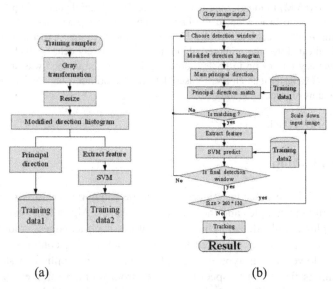

(a)   (b)

Fig. 1. Image-based vehicle detection algorithm. (a) Flowchart of sample training; (b) Flowchart of the vehicle detection algorithm.

The contents of the training sample exerted considerable influence on the detection results. Various front and back images of vehicles were collected to as positive samples, whereas the nonvehicle images were grouped as negative samples. Images of the positive and negative samples were employed as input images (Fig. 2), of which the gradient (G) and angle (θ) of each pixel were calculated (Eqs. 1–3). To reduce complexity, Eq. 4 was employed to categorize θ into nine bins, where a range of 20 degrees constituted a bin (Fig. 3). After the aforementioned calculations had been performed, the directions of each pixel were classified into bins. The input sample was divided into 36 nonoverlapping cells, each of which was 8 × 8 pixels in size (Fig. 4). The direction of each pixel was calculated using Eqs. 1–4; on the basis on the results, the HOG of each cell was determined. The distribution of cell directions within a sample was obtained by the same method (Fig. 5. [b]).

(a)   (b)

Fig. 2. Training sample. (a) positive sample   (b) negative sample

$$G_x = \begin{bmatrix} -1 \\ 0 \\ 1 \end{bmatrix} * I(x,y) \tag{1}$$

$$G_y = \begin{bmatrix} -1 & 0 & 1 \end{bmatrix} * I(x,y) \tag{2}$$

$$\theta(x,y) = \begin{cases} \tan^{-1}(\dfrac{G_y}{G_x}) & , if \quad \tan^{-1}(\dfrac{G_y}{G_x}) < \dfrac{\pi}{2} \\ \tan^{-1}(\dfrac{G_y}{G_x}) - \pi, & else \end{cases} \tag{3}$$

$$Bin(x,y) = \begin{cases} \left\lfloor \dfrac{\theta}{20} \right\rfloor, & if \ 0 \le \theta < 80 \\ \left\lfloor \dfrac{-\theta}{20} \right\rfloor + 5, & if \ -80 \le \theta < 0 \\ 4, & otherwise \end{cases} \tag{4}$$

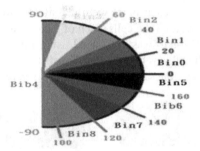

Fig. 3. Schematic of the angle

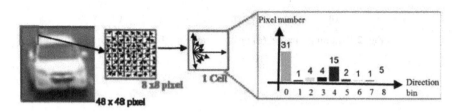

Fig. 4. Schematic of histogram decomposed according to gradient directions

(a)   (b)

Fig. 5. Schematic of histogram decomposed according to gradient directions. (a) Positive sample image input. (b) Gradient directions of each cell

Fig. 5 (b) shows that each cell had eight directions. To reduce calculation, a principal direction was selected from each cell to represent the direction of the entire cell (Fig. 5). However, Fig. 6 shows that the principal direction of most cells fell into the range of Bin 4. An analysis of 555 positive samples revealed that the principal direction of every positive sample was in Bin 4. Therefore, Bin 4 was designated the first feature of vehicles.

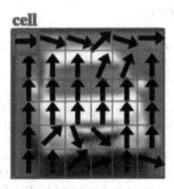

Fig. 6. Principal directions of all cells

After the principal direction was acquired, an SVM was trained to identify the second feature. Fig. 7 shows that every set of four adjacent cells constitute one block (2 × 2 cell). The algorithm defined a block cursor that moved rightward one cell at a time to process the data in all blocks. Because the original image size used 6 × 6 cells, a comprehensive investigation of this image involved 25 incremental updates, from which 900 dimensions (9 bins × 4 cells × 25 blocks) were obtained (Eq. 5). In this study, an SVM was applied for classification. An appropriate hyperplane was trained from 555 positive samples and 1000 negative samples. Vehicle detection was tested upon completion of the training.

$$v_i = [v_1, v_2, v_3, \ldots, v_{900}], \quad i = 1, 2, 3, \ldots, N \tag{5}$$

where N is the amount of the trained samples.

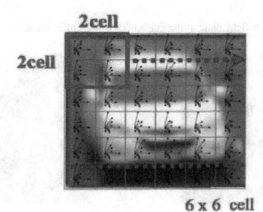

Fig. 7. Schematic of feature extraction from a histogram of gradient directions

3   Experimental Result

Because SVMs are expected to have low execution speeds, a procedure was adopted to eliminate most nonvehicle regions. The SVM was not used until only a small number of regions remained; hence the execution speed of the SVM was higher than typical SVM speeds. Fig. 1(b) demonstrates the execution process of this algorithm. Fig 8 (a) is the input image and Fig. 8 (b) presents the by-pixel direction distribution of the red area in the upper-left corner of the input image. Because the direction distribution is not dominated by Bin 4, this region can be eliminated directly. Similarly, the main gradient direction of the scooter region is Bin 0, therefore this region is eliminated in the first step and does not enter SVM classification (Fig. 9). By comparison, the automobile region is dominated by Bin 4 (Fig. 10). The automobile region enters SVM classification for further verification because it satisfies the condition designated in Step 1. Finally, Fig. 11 is entered as input to a trained SVM classifier; the vehicle region is identified and marked with a green box.

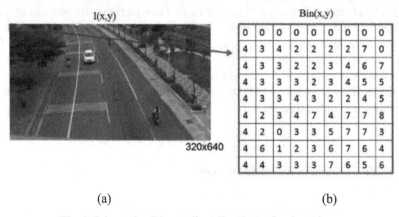

(a)   (b)

Fig. 8. Schematic of the gradient directions of an input image

Fig. 9. Schematic of the gradient directions of a scooter region

 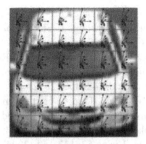

Fig. 10. Schematic of the gradient directions of an automobile region

Fig. 101. Vehicle detection results

4   Conclusion

This study was conducted using the following hardware and software: a custom-assembled personal computer equipped with a 3.2 GHz Intel i5-3470 core processor, 8 GB of DDRIII 800 random access memory, the 64-bit version of the Windows 7 operating environment, and Visual Studio 2010 software. A set of consecutive photographs that had not been used in training supplied the input images to be identified. The vehicle detection algorithm was applied to mark 320 × 640-pixel regions with green boxes.

The empirical results indicated that the proposed algorithm achieved an accuracy rate of 98% at an execution speed of 5.6 frames per second (FPS). By comparison, the conventional HOG method demonstrated an accuracy rate of 96% at an execution speed of 3.16 FPS.

Reference

1. Bing-Fei Wu, Jhy-Hong Juang: Adaptive Vehicle Detector Approach for Complex Environments. IEEE Transactions on Intelligent Transportation Systems 13,817-827 (2012)

2. Wanxin Xu, Meikang Qiu, Zhi Chen, Hai Su: Intelligent Vehicle Detection and Tracking for Highway Driving. IEEE International Conference on Multimedia and Expo Workshops, 67-72 (2012)

3. B. Yogameenaand, S.Md. MansoorRoomi, R. Jyothi Priya, S. Raju, V. Abhaikumar: People/vehicle classification by recurrent motion of skeleton features. IET Journals on Computer Vision 6, 442–450 (2012)

4. N. Dalal and B. Triggs, "Histograms of Oriented Gradients for Human Detection: IEEE Conf. Intell. on Computer Vision and Pattern Recognition 1, 886 – 893 (2005)

5. P. Viola and M. J. Jones, "Robust Real-Time Face Detection: International Journal of Computer Vision 52, 137-154 (2004)

6. Hong Wang, Su Yang, Wei Liao: An Improved PCA Face Recognition Algorithm Based on the Discrete Wavelet Transform and the Support Vector Machines. IEEE International Conference on Computational Intelligence and Security Workshops, 308-311 (2007)

7. Z. Sun,G. Bebis, and R. Miller: Monocular precrash vehicle detection: Features and classifiers. IEEE Trans. Image Processing 15, 2019–2034 (2006)

8. S. Sivaramanand, M. M. Trivedi: A General Active-Learning Framework for On-Road Vehicle Recognition and Tracking. IEEE Trans. Intelligent Transp. System, 267-276 (2010)

9. Jiafu Jiang, Changsha, Hui Xiong, Changsha: Fast Pedestrian Detection Based on HOG-PCA and Gentle AdaBoost. ICCSS. Computer Science and Service System, 1819-1822 (2012)

10. Shunli Zhang, Xin Yu, Yao Sui, Sicong Zhao, Li Zhang: Object Tracking With Multi-View Support Vector Machines. IEEE Trans. on Multimedia 17, 265–278 (2015)

Model-based Vehicle Make and Model Recognition from Roads

Li-Chih Chen

Department of Electrical Engineering, Lee-Ming Institute of Technology.
No. 22, Sec. 3, Tailin Rd., Taishan Dist., New Taipei City, Taiwan.
Email: lcchen@mail.lit.edu.tw

Abstract. In computer vision, the vehicle detection and identification is a very popular research topic. The intelligent vehicle detection application must first be able to detect ROI (Region of Interest) of vehicle exactly in order to obtain the vehicle-related information. This paper uses symmetrical SURF descriptor which enhances the ability of SURF to detect all possible symmetrical matching pairs for vehicle detection and analysis. Each vehicle can be found accurately and efficiently by the matching results even though only single image without using any motion features. This detection scheme has a main advantages that no need using background subtraction method. After that, modified vehicle make and model recognition (MMR) scheme has been presented to resolve vehicle identification process. We adopt a grid division scheme to construct some weak vehicle classifier and then combine such weak classifier into a stronger vehicle classifier. The ensemble classifier can accurately recognize each type vehicle. Experimental results prove the superiorities of our method in vehicle MMR.

Keywords: symmetrical SURF, vehicle make and model recognition (MMR), vehicle classifier

1 Introduction

Vehicle detection and analysis is an important task in various surveillance applications, such as driver assistance systems, self-guided vehicles, electronic toll collection, intelligent parking systems, or in the measurement of traffic parameters such as vehicle count, speed, and flow. In crime prevention or vehicle accident investigation, this task can provide useful information for policemen to search suspicious vehicles from surveillance video.

Faro et al. [5] used a background subtraction technique to subtract possible vehicle pixels from roads and then applied a segmentation scheme to remove partial and full occlusions among vehicle blobs. In [6], Unno et al. integrated motion information and symmetry property to detect vehicles from videos. Jazayeri [7] used HMM to probabilistically model the vehicle motion for vehicle detection. However, this kind of motion feature is no longer usable and available in still images. To treat this problem, this paper will propose a novel vehicle detection scheme to search for areas with a

© Springer International Publishing AG 2017
J.-S. Pan et al. (eds.), *Advances in Intelligent Information Hiding and Multimedia Signal Processing*, Smart Innovation, Systems and Technologies 64,
DOI 10.1007/978-3-319-50212-0_17

high vertical symmetry to locate vehicles in still images or videos by finding pairs of symmetric SURF feature points.

Each vehicle should be detected from images or videos firstly. The most commonly used methods adopted background subtraction to extract possible vehicle candidates from videos. If the environments include various camera vibration and lighting changes, the detection result will be not stable. The background model could not be established successfully in every frame when moving camera is adopted. The background model could not be established when the background is variation in every frames which is the main disadvantage of background subtraction method. Recently, the trained-based scheme is another popular method to detect vehicle by some well-trained classified via SVM or Adaboost. This approach depends heavily on the pre-trained knowledge base. This paper will present a symmetry-based method to detect vehicles from videos without using any motion's information.

The next approach is to identify vehicle make and model after a vehicle candidate was detected. These detection results can be applied to the electronic-toll collection system which can classify each vehicle into its own category. The identification of vehicle make and model offer valuable assistances to the policeman when searching for suspect vehicle candidates. In this paper, we devote our efforts to solve various challenges in vehicle detection and recognition.

2. Related Work

Vehicle detection is a key issue in many surveillance applications such as navigation system, driver assistance system, intelligent parking system, and so on. Exact vehicle detection can promote vehicle recognition accuracy. To develop a robust and effective vision-based vehicle detection and recognition is one of the challenges which result from the variations of vehicle sizes, orientations, shapes, and poses.

2.1 Vehicle Detection

Almost surveillance system equips the camera to detect moving vehicle, the most commonly adopted approach is to extract motion features through background subtraction. However, this technique is not stable when the background includes different varied environment, for example, lighting changes or camera vibrations. In another condition, motion feature is no longer available and usable in still images. Without using any motion features, this section will present a novel approach to detect vehicles on roads by taking advantages of a set of matching pairs of symmetrical SURF points.

Firstly, we describe the flowchart of our vehicle detection scheme briefly shown as Fig 1. Input image had been extracted SURF [1] features which are transformed to mirroring features by our proposed approach. And these SURF features are compared with its mirroring features to produce match pairs. These matching pairs are horizontal symmetrical, then, which are gathered in the histogram bins to dedicate the central line of vehicle candidates. After that, the center lines had been found, the shadow line and hood boundary on the vehicle could be found along these center lines. Finally, we adopt the geometric computation [3] to find the left and right boundary of vehicle candidate.

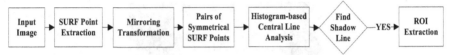

Fig 1. The flowchart of vehicle detection

2.2 Vehicle Recognition

The frontal of vehicle has the significant different features with other brands of vehicle, the front bumper, the front grille and etc., which provide some useful information to recognize the type of vehicle shown as Fig 2. Generally, humans recognize the vehicle model and make is also based on the frontal significant characteristics.

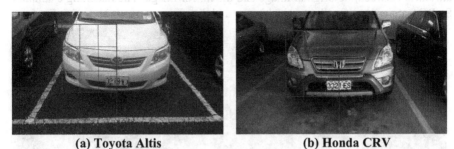

(a) Toyota Altis   (b) Honda CRV

Fig 2.The significant different features on the frontal of vehicle, the cyan-blue points are the SURF features of image.

After the front vehicle is extracted, different features will be extracted from by our vehicle MMR scheme. In order to satisfy various markets' requirements, the manufacturer of vehicle will modify the vehicle's shape. In this paper, a novel classification scheme will be presented to classify vehicles into different model categories. The flowchart of vehicle's make and model recognition is shown in Fig. 3. This system divides the vehicle's ROI into several grids. The vehicle's features will be extracted from each grid and the grid's features are trained as weak classifiers. After that, these weak classifiers are been assembled as a strong classifier to recognize vehicle's make and model. In order to promote the recognition performance, we collect the features in the left or right half-frontal of vehicle region shown as Fig 4.

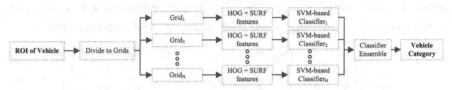

Fig 3. The flowchart of vehicle's make and model recognition

Fig 4. Two grid-based methods for vehicle type recognition

3. Symmetrical SURFs

The basic idea of our approach is that the matched points have some symmetric attributes between the original image and the mirror image. The matching example of SURF features between the original image and the mirror image is shown as Fig 5.

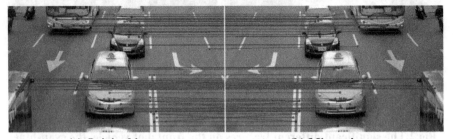

(a) Original image   **(b) Mirror image**
Fig 5. The matching example between the original image and the mirror image

The method [2] provides the transformation matrix M to convert the SURF [1] features in the same image and then to calculate its' similarity between the original SURF features and the symmetrical SURF features. However, the SURF lacks the capability of finding symmetrical pairs of feature points. In real word, the symmetrical pair is common to find objects with various symmetry properties. In real applications, the horizontal and vertical reflection symmetries appear most frequently than other symmetry.

The original definition of SURF [1] lacks of the supports to match two points if they are symmetrical. To provide this ability, the relations of SUFR descriptors between two symmetrical points should be derived. Let $B_{original}$ denote the original square extracted from an interest point and B_{mirror} be its horizontally mirrored version. Then, we can divide $B_{original}$ and B_{mirror} form 8×8 square to 4×4 sub-regions as

$$B_{original} = \begin{bmatrix} B_{00} & B_{01} & B_{02} & B_{03} \\ B_{10} & B_{11} & B_{12} & B_{13} \\ B_{20} & B_{21} & B_{22} & B_{23} \\ B_{30} & B_{31} & B_{32} & B_{33} \end{bmatrix}, B_{mirror} = \begin{bmatrix} B_{03}^{m} & B_{02}^{m} & B_{01}^{m} & B_{00}^{m} \\ B_{13}^{m} & B_{12}^{m} & B_{11}^{m} & B_{10}^{m} \\ B_{23}^{m} & B_{22}^{m} & B_{21}^{m} & B_{20}^{m} \\ B_{33}^{m} & B_{32}^{m} & B_{31}^{m} & B_{30}^{m} \end{bmatrix} \quad (1)$$

For each sub-region B_{ij}, its sums of wavelet responses can be calculated by the form:

$$f_{ij} = (\sum_{b \in B_{ij}} dx(b), \sum_{b \in B_{ij}} dy(b), \sum_{b \in B_{ij}} |dx(b)|, \sum_{b \in B_{ij}} |dy(b)|) \tag{2}$$

where $dx(b) = b_{y,x+1} - b_{y,x}$ and $dy(b) = b_{y+1,x} - b_{y,x}$. We use $d_{i,j}^x$, $d_{i,j}^y$, $|d_{i,j}^x|$, and $|d_{i,j}^y|$ to denote the sums of wavelet responses, *i.e.*,

$$d_{i,j}^x = \sum_{b \in B_{i,j}} dx(b),\ d_{i,j}^y = \sum_{b \in B_{i,j}} dy(b),\ |d_{i,j}^x| = \sum_{b \in B_{i,j}} |dx(b)|,\ and\ |d_{i,j}^y| = \sum_{b \in B_{i,j}} |dy(b)| \tag{3}$$

According to the definitions of Eq.(2) and Eq.(3), we have:

$$d_{i,j}^x = b_{2i,2j+1} + b_{2i+1,2j+1} - b_{2i,2j} - b_{2i+1,2j},\ d_{i,j}^y = b_{2i+1,2j} + b_{2i+1,2j+1} - b_{2i,2j} - b_{2i,2j+1} \tag{4}$$

With Eq. (3) and (4), the SURF descriptors of B_{ij} and B_{ij}^m can be extracted, respectively, by

$$f_{i,j} = (d_{i,j}^x, d_{i,j}^y, |d_{i,j}^x|, |d_{i,j}^y|), f_{i,j}^m = (-d_{i,j}^x, d_{i,j}^y, |d_{i,j}^x|, |d_{i,j}^y|) \tag{5}$$

Let $B_i = \begin{bmatrix} B_{i0} & B_{i1} & B_{i2} & B_{i3} \end{bmatrix}$ and $B_i^m = \begin{bmatrix} B_{i3}^m & B_{i2}^m & B_{i1}^m & B_{i0}^m \end{bmatrix}$. From B_i, a new feature vector f_i can be constructed, *i.e.*, $f_i = (f_{i,0}, f_{i,1}, f_{i,2}, f_{i,3})$ or

$$f_i = (d_{i,0}^x, d_{i,0}^y, |d_{i,0}^x|, |d_{i,0}^y|, d_{i,1}^x, d_{i,1}^y, |d_{i,1}^x|, |d_{i,1}^y|, d_{i,2}^x, d_{i,2}^y, |d_{i,2}^x|, |d_{i,2}^y|, d_{i,3}^x, d_{i,3}^y, |d_{i,3}^x|, |d_{i,3}^y|) \tag{6}$$

Similarly, from B_i^m, another feature vector f_i^m can be constructed:

$$f_i^m = (-d_{i,3}^x, d_{i,3}^y, |d_{i,3}^x|, |d_{i,3}^y|, -d_{i,2}^x, d_{i,2}^y, |d_{i,2}^x|, |d_{i,2}^y|, -d_{i,1}^x, d_{i,1}^y, |d_{i,1}^x|, |d_{i,1}^y|, -d_{i,0}^x, d_{i,0}^y, |d_{i,0}^x|, |d_{i,0}^y|) \tag{7}$$

With f_i and f_i^m, the SURF descriptors $f_{original}$ and f_{mirror} of $B_{original}$ and B_{mirror} can be constructed, respectively, as follows:

$$f_{original} = [f_0\ f_1\ f_2\ f_3]^t\ and\ f_{mirror} = [f_0^m\ f_1^m\ f_2^m\ f_3^m]^t \tag{8}$$

The transformation between $f_{original}$ and f_{mirror} can be easily built by converting each row feature f_i to f_i^m using the relations between Eq.(6) and (7). In Eq.(8), the dimensions of $f_{original}$ and f_{mirror} are 4×16. Then, given two SURF descriptors f^p and f^q, their distance is defined as:

$$\xi_{SURF}(f^p, f^q) = \sum_{m=1}^{4} \sum_{n=1}^{16} [f^p(m,n) - f^q(m,n)]^2 \tag{9}$$

In Eq.(9), the sums of wavelet responses are computed from 2×2 samples. However, in the original SURF descriptor, the wavelet responses $d_{i,j}^x$ and $d_{i,j}^y$ are summed up over 5×5 samples. Under this condition, there is no exact symmetric transformation between $f_{original}$ and f_{mirror}. To treat this problem, it is suggested that each sub-region is with 4×4 samples. Then, the sub-region B_{ij} in $B_{original}$ can be expressed as:

$$B_{ij} = \begin{bmatrix} b_{4i,4j} & b_{4i,4j+1} & b_{4i,4j+2} & b_{4i,4j+3} \\ b_{4i+1,4j} & b_{4i+1,4j+1} & b_{4i+1,4j+2} & b_{4i+1,4j+3} \\ b_{4i+2,4j} & b_{4i+2,4j+1} & b_{4i+2,4j+2} & b_{4i+2,4j+3} \\ b_{4i+3,4j} & b_{4i+3,4j+1} & b_{4i+3,4j+2} & b_{4i+3,4j+3} \end{bmatrix} \tag{10}$$

B_{ij} is further divided to 2×2 sub-grids $g_{m,n}$ which consists of 2×2 samples, where

$$B_{ij} = \begin{bmatrix} g_{2i,2j} & g_{2i,2j+1} \\ g_{2i+1,2j} & g_{2i+1,2j+1} \end{bmatrix} \text{ and } g_{m,n} = \begin{bmatrix} b_{2m,2n} & b_{2m,2n+1} \\ b_{2m+1,2n} & b_{2m+1,2n+1} \end{bmatrix} \qquad (11)$$

Then, the mean value $\bar{g}_{m,n}$ of $g_{m,n}$ is calculated by

$$\bar{g}_{m,n} = \frac{1}{4}(b_{2m,2n} + b_{2m,2n+1} + b_{2m+1,2n} + b_{2m+1,2n+1}) \qquad (12)$$

Based on $\bar{g}_{m,n}$, the wavelet responses $d_{i,j}^x$ and $d_{i,j}^y$ of B_{ij} can be summed up with the form

$$\begin{aligned} d_{i,j}^x &= \bar{g}_{2i,2j+1} + \bar{g}_{2i+1,2j+1} - \bar{g}_{2i,2j} - \bar{g}_{2i+1,2j} \\ d_{i,j}^y &= \bar{g}_{2i+1,2j} + \bar{g}_{2i+1,2j+1} - \bar{g}_{2i,2j} - \bar{g}_{2i,2j+1} \end{aligned} \qquad (13)$$

With Eq.(13), the SURF descriptor $f_{i,j}$ of B_{ij} can be constructed by using Eq.(11). The descriptor $f_{i,j}^m$ of B_{ij}^m is calculated similarly. With $f_{i,j}$ and $f_{i,j}^m$, the features f_i and f_i^m can be obtained by using Eq.(6) and (7). Then, the symmetrical transformation between $f_{original}$ and f_{mirror} is the same to Eq.(8). Actually, even though the wavelet responses $d_{i,j}^x$ and $d_{i,j}^y$ are summed up over 5×5 samples, Eq.(8) still works well for real cases. Fig 6 shows an example to match two symmetric points when the SURF descriptors are extracted by summing up the wavelet responses $d_{i,j}^x$ and $d_{i,j}^y$ from 5×5 samples. Fig 6(a) shows the matching result of these two points without using the proposed transformation. Fig 6 (b) is the matching result after the symmetric transformation. Clearly, after converting, the transformed SURF descriptor (denoted by a red line) is very similar to the blue one.

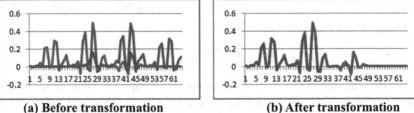

(a) Before transformation   (b) After transformation
Fig 6. Matching results of two symmetrical SURF points before/after the symmetric transformation. Lines with different colors denote the features of two symmetrical SURF points.

Given a frame I_t, a set F_{I_t} of SURF points is first extracted. For an interest point p in F_{I_t}, we can extract its SURF descriptor $f_{original}^p$ and obtain its mirrored version f_{mirror}^p by using Eq.(8). Let Min_r denote the minimum distance between an interest point r and other points in F_{I_t}, i.e.,

$$Min_r = \min_{s \in F_{I_t}, s \neq r} \xi_{SURF}(f^r, f^s) \tag{14}$$

In addition, let Min_r^{Sec} denote the distance of the best second match of r to other points in F_{I_t}. For two points p and q in F_{I_t}, they form a symmetrical match if they satisfy

$$\xi_{SURF}(f_{original}^p, f_{mirror}^q) = Min_p \text{ and } \frac{Min_p}{Min_p^{Sec}} < 0.65 \tag{15}$$

Fig 7 shows an example of horizontal symmetry matching from a vehicle by using Eq.(15).

Fig 7. Horizontal symmetry matching example

4. Experimental Results

To evaluate the performances of our proposed system, an automatic system for vehicle classification was implemented in this paper. Twenty-nine vehicle types were collected in this paper for performance evaluation. To train the HOG-classifier and the SURF-classifier, a database containing 2,840 vehicles was collected. For training, the SVM library was used from [4]. The radial basis function (RBF) was selected as the kernel function with the gamma value 0.08125 and the parameter C of nu-SVC is 12.0. In addition, another testing database with 4,090 vehicles was also collected. This vehicle database was collected by the VBIE (Vision-Based Intelligent Environment) project [8]. Fig 8 shows the two results of vehicle detection. And the performance analysis of our method is shown in **Table 1**.

(a) Toyota Altis, 2010   (b) Toyota Camry, 2008
Fig 8. Results of front vehicle detection.

Table 1: Performance analyses of MMR among different vehicle types.

Types / Analyses	Toyota Altis	Toyota Camry	Toyota Vios	Toyota Wish	Toyota Yaris	Toyota Previa	Toyota Innova	Toyota Surf	Toyota Tercel	Toyota Rav4
Correct	701	404	309	217	200	54	42	38	74	56
Total	735	441	319	220	205	58	44	43	84	56
Accuracy	95.37%	91.61%	96.87%	98.64%	97.56%	93.10%	95.45%	88.37%	88.10%	100%
Types / Analyses	Honda CRV	Honda Civic	Civic FIT	Nissan March	Nissan Livna	Nissan Teana	Nissan Sentra	Nissan Cefiro	Nissan Xtrail	Nissan Tiida
Correct	340	291	81	127	172	58	39	31	84	152
Total	351	309	82	139	177	60	41	38	99	179
Accuracy	96.87%	94.17%	98.78%	91.37%	97.18%	96.67%	95.12%	81.58%	84.85%	84.92%
Types / Analyses	Mitsubishi Zinger	Mitsubishi Outlander	Mitsubishi Savrin	Mitsubishi Lancer	Suzuki Solio	Ford Liata	Ford Escape	Ford Mondeo	Ford Tierra	Total
Correct	20	68	22	39	86	11	75	27	29	3,847
Total	22	71	22	40	90	11	88	35	31	4,090
Accuracy	90.91%	95.77%	100%	97.50%	95.56%	100%	85.23%	77.14%	93.55%	94.06%

5. Conclusions

This paper has proposed a novel symmetric SURF descriptor and applied to vehicle MMR system to detect vehicles and recognize their makes, models extremely accurately. The major contributions of this paper are noted as follows: A new vehicle detection scheme is proposed to detect vehicles from moving cameras. The advantage of this scheme is no need of background modeling and subtraction. In addition, it is suitable for real-time applications because of its extreme efficiency. A new grid-based scheme is proposed to recognize vehicles. The different weak classifiers are integrated to build a strong ensemble classifier so as to recognize the vehicle models efficiently and accurately.

Experimental results have proved the superiority of our proposed analysis system of vehicle information using symmetrical SURF. In the future, the optimization of our program should be considered to gain more efficiency of the process. How to get vehicle ROI more quickly is also a very important issue. In the spirit of research is the endless pursuit of accuracy and rapidity of the system. I will lift the spirit of research and continue our efforts in the pursuit of higher system performance.

6. References

[1] H. Bay, A. Ess, T. Tuytelaars, and L. Van Gool, "Speeded-Up Robust Features (SURF)," Computer Vision and Image Understanding, vol. 110, no. 3, pp. 346-359, 2008.

[2] Li-Chih Chen, Jun-Wei Hsieh, Yilin Yan and Duan-Yu Chen, "Vehicle make and model recognition using sparse representation and symmetrical SURFs," Pattern Recognition, vol. 48, pp. 1979-1998, Jan. 2015.

[3] Sin-Yu Chen and Jun-Wei Hsieh, "Jointing Edge Labeling and Geometrical Constraint for Lane Detection and its Application to Suspicious Driving Behavior Analysis," Journal of Information Science and Engineering, vol. 27, no.2, pp. 715-732, 2011.

[4] C.-C. Chang and C.-J. Lin, LIBSVM: a library for support vector machines, 2001. Software available at http://www.csie.ntu.edu.tw/~cjlin/libsvm.

[5] A. Faro, D. Giordano, and C. Spampinato, "Adaptive background modeling integrated with luminosity sensors and occlusion processing for reliable vehicle detection," IEEE Transactions on Intelligent Transportation Systems, vol. 12, no.4, pp.1398-1412, 2011.

[6] H. Unno, K. Ojima, K. Hayashibe, and H. Saji, "Vehicle Motion Tracking Using Symmetry of Vehicle and Background Subtraction," *IEEE Intelligent Vehicles Symposium*, 2007.

[7] A. Jazayeri, H.-Y Cai, J.-Y. Zheng, and M. Tuceryan, "Vehicle detection and tracking in car video based on motion model," *IEEE Transactions on Intelligent Transportation Systems*, vol. 12, no.2, pp.583-595, 2011.

[8] http://vbie.eic.nctu.edu.tw/en/introduction

Frames Motion Detection of Quantum Video

Shen Wang*

School of Computer Science and Technology, Harbin Institute of Technology,
Harbin, 150080, China
shen.wang@hit.edu.cn

Abstract. A novel frames motion detection scheme for quantum video
is proposed. To effectively demonstrate the motion detection scheme,
a novel quantum video representation based on NEQR quantum image
is proposed at first. Then, the new strategy comprises quantum video
frame blocking, comparison of frame blocks and residual calculation for
quantum video frames motion detection. Experimental simulations conducted on a simple video demonstrate that significant improvements in
the results are in favor of the proposed approach.

Keywords: quantum computation, quantum video representation, frames
motion detection

1   Introduction

Quantum computer era is an inevitable stage in the development process of computer. Taken quantum mechanics as a theoretical foundation, quantum computer
encodes quantum state used in computer, implements computation tasks that is
transforming or involving quantum states according to quantum mechanics rules
and extracts computational results using quantum measurements. Quantum s-
tate has the coherent superposition property, especially the quantum entanglement property which is not in the classical physics, which makes the computing
ability of the quantum computer far higher than the classical computer [1].
Therefore, Quantum information, especially quantum media information, will
have important researching value in the era of quantum computer.

Compared with text media, audio media and image media, video media, as
the most closely information carrier related to people's sensory, contains a greater
amount of information, and more intuitive. Therefore, video plays a greater role
in the information representation. Quantum video processing, an area focusing
on extending conventional video processing tasks and operations to the quantum
computing framework, will be a key issue in quantum information processing
field[2].

Generally speaking, video composes of a series of images, accordingly, quantum video representation and processing can not be separated from the quantum

---

* This work is supported by the National Natural Science Foundation of Chi-
na(61301099,61501148).

© Springer International Publishing AG 2017   145
J.-S. Pan et al. (eds.), *Advances in Intelligent Information Hiding
and Multimedia Signal Processing*, Smart Innovation, Systems and Technologies 64,
DOI 10.1007/978-3-319-50212-0_18

image representations. At present, the quantum image representation methods can be divided into the following categories: lattice method based[3][4][5], entanglement based[6], vector based[7][8], FRQI (Flexible Representation of Quantum Images) method[9] and NEQR (Novel Enhanced Quantum Representation) method[10]. The two methods of FRQI and NEQR are widely used in the research of quantum image processing and security. In FRQI, the gray level and the position of the image are expressed as a normalized quantum superposition state. NEQR quantum image indicates an increase in the number of quantum bits required for representing the gray level of image. However, due to the representation of the gray value is similar to the bit plane coding in the classical image, NEQR method is more easy to design quantum image processing algorithm.

Until now, quantum video representations [11][12][13] are mainly based on FRQI[9]. Inspired the advantage of NEQR, in this paper, a novel quantum video representation based on NEQR (QVNEQR) is introduced at first. With the intelligent video surveillance systems used widely and the monitoring conditions become more and more complicated, moving target detection plays an important role for computer vision, image applications and so on. Then, a quantum frames motion detection algorithms for QVNEQR is proposed which can detect target from adjacent frames. A simple simulation verifies the efficiency of this scheme. Compared with other moving target detection methods, the proposed scheme realizes the detection aim but has no need for any measurement operation, therefore, it can not destroy quantum state[14][15].

The rest of the paper is organized as follows. Section 2 gives a new quantum video representation method based on NEQR image presentation. In Section 3, the quantum video frames motion detection algorithm is proposed. Section 4 is devoted to the simulation results and result analyses. Finally, Section 5 concludes the paper.

2   Quantum Video Representation based on NEQR

Backgroud about NEQR is introduced at first and then the quantum video based on NEQR (QVNEQR) is proposed.

2.1   Novel Enhanced Quantum Representation(NEQR)

A novel enhanced quantum representation for images is proposed by Zhang [10]. In this new model, the classical image color information is represented by the basis states of the color qubit. Therefore, color information qubit sequence and the position information qubit sequence are entangled in NEQR representation to store the whole image. Suppose that the gray range of the image is 2^q. Gray-scale value of the corresponding pixel is encoded by binary sequence $c_{q-1}^{YX} c_{q-2}^{YX} \cdots c_0^{YX}$, $c_m^{YX} \in [0,1] \, (m = 0, 1, \cdots q - 1)$, $f(Y, X) \in [0, 2^q - 1]$. The new representation model of a quantum image for a $2^n \times 2^n$ image is described as follows:

$$|I\rangle = \frac{1}{2^n} \sum_{Y=0}^{2^n-1} \sum_{X=0}^{2^n-1} |f(Y, X)\rangle |Y\rangle |X\rangle \tag{1}$$

Position information includes the vertical information and the horizontal information.

$$|Y\rangle|X\rangle = |Y_{n-1}Y_{n-2}\cdots Y_0\rangle|X_{n-1}X_{n-2}\cdots X_0\rangle \tag{2}$$

$|Y\rangle$ encodes the vertical information and $|X\rangle$ encodes the horizontal information. NEQR uses the basis qubit to store the gray-scale information for each pixel in an image, some digital image-processing operations, for example certain complex color operations, can be done on the basis of NEQR. And partial color operations and statistical color operations can be conveniently performed based on NEQR.

2.2   quantum video based on NEQR (QVNEQR)

Inspired by the concept of strip, m quantum circuits representing time axis are added to the 2^m frames NEQR quantum images and generate a quantum video QVNEQR.

$$|V\rangle = \frac{1}{2^{m/2}}\sum_{j=0}^{2^m-1}|I_j\rangle \otimes |j\rangle \tag{3}$$

$$|I_j\rangle = \frac{1}{2^n}\sum_{i=0}^{2^{2n}-1}|c_{j,i}\rangle \otimes |i\rangle = \frac{1}{2^n}\sum_{i=0}^{2^{2n}-1}\left|c_{q-1}^{j,i}c_{q-2}^{j,i}...c_0^{j,i}\right\rangle|i\rangle \tag{4}$$

$$c_s^{j,i} \in \{0,1\}, s = 0,1,\ldots,q-1, q = 24$$

herein, every frame is an NEQR quantum color image size of $2^n \times 2^n$ and includes R, G, B three color information. Every color channel uses 8 qubits to encode its intensity. Besides, the QVNEQR state is a normalized state, i.e.,

$$\||V\rangle\| = \sqrt{(\frac{1}{2^{m/2+n}})^2 \cdot 2^{m+2n}} = 1 \tag{5}$$

3   Frames Motion Detection of QVNEQR

Th motion of quantum video can be calculated via the change between the corresponding blocks come from two adjacent video frames. The concrete computation process divides into three steps i.e., frame blocking, comparison of frame blocks and residual calculation. In bellows, the three steps are described in detail.

Frame Blocking

The $2^n \times 2^n$ size frames of quantum video shown in Eq. (3) use n qubits coding x axis and y axis, respectively. Suppose that the size of each frame is 8×8 , thus 2^{2n-6} image blocks can be obtained. Encoding these blocks need $2n-6$ qubits and x axis and y axis need $n-3$ qubits, respectively. Moreover, the binary expression of the block's serial number has the one-to-one correspondence with the codes of these blocks.

To finish the frame blocking process, implementing the following operations to video frame $|I_j\rangle$ in (3):

$$|I_j\rangle = \frac{1}{2^n}\sum_{i=0}^{2^{2n}-1}|c_{j,i}\rangle \otimes |i\rangle = \frac{1}{2^n}\sum_{y=0}^{2^n-1}\sum_{x=0}^{2^n-1}\left|c_{q-1}^{j,i}c_{q-2}^{j,i}\cdots c_0^{j,i}\right\rangle|y\rangle\,|x\rangle$$

$$=\frac{1}{2^n}\sum_{y_i\in\{0,1\},x_i\in\{0,1\}}\left|c_{q-1}^{j,i}c_{q-2}^{j,i}\cdots c_0^{j,i}\right\rangle\left|\underbrace{000\cdots0}\right\rangle|y_2y_1y_0\rangle\left|\underbrace{000\cdots0}\right\rangle|x_2x_1x_0\rangle$$

$$+\frac{1}{2^n}\sum_{y_i\in\{0,1\},x_i\in\{0,1\}}\left|c_{q-1}^{j,i}c_{q-2}^{j,i}\cdots c_0^{j,i}\right\rangle\left|\underbrace{000\cdots0}\right\rangle|y_2y_1y_0\rangle\left|\underbrace{000\cdots1}\right\rangle|x_2x_1x_0\rangle$$

$$+\cdots$$

$$+\frac{1}{2^n}\sum_{y_i\in\{0,1\},x_i\in\{0,1\}}\left|c_{q-1}^{j,i}c_{q-2}^{j,i}\cdots c_0^{j,i}\right\rangle\left|\underbrace{111\cdots1}\right\rangle|y_2y_1y_0\rangle\left|\underbrace{111\cdots0}\right\rangle|x_2x_1x_0\rangle$$

$$+\frac{1}{2^n}\sum_{y_i\in\{0,1\},x_i\in\{0,1\}}\left|c_{q-1}^{j,i}c_{q-2}^{j,i}\cdots c_0^{j,i}\right\rangle\left|\underbrace{111\cdots1}\right\rangle|y_2y_1y_0\rangle\left|\underbrace{111\cdots1}\right\rangle|x_2x_1x_0\rangle$$

In the above equation, the number of qubits using symbol '$\underbrace{\quad\cdot\quad}$' is $n-3$. After recoding the quantum video state, every item in the above equation is a 8×8 image block.

Comparison of Frame Blocks

The purpose of comparison of frame blocks is retrieving the similar blocks between two adjacent video frames. Suppose the image block sets of video frame $|I_1\rangle$ and $|I_2\rangle$ are $\{|B_i^1\rangle, i=1,2,...,2^{2n-6}\}$ and $\{|B_i^2\rangle, i=1,2,...,2^{2n-6}\}$, respectively. $|p_s^{u,v}\rangle$, $s=1,...,64, u=0,1,\cdots,2^m-1$, $v=0,1,\cdots,2^{2n-6}-1$ denotes the sth pixel of vth block from the uth frame. The comparison of frame blocks uses the qubit comparator shown in Fig. 1 Starting from the highest qubit, the comparisons are implemented on each qubit in turn until the lowest qubit. It can be computed that the maximal range of the difference between $|p_s^{u,v_1}\rangle$ and $|p_s^{u+1,v_2}\rangle$ is $2^{q-1}+2^{q-2}+...+1=2^q-1$. The evaluation of the pixel difference can be calculated through the following equation:

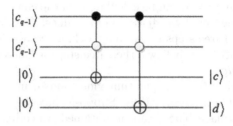

Fig. 1. Qubits comparator

$$|c'_{q-1} - c_{q-1}| \cdot 2^{q-1} + |c'_{q-2} - c_{q-2}| \cdot 2^{q-2} + \cdots |c'_0 - c_0| \cdot 2^0$$
$$\leq 2^{q-1} + 2^{q-2} + \ldots + 1$$
$$= 2^q - 1$$

The complexity of the qubit comparator is $O(N)$, thus the complexity of the comparison of video frame blocks is $2^{2n-6} \cdot 64 \cdot 2O(N) = 2 \cdot 2^{2n}O(N)$, which locates in the acceptable complexity range.

residual calculation

Suppose the adjacent video frames are $|I_1\rangle$ and $|I_2\rangle$, the tth block $|B_t^k\rangle$ of video frame $|I_k\rangle$, $k = 1, 2$ has the following form:

$$|B_t^k\rangle = \frac{1}{2^n} \sum_{y_i \in \{0,1\}, x_i \in \{0,1\}} |c_{q-1}^{k,t} c_{q-2}^{k,t} \cdots c_0^{k,t}\rangle |t_{b_y}\rangle |y_2 y_1 y_0\rangle |t_{b_x}\rangle |x_2 x_1 x_0\rangle$$

herein, $|t_{b_y}\rangle$ and $|t_{b_x}\rangle$ express the needed qubits of y-axis and x-axis for the binary coding of t. Thus, the residual of video frame block $|B_{t_1}^1\rangle$ and $|B_{t_2}^1\rangle$ can be gained as below:

$$|\Delta B\rangle = \frac{1}{2^n} \sum_{y_i \in \{0,1\}, x_i \in \{0,1\}} |C^{1,t_1} - C^{2,t_2}\rangle |t_{b_y}\rangle |y_2 y_1 y_0\rangle |t_{b_x}\rangle |x_2 x_1 x_0\rangle$$

where $|C^{1,t_1}\rangle = |c_{q-1}^{2,t_1} c_{q-2}^{2,t_1} \cdots c_0^{2,t_1}\rangle$ and $|C^{2,t_2}\rangle = |c_{q-1}^{2,t_2} c_{q-2}^{2,t_2} \cdots c_0^{2,t_2}\rangle$.

The quantum circuit of the residual $|\Delta B\rangle$ calculation can be realized by the following two stages:

stage 1 Designing the complement quantum circuit and taking the complement operation to quantum state $|C^{2,t_2}\rangle$. The specific quantum circuit is shown in Fig.2 and the output state is denoted as $|\overline{C^{2,t_2}}\rangle$.

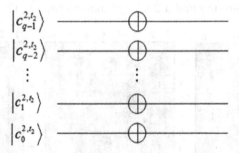

Fig. 2. Quantum circuit of complement

stage 2 Computing the sum of $|C^{2,t_1}\rangle$ and $|\overline{C^{2,t_2}}\rangle$ by means of q-ADD quantum circuit, which is shown in Figs.3 and 4.

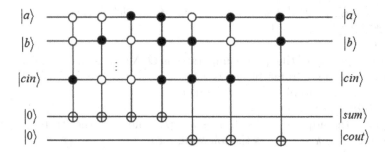

Fig. 3. Quantum circuit of 1-ADD

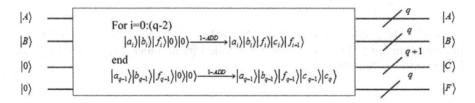

Fig. 4. Quantum circuit of q-ADD

q-ADD quantum circuit can realize $|C\rangle = |A+B\rangle$. Set $|A\rangle = |C^{2,t_1}\rangle$ and $|B\rangle = |\overline{C^{2,t_2}}\rangle$ can finish the calculation of residual.

4   Experiments

In this section, a simple experiment to demonstrate the execution of the proposed scheme. A three-frame QVNEQR color quantum video has been produced as shown in Fig.5. Each frame is an 8×8 color NEQR quantum image with a common blue background and a 2×2 red moving target. Through the proposed

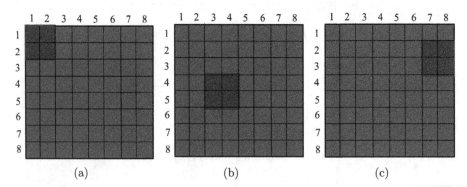

Fig. 5. A quantum video with three frames representing a target moving

three step, it can be seen that the frames motion detection in QVNEQR video is feasible.

5   Conclusions

In this paper, a novel quantum video frames motion detection scheme based on a novel quantum video representation has been proposed. The proposed strategy comprises of three stages: frame blocking, comparison of frame blocks and residual calculation. Compared with other moving target detection methods, the proposed scheme realizes the detection aim but has no need for any measurement operation, therefore, it can not destroy quantum state. This is a meaningful attempt to introduce information processing techniques into quantum video scenarios.

References

1. D.Deutsch, Quantum theory, the Church-Turing principle and the universal quantum computer, Proc.R. Soc. London A400,97-117 (1985)
2. F. Yan, A.M.Iliyasu, S.E.Venegas-Andraca: A survey of quantum image representations. Quantum Inf. Process. 15, pp.1-35 (2016)
3. Venegas-Andraca S E, Bose S. Storing, processing, and retrieving an image using quantum mechanics//Quantum Information and ComputationProceedings of SPIE. Orlando, FL, United states: International Society for Optics and Photonics, 5105:137-147(2003).
4. Li H S, Qing X Z, Lan S, et al. Image storage, retrieval, compression and segmentation in a quantum system. Quantum information processing,12(6):2269-2290(2013).
5. Yuan S, Mao X, Xue Y, et al. SQR: a simple quantum representation of infrared images. Quantum Information Processing. 13(6), pp.1353-1379(2014).
6. Venegas A S, Ball J. Processing images in entangled quantum systems[J]. Quantum Information Processing,9(1):1-11(2010).
7. Latorre J I. Image compression and entanglement[J]. arXiv preprint quantph/ 0510031, 2005.
8. Hu B Q, Huang X D, Zhou R G, et al. A theoretical framework for quantum image representation and data loading scheme[J]. SCIENCE CHINA Information Sciences, 57: 032108:1-11(2014).
9. Le, P.Q., Dong, F., Hirota, K.: A flexible representation of quantum images for polynomial preparation, image compression and processing operations. Quantum Inf. Process. 10(1), pp.63-84 (2010)
10. Y. Zhang, K. Lu, Y. H. Gao, M. Wang: NEQR: a novel enhanced quantum representation of digital images, Quantum Inf. Process,12(8), pp.2833-2860 (2013)
11. B. Sun, P. Q. Le, A.M. Iliyasu, J. Adrian Garcia, F. Yan, J., F. Dong, and K. Hirota: A multi-channel representation for images on quantum computers using the RGBα color space. Proceedings of the IEEE 7th International Symposium on Intelligent Signal Processing, pp.160-165 (2011)
12. Iliyasu, A.M., Le, P.Q., Dong, F., Hirota, K.: A framework for representing and producing movies on quantum computers[J]. International Journal of Quantum Information, 2011, 9(6) :1459C1497.
13. Yan, F., Iliyasu, A., Venegas-Andraca, S., Yang, H.: Video encryption and decryption on quantum computers[J]. Int. J. Theor. Phys. 2015, 54(8), 2893C2904.
14. Yan, F., Iliyasu, A., Khan, A., Yang, H. Moving target detection in multi-channel quantum video. In: IEEE 9th International Symposium on Intelligent Signal Processing (WISP),2015, 1C5.
15. Yan, F., Iliyasu, A., Khan, A. Measurements-based moving target detection in quantum video[J]. Int. J. Theor. Phys. 2015, DOI 10.1007/s10773-015-2855-0

Part II
Multisignal Processing Techniques

Blind Quantum Computation with Two Decoy States

Qiang Zhao Qiong Li*

*School of Computer Science and Technology,
Harbin Institute of Technology, Harbin, China
qiong.li@hit.edu.cn,

Abstract. The Universal Blind Quantum Computation(UBQC) protocol allows a client to perform quantum computation on a remote server.
In the UBQC with Weak Coherent Pulses(WCP), the Remote Blind
qubit State Preparation(RBSP) protocol requires larger scale number of
pulses for preparing one single qubit, especially the long-distance communication. In the paper, we present a modified RBSP protocol with two
decoy states to reduce the required number of pulses for the client. Both
theoretical analysis and simulation results demonstrate that decoy state
can improve the qubit preparation efficiency of UBQC protocol greatly
and two decoy states perform even better than one decoy state. What is
more,the advantage of two decoy states become more highlighted with
the increasing of distance.

Keywords: Universal blind quantum compuation; Weak coherent pluses; Remote blind qubit state preparation; Decoy state; Preparation efficiency

1 Introduction

In recent years, although modern advances in quantum information have making
stride towards scalable quantum computers, it is still distant to the dream of
small and privately owned quantum computers. In the near future, the realistic
mode is to use the large-scale quantum computer as the computing center in
the cloud application framework. The classical clients, using their home-based
simple devices, can realize their desired quantum computation in a rental fashion. Consequently, some security problems arise, such as how to guarantee the
privacy of the client, named as Alice, when she interacts with the quantum server, named as Bob. These related questions have been studied. In the classical
field, Feigenbaum introduced the notion of computing with encrypted data to
guarantee the client's privacy[1]. After then, Abadi, Feigenbaum, Kilian gave
an negative result: no NP-hard function can be computed with encrypted data
(even probabilistically and with polynomial interaction), unless the polynomial
hierarchy collapses at the third level[2]). Another solution to that problem is
known as fully homomorphic encryption, but they depend on the computational
assumptions[3]. In the quantum world, various protocols have been presented to

© Springer International Publishing AG 2017
J.-S. Pan et al. (eds.), *Advances in Intelligent Information Hiding
and Multimedia Signal Processing*, Smart Innovation, Systems and Technologies 64,
DOI 10.1007/978-3-319-50212-0_19

realize secure delegated quantum computing demanding different requirements on the client or implementing different security levels. So far, the optimal scheme is Universal Blind Quantum Computing(UBQC) protocol, which is proposed by Broadbent, Fitzsimons and Kashefi[4]. The ideal UBQC protocol can realize the unconditional privacy, i.e. the server cannot eavesdrop anything about the client's computation, inputs and outputs. What is more, the quantum requirement of the protocol on the client is the lowest, i.e., only the ability of preparing single photons is needed. The client sends the initial quantum states, which are carried by some modulated single photons, to server through a quantum channel. After that, Alice sends measurement orders to Bob through a two-way classical channel, which can guide Bob to execute the corresponding quantum computations.

In UBQC protocol[4], the only non-classical requirement on the client is the ability of preparing single qubits. Comparing to other protocols[5], the burden of client is greatly reduced since there is no need of any other quantum resource. The blindness of the UBQC protocol only stands in the ideal case where the client prepares perfect qubits. In the practical implementation, the polarization of single photons can be used to encode the qubits. However, any exiting physical device is inevitably imperfect, because it is almost impossible to completely avoid to send two or more identically polarized photons instead of one to the server. Such event would destroy the perfect preparation, which directly affects the perfect privacy of the client's information. So the perfect security of the UBQC protocol is unlikely to be achieved in practice. On the other hand, the number of the photons emitted by a single photon source satisfies the Poisson distribution, and the probability of generated single photons is very low. Therefore, Dunjko, Kashefi, and Leverrier present a Remote Blind Single qubit Preparation(RBSP) protocol with Weak Coherent Pluses(WCP) to prepare qubits in UBQC[6]. In RBSP protocol, the client merely needs to send WCP to the server. The server is responsible to prepare the quantum states which are arbitrarily close to the perfect single qubit. This allows us to achieve $\epsilon - blind$ UBQC for any $\epsilon > 0$. Meanwhile, this paper gave a rough lower bound of the number of required pulses for each prepared qubit. But the lower bound is not efficient when the communication distance are long(the small transmittance of the channel). In the paper, we present a modified RBSP protocol with two decoy WCP to decrease the number of required pluses. In fact, the decoy state is the WCP on different intensity, which has been discussed widely in the QKD system[7–11]. In QKD, If the decoy states marked are sent to Bob, Bob can estimate the lower bound of gain of received single photons and the upper bound of error rate of received single photons to gain a higher secure key rate and longer communication distance[12–14]. Recently, Lo and Xu present a modified RSBP protocol with one decoy WCP[15], which can reduce the scaling of the required number of pulses from $O\left(1/T^4\right)$ to nearly $O\left(1/T\right)$(T is the transmittance of the channel). In our paper, we present a modify the RBSP protocol with two decoy states to further reduce the lower bound of the number of required pulses. In particular, the advantages of the modified RBSP protocol with two decoy states are more highlighted than one decoy with the increase of disantanc.

The rest of this paper is organized as follows: in Section II, technical preliminaries are introduced. In Section III, theoretical analysis shows that the modified RBSP protocol with two decoy WCP reduces the number of required pluses greatly. In Section IV, the numeric simulations are introduced and the results verify the theoretical analysis. In Section V, conclusions are drawn.

2   Technical Preliminaries

We give a brief recap of the UBQC protocol[4], as follows: the UBQC protocol is set in the framework of measurement-based quantum computation(MBQC)[16–18]. The crux of MBQC is a generic brickwork state, which is a multipartite entangled quantum state, and the computation is executed by performing measurements on its subsystems. The MBQC can be conceptually separated as classical and quantum part. These measurement angles which are used to guide the computation are generated by the classical client, as the classical controller unit. The preparation and measurements of the brickwork state are executed by the quantum server, as the quantum unit.

The blindness of UBQC only holds if the client can prepare the needed qubits perfectly. However, in a practical implementation, imperfection is inevitable and perfect blindness cannot be achieved. For the reason, the Remote Blind Single qubit Preparation(RBSP) protocol is proposed by the Dunjko etc. Due to quantum states arbitrarily are close to perfect random single qubit states, this allows us to efficiently achieve ϵ-blind UBQC for any $\epsilon > 0$, even if the channel between the client and the server is arbitrarily lossy. Then, we can achieve the number of required pulses for preparing one single qubit, as follows

Theorem 1[6]. A UBQC protocol of computation size S, where the clients preparation phase is replaced with S calls to the coherent state RBSP protocol, with a lossy channel connecting the client and the server of transmittance no less than T, is correct, ϵ-robust and ϵ-blind for a chosen $\epsilon > 0$ if the parameter N of each instance of the RBSP protocol called is chosen as follows:

$$N \geq \frac{18 \ln (S/\epsilon)}{T^4}.$$ (1)

3   Two Decoy States RBSP Protocol

Firstly, the modified RBSP protocol with two decoy states is presented as follows:
   1. *Client's preparation*
1.1 The client generates N weak coherent pulses which contain signal states and two decoy states with mean photon number μ, v_1, v_2 respectively. A random polarisation can be described as σ_l(for $l = 1, ..., N$). These signal states are described by

$$\rho_{\mu}^{\sigma_l} = e^{-\mu} \sum_{k=0}^{\infty} \frac{\mu^k}{k!} |k\rangle\langle k|_{\sigma_l}$$

The two decoy states denoted as $\rho_{v_1}^{\sigma_l}, \rho_{v_2}^{\sigma_l}$, is similar to the above formula except that the different mean photon numbers. The polarisation angles σ_l are chosen at random in $\{k\pi/4 : 0 \leq k \leq 7\}$. The client stores the sequence $(\sigma_1, ..., \sigma_N)$.

1.2 The client sends these signal states $\rho_\mu^{\sigma_l}$ and two decoy states $\rho_{v_1}^{\sigma_l}, \rho_{v_2}^{\sigma_l}$ to the server.

2. *Server's preparation*

2.1 For each state he receives, the server reports which pulses are received to the client.

3. *Client-Server's interaction*

3.1 The client verifies the gain of signal and two decoy states$(Q_\mu, Q_{v_1}, Q_{v_2})$ from the statistics reported by the server based on overall transmittance T. The client aborts the protocol if this number is larger than her preset threshold, otherwise the protocol continues. The client then declares the position of the decoy states and the quantum computation scale S.

3.2 The server discards the decoy states, and the total remaining number of qubits is M_μ. The server then randomly divides these M_μ signal states into S groups, and performs the I1DC computation[6]. He can obtain the resulting s-tate $|+_\theta\rangle$, and reports the measurement information to client.

3.3 Using her knowledge about the angles σ_l of the qubits used in the I1DC procedure by the server and the received measurement outcomes, the client computes θ based on I1DC computation.

In [6], Dunjko etc. pointed out the crux of the security of UBQC protocol: each I1DC subroutine is such that if the server is totally ignorant about the polarization of at least one photon in the 1D cluster, then he is also totally ignorant about the final qubit. In order to exploit this property, the server should make sure that each preparation group must have at least a single photon for generating a single qubit. Based on the above principle, we refer to the literature [15] to establish the mathematical model.

After the server discarding these decoy states, the total remaining number of the signal states is left with M_μ, the number of the single photons is denoted by M_1, and we can describe the proportion of these single photons in the signal states is $p_1 = M_1/M_\mu$. If the computation scale of the UBQC protocol is S, the server then divides the signal states into S groups, and the number of the signal states in the every group is described by $m = M_\mu/S$. If there is no single photon in a measurement group, the group is a failure measurement, and the probability of failure is denoted as p_{fail}.

$$p_{fail} = \frac{\binom{m}{M_\mu - M_1}}{\binom{m}{M_\mu}} \leq \left(\frac{M_\mu - M_1}{M_\mu}\right)^m = (1 - p_1)^m. \tag{2}$$

If there is a failure measurement group in the S groups, the total RBSP protocol will fail, and the total probability of preparation failure is described as P_{fail}.

$$P_{fail} = S p_{fail} \leq S(1 - p_1)^m. \tag{3}$$

If the security of the UBQC protocol is $\epsilon - secure$, the probability of the preparation failure P_{fail} must satisfy $P_{fail} \leq \epsilon$. According to (2) and (3), we can estimate the lower bound of required pulses for preparing a single qubit.

$$m \geq \frac{\ln(\epsilon/S)}{\ln(1-p_1)}. \tag{4}$$

The probabilities of the signal states and the two decoy states chosen from the client are defined as p_μ, p_{v_1}, p_{v_2} respectively, and $p_\mu + p_{v_1} + p_{v_2} = 1$. The total number of required pulses for computation scale S is denoted as N, and its lower bound is estimated by

$$N = \frac{M_\mu}{p_\mu Q_\mu} = \frac{mS}{p_\mu Q_\mu} \geq \frac{S}{p_\mu Q_\mu} \frac{\ln(\epsilon/S)}{\ln(1-p_1)}. \tag{5}$$

In (5), we can know that the proportion of the single photons p_1 is proportional to the total number N, then we only need to estimate the lower bound of p_1. In [9] and [11], Lo and Ma etc. gave the estimation formula of the gain of single photons Q_1 and the gain of signal states Q_μ. So we can estimate the lower bound of p_1, i.e.

$$p_1 = \frac{Q_1}{Q_\mu} \geq \frac{Y_1^{L,v_1,v_2} \mu e^{-\mu}}{Q_\mu}$$

$$= \frac{\mu^2 e^{-\mu}}{\mu v_1 - \mu v_2 - v_1^2 + v_2^2} \left[\frac{Q_{v_1}}{Q_\mu} e^{v_1} - \right. \tag{6}$$

$$\left. \frac{Q_{v_2}}{Q_\mu} e^{v_2} - \frac{v_1^2 - v_2^2}{\mu^2 Q_\mu} \left(Q_\mu e^\mu - Y_0^{L,v_1,v_2} \right) \right].$$

The dark count rate is described as Y_0. According to the literature [11], we can get its lower bound in a real run of protocol, i.e.

$$Y_0 \geq Y_0^{L,v_1,v_2} = \max \left\{ \frac{v_1 Q_{v_2} e^{v_2} - v_2 Q_{v_1} e^{v_1}}{v_1 - v_2}, 0 \right\}. \tag{7}$$

According to (6) and (7), we can achieve the approximate evaluation of the lower bound of p_1. Then, based on (5), we can estimate the total number of required pulses for the computation scale S.

4 Simulation

To estimate the total number of required pulses, we need to evaluate the lower bound of p_1. In (6), the gain of signal states and the gain of decoy states Q_μ, Q_{v_1} and Q_{v_2} can be obtained in a real system. To get a taste of the efficiency of this result, we consider the asymptotic case where the Q_μ, Q_{v_1} and Q_{v_2} are replaced with the expectation values $\widetilde{Q_\mu}, \widetilde{Q_{v_1}}$ and $\widetilde{Q_{v_2}}$.

$$\widetilde{Q_\mu} = Y_0 + 1 - \exp(-T\mu) \cong Y_0 + T\mu,$$
$$\widetilde{Q_{v_1}} \cong Y_0 + Tv_1, \tag{8}$$
$$\widetilde{Q_{v_2}} \cong Y_0 + Tv_2.$$

T is the overall transmittance and detection efficiency between client and server, and can be given by

$$T = 10^{-\alpha/10} \times t_S \times \eta_S, \tag{9}$$

where α is the loss coefficient measured in dB/km, L is the length of the fiber in km, t_S denotes for the internal transmittance of optical components in the server's side, and η_S is detector efficiency in the server's side. According to (5),(6) and (8), we know that our new protocol allow us to reduce the scaling N from $O(1/T^4)$ to nearly $O(1/T)$, shown as Table.1 We assume that α is 0.2dB/km,

Table 1. The relationship between N and T

Protocol	The total number of required pulses(N)
UBQC with weak coherent pulses	$O(1/T^4)$
our new protocol	$O(1/T)$

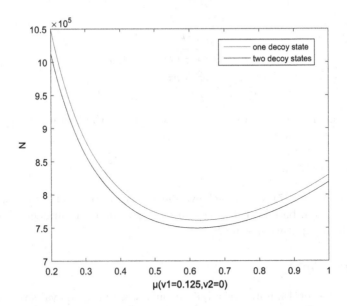

Fig. 1. The relationship between the number of pulses N and the mean photon numbers of signal states μ.

L is 25km, t_S is 0.45, η_S is 0.1, and Y_0 is 6×10^{-5}. The mean photon number of decoy states v_1 is 0.125 and v_2 is 0. The chosen probability of the signal states p_μ is 0.9. The computation scale S is 1000. The required security of the UBQC is 10^{-10}. According to (9) and (8), we can achieve the expectation value

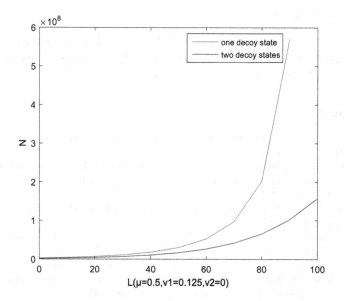

Fig. 2. The relationship between the number of pulses N and the communication distance L.

$\widetilde{Q_\mu}, \widetilde{Q_{v_1}}$ and $\widetilde{Q_{v_2}}$. And then, the real dark count rate Y_0 can be estimated by (7). Combining the formula (5) and (6),we simulate the relationship figure between N and μ, as shown Fig.1. In Fig.1, we can see that N does not change too much for $0.5 < \mu < 0.6$. The two decoy states case can achieve better optimal value of N than the one decoy state. If the mean photon number of μ is set to 0.5, we achieve the relationship between the number of pulses N and the transmission distance L, as shown Fig.2. We can know the advantages of two decoy protocol are highlighted with the increasing distance.

5   Conclusion

In order to improve the efficiency of UBQC, we utilize two decoy states into preparation protocol, and present a modified RBSP protocol with two decoy states. In [6], we know that the total number of required pulses is $O(1/T^4)$ according to (1). Our new protocol allow us to reduce the scaling N from $O(1/T^4)$ to nearly $O(1/T)$, and the two decoy states case can achieve a better value of N than one decoy state, significantly increasing the efficiency. Meanwhile, our results also show that the advantages of two decoy states are more highlighted with the increasing distance.

Acknowledgment

This work is supported by the National Natural Science Foundation of China (Grant Number: 61471141, 61361166006, 61301099), and Basic Research Project of Shenzhen, China (grant Number: JCYJ20150513151706561).

References

1. J. Feigenbaum, "Encrypting problem instances," in *Advances in Cryptology-CRYPTO'85 Proceedings*. Springer, 1985, pp. 477–488.
2. M. Abadi, J. Feigenbaum, and J. Kilian, "On hiding information from an oracle," in *Proceedings of the nineteenth annual ACM symposium on Theory of computing*. ACM, 1987, pp. 195–203.
3. C. Gentry *et al.*, "Fully homomorphic encryption using ideal lattices." in *STOC*, vol. 9, 2009, pp. 169–178.
4. A. Broadbent, J. Fitzsimons, and E. Kashefi, "Universal blind quantum computation," in *Foundations of Computer Science, 2009. FOCS'09. 50th Annual IEEE Symposium on*. IEEE, Conference Proceedings, pp. 517–526.
5. P. Arrighi and L. Salvail, "Blind quantum computation," *International Journal of Quantum Information*, vol. 4, no. 05, pp. 883–898, 2006.
6. V. Dunjko, E. Kashefi, and A. Leverrier, "Blind quantum computing with weak coherent pulses," *Physical review letters*, vol. 108, no. 20, 2012.
7. W.-Y. Hwang, "Quantum key distribution with high loss: toward global secure communication," *Physical Review Letters*, vol. 91, no. 5, 2003.
8. D. Gottesman, H.-K. Lo, N. Lütkenhaus, and J. Preskill, "Security of quantum key distribution with imperfect devices," in *Information Theory, 2004. ISIT 2004. Proceedings. International Symposium on*, 2004, p. 136.
9. H.-K. Lo, X. Ma, and K. Chen, "Decoy state quantum key distribution," *Physical review letters*, vol. 94, no. 23, 2005.
10. X.-B. Wang, "Beating the photon-number-splitting attack in practical quantum cryptography," *Physical review letters*, vol. 94, no. 23, 2005.
11. X. Ma, B. Qi, Y. Zhao, and H.-K. Lo, "Practical decoy state for quantum key distribution," *Physical Review A*, vol. 72, no. 1, 2005.
12. Y. Zhao, B. Qi, X. Ma, H.-K. Lo, and L. Qian, "Experimental quantum key distribution with decoy states," *Physical review letters*, vol. 96, no. 7, 2006.
13. T. Schmitt-Manderbach, H. Weier, M. Fürst, R. Ursin, F. Tiefenbacher, T. Scheidl, J. Perdigues, Z. Sodnik, C. Kurtsiefer, J. G. Rarity *et al.*, "Experimental demonstration of free-space decoy-state quantum key distribution over 144 km," *Physical Review Letters*, vol. 98, no. 1, 2007.
14. Y. Liu, T.-Y. Chen, J. Wang, W.-Q. Cai, X. Wan, L.-K. Chen, J.-H. Wang, S.-B. Liu, H. Liang, L. Yang *et al.*, "Decoy-state quantum key distribution with polarized photons over 200 km," *Optics Express*, vol. 18, no. 8, pp. 8587–8594, 2010.
15. K. Xu and H.-k. Lo, "Blind quantum computing with decoy states," *arXiv preprint arXiv:1508.07910*, 2015.
16. R. Raussendorf and H. J. Briegel, "A one-way quantum computer," *Physical Review Letters*, vol. 86, no. 22, 2001.
17. R. Raussendorf, D. E. Browne, and H. J. Briegel, "Measurement-based quantum computation on cluster states," *Physical review A*, vol. 68, no. 2, 2003.
18. R. Jozsa, "An introduction to measurement based quantum computation, 2005," *arXiv preprint quant-ph/0508124*.

A Bayesian based finite-size effect analysis of QKD

Xuan Wen, Qiong Li, Hongjuan Wang, Hui Yang, and Hanliang Bai

School of Computer Science and Technology,
Harbin Institute of Technology,
Harbin, China

Abstract. The finite-size of pulses processed in practical quantum key distribution system result in statistical fluctuations. Here, we analyze what parameters are fluctuating and respectively estimate them based on Bayesian estimation theory, then the formula for secret key rate of two decoy states protocol in finite case is obtained in this paper. Moreover, we simulate the secret key rate by using data from a proposed finite experiment of four different transmission distances quantum key distribution systems. Our simulation results show that the secret key rate based on Bayesian estimation theory we used is much higher than existing results.

Keywords: Finite-size, Two decoy states, Quantum key distribution, Secret key rate, Statistical fluctuation, Bayesian estimation theory

1 Introduction

Quantum key distribution (QKD) [1] provides a means of sharing secret keys between two parties (Alice and Bob) in the presence of an eavesdropper (Eve). The ideal security proof for the BB84 protocol was given when a single-photon source was assumed. However, due to the limitation of the technology, real implementations of QKD, which is differ from the ideal models in security proof, often adopt highly attenuated laser source which produces multi-photon states obeying Poisson distribution. Multi-photon states opens the door of photon-number-splitting (PNS) attack [2] for Eve.

Gottesman, Lo, Lütkenhaus, and Preskill (GLLP) [3] first proposed formula for key generation rate of QKD system under PNS attack. GLLP roughly discuss the gain rate lower bound and quantum bit error rate (QBER) upper bound of single-photon under the worst assumption: All the loss and error of signal are attributed to single-photon. GLLP security analysis theory indicates that PNS attack severely restrict the secret key rate and transmission distance of practical QKD systems. Hwang [4] proposed the innovative decoy state method to combat PNS attack. Hwangs idea was highly. However, his security analysis was heuristic. Lo [5] presented a rigorous security analysis of the decoy state protocol. They use decoy state to estimate the gain and error rate of single-photon states and

© Springer International Publishing AG 2017

J.-S. Pan et al. (eds.), *Advances in Intelligent Information Hiding and Multimedia Signal Processing*, Smart Innovation, Systems and Technologies 64, DOI 10.1007/978-3-319-50212-0_20 achieved a tighter formula for secret key generation rate by combining the idea GLLP.

In [3,5], security analysis are given by assuming system with infinite resources. However, the pulse number processed in practical QKD system are necessarily of finite length, which will inevitably cause statistical fluctuationss between parameters' measured value and their true value. It's imperative to take the effects of finite resource into account when analyze the secret key rate of QKD system.

In finite resources case, Scarani and Renner [6] proved the security of the ideal BB84 (experimental devices are perfect) protocol with finite resources. Cai and Scarani [7] estimated statistical fluctuationss of parameters in finite resource by using the law of large numbers theorem and derived bounds of the secret key rate. Later in [8,9,10,11,12,13,14], many scholars also analyzed the security of finite resources QKD system by using parameter estimation method in [7]. However, in [7] only considered the confidence level $1-\varepsilon$ and sample capacity N in estimating statistical fluctuations, but didn't considered the sample information which is valuable for estimation.

In this paper, taking the sample information into account, we estimate the statistical fluctuations based on Bayesian estimation theory and compared the secret key rate of finite resources decoy states QKD system in different estimation methods of this paper and [8] by simulating data from a finite experiment [15].

The rest of this paper is arranged as follows: In Section 2, we will give a brief discussion of Bayesian estimation theory. In Section 3, we will discuss what parameters are fluctuating in finite resources QKD system with weak+vacuum decoy states and respectively estimate the fluctuating parameters based on Bayesian estimation theory, then give formula for secret key generation rate by combining [16]. In Section 4, numerical simulation results for different transmission distances based on the existing experimental data will be shown. Finally, conclusions are drawn in Section 5.

2   Review of infinite-size two decoy states protocol

Two decoy states protocol [17] add two different quantum states on the basis of BB84 protocol. In process of two decoy states, Alice send three kinds of quantum states with different average number of photons to Bob. Ma combined GLLP theory with the decoy method and achieved key generation rate of two decoy states [16]:

$$R \geq q\left\{-Q_\mu f(E_\mu)H(E_\mu) + Q_1[1 - H(e_1)]\right\}, \tag{1}$$

which

$$Q_1 = \frac{\mu^2 e^{-\mu}}{\mu\nu_1 - \nu_1{}^2}(Q_{\nu_1}e^{\nu_1} - \frac{\nu_1{}^2}{\mu^2}Q_\mu e^\mu - \frac{\mu^2 - \nu_1{}^2}{\mu^2}Q_{\nu_2}),$$

$$e_1 = \frac{(\mu\nu_1 - \nu_1^2)(E_\mu Q_\mu e^\mu - E_{\nu_2}Q_{\nu_2})}{\mu^2 Q_{\nu_1}e^{\nu_1} - Q_\mu e^\mu \nu_1^2 - Q_{\nu_2}(\mu^2 - \nu_1^2)},$$

and q depends on the implementation (1/2 for the BB84 protocol due to the fact that half of the time Alice and Bob disagree with the bases), μ , ν_1 , ν_2 respectively denote the intensities of signal states and two decoy states. Q_μ is the gain of signal states, E_μ is the overall quantum bit error rate, Q_1 is the gain of single-photon states, e_1 is the error rate of single-photon states, $f(x)$ is the bidirectional error correction efficiency (see, for example, [18]) as a function of error rate, normally $f(x) \geq 1$ with Shannon limit $f(x)=1$, and $H(x)$ is the binary Shannon entropy function, given by $H(x) = -x\log_2(x) - (1-x)\log_2(1-x)$.

3 Secret key rate estimation based on Bayesian estimation theory

Recall that from (1), the secret key rate is actually a function of eight parameters: μ, ν_1, ν_2, Q_μ, Q_{ν_1}, Q_{ν_2}, E_μ, E_{ν_2}. Among them, the intensities μ, ν_1, ν_2 of three lasers used by Alice are system parameters set up before the experiment is run. Although the three intensities will suffer from small statistical fluctuationss, it is difficult to pinpoint the extent of their fluctuations without hard experimental data. To simplify our analysis, we will ignore their fluctuations in this paper. The gain and QBER of three quantum states Q_μ, Q_{ν_1}, Q_{ν_2}, E_μ, E_{ν_2} are directly measured from experiment, there must exist statistical fluctuationss from their real value because of finite resource. So it is indispensable to estimate the five statistical parameter before calculating secret key rate. In two decoy states protocols, the optimal and widely used case is weak+vacuum decoy state protocol, of which decoy state with intensity $\nu_1 \ll \mu$ and $\nu_2=0$. Specially, we use ν note ν_1 and ϕ note $\nu_2=0$ in rest of this paper. We mainly analyze the security and statistical fluctuationss of weak+vacuum decoy states protocol QKD system with finite resources in this paper. So our main goal is to estimate Q_μ, Q_{ν_1}, Q_{ν_2}, E_μ, E_{ν_2}.

If regarding the gain case of a pulse signal as a random variable A, obviously A obey binomial distribution $b(1,Q)$ (since the exchanged pulse can be divided into count pulse $A=1$ and pulse that have been lost $A=0$), where Q is the probability of count pulse. According to this view, the count pulses number X of N exchanged pulses in practical system satisfy $X \sim b(N,Q)$ as follow

$$P(X = x|Q) = \binom{N}{x} Q^x(1-Q)^{N-x}, x = 0, 1, 2, \cdots, N. \qquad (2)$$

Our purpose is to estimate the unknown parameter Q. In Bayesian estimation theory, any unknown parameter can be seen as a random variable, and we need to assume Q obey a certain probability distribution before estimating. The performance will be dissimilar for different experimental systems, this means the experimental data can't be shared for different systems. Without loss of generality, there is no priori information can be provided before experiment. For this situation, we assume Q is uniform distribution on codomain $[0,1]$ according to suggestion of Bayesian estimation theory,

$$\pi(Q) = \begin{cases} 1, 0 < Q < 1 \\ 0, else \end{cases}. \qquad (3)$$

Then, the marginal probability distribution of X can be derived as follow

$$m(x) = \binom{N}{x} \int_0^1 Q^x (1-Q)^{N-x} dQ = \binom{N}{x} \frac{\Gamma(x+1)\Gamma(N-x+1)}{\Gamma(N+2)}. \qquad (4)$$

Hence, combining (2),(3),(4), we can obtain the posterior distribution of Q,

$$\pi(Q|x) = \frac{p(x|Q)\pi(Q)}{m(x)} = Be(x+1, N-x+1), \qquad (5)$$

where $\Gamma(x)$, $Be(x,y)$ are Gamma and Beta function respectively.

Beta function $Be(x,y)$ is unimodal function achieving maximum value at point $\frac{x-1}{x+y-2}$, that is to say maximum point of $\pi(Q|x)$ is frequency value $p=\frac{x}{N}$ of random variable A. It further indicates that the posterior distribution based on Bayesian estimation theory well reflected sample information from experiment.

To make the results more compact and accurate, the length of confidence interval should be as short as possible. This requires the posteriori distribution function values of points inside of the confidence interval are entirely larger than that outside of confidence interval. Interval getting by this way is Highest Posterior Density (HPD) confidence interval. Because the expression of Beta function is too complex, finding HPD confidence interval becomes very difficult. In this paper, we use a neighborhood C of point $Q=p$ with confidence level $1-\varepsilon$ to approximatively instead of HPD confidence interval. Our main goal translate into using $\pi(Q|x) = Be(x+1, N-x+1) = Be(pN+1, N-pN+1)$ to find an interval C satisfy

$$\int_C Be(pN+1, N-pN+1)dQ = 1 - \varepsilon, \qquad (6)$$

where p is frequency value of A, it is also the measured value of Q.

Method of finding confidence interval C proposed in this paper is detailedly described as follow:

1. If $\int_{p-D}^{p+D} Be(x+1, N-x+1)dQ \geq 1 - \varepsilon$, where $D = \min\{p, 1-p\}$.
   Finding a r makes $C=[p-r, p+r]$ satisfy (6) by using binary search in $[0, D]$, then C is the confidence interval of Q we need.
2. If $\int_{p-D}^{p+D} Be(x+1, N-x+1)dQ < 1 - \varepsilon$ and $p \geq D$.
   Finding a r makes $C=[p-r, 1]$ satisfy (6) by using binary search in $[D, p]$, then C is the confidence interval of Q we need.
3. If $\int_{p-D}^{p+D} Be(x+1, N-x+1)dQ < 1 - \varepsilon$ and $p < D$.
   Finding a r makes $C=[0, p+r]$ satisfy (6) by using binary search in $[p, D]$, then C is the confidence interval of Q we need.

In the similar principle of estimating gain, the error case of an exchanged pulse signal also obey binomial distribution. Next, we respectively estimate the gain and QBER of three quantum states Q_μ, Q_{ν_1}, Q_{ν_2}, E_μ, E_{ν_2} based on the above analysis.

1. Q_μ : In practical experiment, Q_μ is measured through observing the count situation of all signal states Alice send to Bob. Then the sample capacity for measuring Q_μ is the total number N_μ of signal states sent by Alice. Therefore, the posterior distribution of Q_μ is

$$Be(N_\mu \tilde{Q}_\mu + 1, N_\mu(1 - \tilde{Q}_\mu) + 1)$$

where $\tilde{Q}_\mu$ is the measured value of Q_μ, we use $\tilde{X}$ note measured value of X in this paper.
Similar with Q_μ, we can get the posterior distribution of Q_ν, Q_ϕ as follow respectively

$$Be(N_\nu \tilde{Q}_\nu + 1, N_\nu(1 - \tilde{Q}_\nu) + 1)$$
$$Be(N_\phi \tilde{Q}_\phi + 1, N_\phi(1 - \tilde{Q}_\phi) + 1)$$

Where N_ν is the total number of weak state sent by Alice, N_ϕ is the total number of vacuum state. Let $[Q_\mu^L, Q_\mu^U]$, $[Q_\nu^L, Q_\nu^U]$, $[Q_\phi^L, Q_\phi^U]$ be the confidence interval of Q_μ, Q_ν, Q_ϕ derived according to Bayesian estimation theory.

2. E_μ : In practical experiment, we randomly select partial bits from the counting signal state generated by identical base at a certain ratio w when measure E_μ. Then the sample capacity for measuring E_μ is $N_\mu^E = wqQ_\mu N_\mu$. Therefore, the posterior distribution of Q_μ is

$$Be(N_\mu^E \tilde{E}_\mu + 1, N_\mu^E(1 - \tilde{E}_\mu) + 1)$$

Let $[E_\mu^L, E_\mu^U]$ is the confidence interval of E_μ.

3. E_ϕ :In practical experiment, Alice didn't generate any pulse for vacuum state, so the QBER of vacuum state is equal to that of dark count. We measure it by comparing the counting vacuum state. The sample capacity for measuring E_ϕ is $N_\phi^E = N_\phi Q_\phi$. Therefore, the posterior distribution of E_ϕ is

$$Be(N_\phi^E \tilde{E}_\phi + 1, N_\phi^E(1 - \tilde{E}_\phi) + 1)$$

Let $[E_\phi^L, E_\phi^U]$ is the confidence interval of E_ϕ.

We should always consider the worst case when analysis security of QKD. That is to say that it's necessary to compute the minimum value of R. So we need use lower bounds of parameters in direct proportion to R and use upper bounds of parameters in inverse proportion to R. We use $f(E_\mu) = f^U = 1.22$, which is the upper bound of $f(x)$ in secure distance [18]. Combining (1) and the estimation results above, the final secret key rate of weak+vacuum decoy states protocol QKD system with finite resources is characterized by the following formula

$$R \geq -qQ_\mu^U f^U H(E_\mu^U) + qQ_1^L[1 - H(e_1^U)] \tag{7}$$

which

$$Q_1^L = \frac{\mu^2 e^{-\mu}}{\mu\nu - \nu^2}(Q_\nu^L e^\nu - \frac{\nu^2}{\mu^2}Q_\mu^U e^\mu - \frac{\mu^2 - \nu^2}{\mu^2}Q_\phi^U),$$

$$e_1^U = \frac{(\mu\nu - \nu^2)(E_\mu^U Q_\mu^U e^\mu - E_\phi^L Q_\phi^L)}{\mu^2 Q_\nu^L e^\nu - Q_\mu^U e^\mu \nu^2 - Q_\phi^U(\mu^2 - \nu^2)}.$$

4   Simulation and Analysis

In practical QKD system, the dark count rate of detector is usually equal to about 10^{-5}, we fix Q_ϕ by 10^{-5} in simulation. The bits from dark count is random, which with QBER about 0.5. So we make E_ϕ=0.5. If the failure probability of each parameter estimation is ε then the confidence level of quantity contains n parameters is at least $(1-\varepsilon)^n \simeq 1 - n\varepsilon$. We use $\varepsilon=10^{-6}$ for each parameters in simulation, then the final secret key rate with confidence level at least $5\times10^{-6} \simeq 10^{-5}$, that is to say the security parameter of QKD system is 10^{-5} (equal to security parameter in [7]). We use $q=1/2$ in our simulation. For the rest of the parameters, our simulation is based on experimental results reported by Yin [15] as shown in Table 1.

Table 1. All of the values are measured from experiment with μ=0.6 and ν=0.2

Length	Q_μ	E_μ	Q_ν	E_ν
49.2 km	8.6×10^{-4}	0.0103	2.9×10^{-4}	0.02
62.1 km	2.88×10^{-4}	0.0108	1.08×10^{-4}	0.0225
83.7 km	1.57×10^{-4}	0.0145	5.28×10^{-5}	0.019
108 km	7.1×10^{-5}	0.016	2.52×10^{-5}	0.027

Combining (7) and the experimental data above, we simulate secret key rate based on the two different estimation methods and get results in Fig.1 and Fig.2.

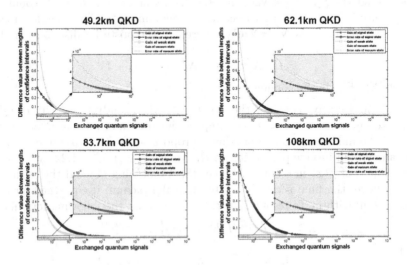

Fig. 1. Difference value between the lengths of confidence intervals estimated by two different estimation.

According to Fig.1, For the same confidence level $1-\varepsilon$ and sample informa-tion, the difference values of five parameters are always positive. That is to say

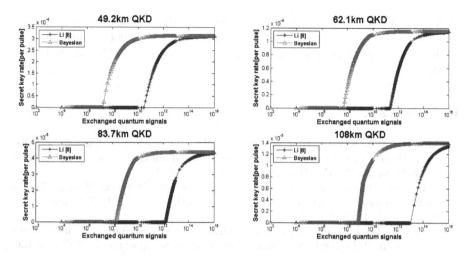

Fig. 2. Secret key rate for different exchanged quantum signals with different transmission distances.

that the confidence intervals we obtained are entirely shorter than that by using estimation method in [8], especially for E_μ and E_ϕ. It indicates that Bayesian estimation theory estimate more accurate than method in [8].

From Fig.2, we can clearly see: In the four practical QKD systems with different transmission distances, for the same security parameters ε and total number of pulses N, the secret key rate we obtained are higher than method in [8]. The minimal number of the exchanged quantum signals for generating secret key become much smaller and the secret key rate reach the limit value more quickly when we use Bayesian estimation theory. For example, in the case of transmission distance is 49.2 km, estimation method in [8] obtain a non-negligible secret key rate for N as small as 10^{11}, while our method do not obtain any key for $N < 10^7$ signals. All phenomena above indicates that Bayesian estimation theory we used estimate more accurate than estimation method in [8] and greatly improved the secret key rate of QKD system. On the other hand, our work greatly reduces the performance requirement of laser in QKD system.

5   Conclusion

We estimate the statistical fluctuations of finite-size QKD system based on Bayesian estimation theory. Then we provide explicit security bounds for the two decoy states QKD protocol in the finite-size regime. By simulating, it shows that our estimation result is more accurate than that of method in [7]. Therefore our secret key rate in finite-size based on Bayesian estimation theory is much higher and tighter than that of existing literature.

Acknowledgments. This work is supported by the National Natural Science Foundation of China (Grant Number: 61471141, 61361166006, 61301099), and Basic Research Project of Shenzhen, China (grant Number: JCYJ201505131517 06561).

References

1. C.H., G.:Public Key Distribution and Coin Tossing. In: Proceedings of the IEEE International Conference on Computers Systems and Signal Processing, IEEE, Bangalore, India, pp.175-179. IEEE Press, New York (1984)
2. Lütkenhaus, Jahma: Quantum Key Distribution with Realistic States: Photon-Number Statistics in the Photon-Number Splitting Attack. J. New Journal of Physics. 4(1): 44 (2002)
3. Gottesman, Lo, Lütkenhaus: Security of Quantum Key Distribution with Imperfect Devices. In: ISIT 2004 Proceedings International Symposium on IEEE, pp.136. IEEE Press, New York (2004)
4. Hwang: Quantum Key Distribution with High Loss: Toward Global Secure Communication. J. Physical Review Letters. 91(5): 057901 (2003)
5. Lo , Ma , Chen: Decoy State Quantum Key Distribution. J. Physical Review Letters. 94(23): 230504 (2005)
6. Scarani, Renner: Security Bounds for Quantum Cryptography with Finite Resources. M, Theory of Quantum Computation, Communication, and Cryptography. Springer Berlin Heidelberg, pp. 83-95 (2008)
7. Cai, Scarani: Finite-Key Analysis for Practical Implementations of Quantum Key Distribution. J. New Journal of Physics. 11(4): 045024 (2009)
8. L.H.W, Z.Y.B, Y.Z.Q: Security of Decoy States QKD with Finite Resources against Collective Attacks. J. Optics Communications. 282(20):4162-4166 (2009)
9. Sheridan, Le, Scarani: Finite-key Security against Coherent Attacks in Quantum Key Distribution. J. New Journal of Physics. 12(12):128-134 (2010)
10. Mertz, Kampermann, Shadman: Quantum Key Distribution with Finite Resources: Taking Advantage of Quantum Noise. J. Physical Review A. 52(87):333-43 (2011)
11. Tomamichel, Lim, Gisin: Tight Finite-Key Analysis for Quantum Cryptography. J. Nature Communications. 3(48):19596-19600 (2012)
12. Salas: Security of Plug-and-Play QKD Arrangements with Finite Resources. J. Quantum Information and Computation. 13(9-10):861-879 (2013)
13. Curty, X., Cui: Finite-Key Analysis for Measurement-Device-Independent Quantum Key Distribution. J. Nature Communications. 5(4):643-648 (2014)
14. T.T.S, S.J.Q, Q.Y.W: Finite-Key Security Analyses on Passive Decoy-State QKD Protocols with Different Unstable Sources. J. Scientific Reports. (2015)
15. Y.Z.Q, H.Z.F, C.W: Experimental Decoy Quantum Key Distribution up to 130km Fiber. J . arXiv Preprint arXiv. 0704.2941 (2007)
16. Ma, B.Q, Y.Z, Lo: Practical Decoy State for Quantum Key Distribution. J. Phys Rev A. 72: 012326 (2005)
17. X.B . Wang: Beating the PNS Attack in Practical Quantum Cryptography. J. Phys Rev Lett. 94230503 (2005)
18. Brassard, Salvail: Secret-Key Reconciliation by Public Discussion. J. Lecture Notes in Computer Science. 765:410–423 (1998)

A Novel Aeromagnetic Compensation Method Based on the Improved Recursive Least-Squares

Guanyi Zhao[1], Yuqing Shao[1], Qi Han[1], and Xiaojun Tong[2]

[1] School of Computer Science and Technology, Harbin Institute of Technology,
Harbin, 150001, China
qi.han@hit.edu.cn
[2] School of Computer Science and Technology, Harbin Institute of Technology,
Weihai, 264209, China

Abstract. Aeromagnetic compensation is a significant technique in airborne magnetic surveys which are usually applied in mineral prospection. In this paper, a novel method for aeromagnetic compensation based on the improved recursive least-squares (RLSQ) is studied. The method will reduce the errors caused by the trend of the geomagnetic variation, which will remain in the band-pass filtered signal of the geomagnetic field. Simulation results demonstrate its higher performance. Also, the ability and the limitation of the proposed method are discussed.

Keywords: Magnetic survey · Aeromagnetic compensation · Recursive least-squares · Geomagnetic field

1 Introduction

Airborne magnetic surveys are performed in order to detect mineral deposits or ferromagnetic targets [1]. These substances cause abnormal changes in the local magnetic field. These changes are too weak to be visible in the background signal. The background, or the external noise, can be classified into two groups [2], the interference unrelated to the aircraft maneuvers and the interference related to the aircraft maneuvers.

Aeromagentic compensation is aimed at removing the magnetic interference associated to the aircraft maneuvers. The target field signal can be detected in the compensated total field signal. There is a lot of research around the compensation methods in recent decades as [2]-[9]. Most of the compensation algorithms rely on the T-L model, which was presented by Tolles and Lawson in 1950 [5]. The model separates the interference field related to the aircraft maneuvers into three sources. The first one is the permanent field, which can be described as a constant vector in the aircraft reference frame. The second source is the induced field, which is produced by magnetizing of some aircraft's structures in the environmental magnetic field. The third source is the eddy-current field, which is caused by the eddy currents through aircraft's skin and some other structures. Herein, the T-L model can be transformed into a parameter estimation question.

© Springer International Publishing AG 2017

J.-S. Pan et al. (eds.), *Advances in Intelligent Information Hiding and Multimedia Signal Processing*, Smart Innovation, Systems and Technologies 64,
DOI 10.1007/978-3-319-50212-0_21

The least-squares (LS) estimation was first applied in [6]. But the ill-conditioned nature of the coefficients matrix makes the solution unstable. So some other methods such as stepwise regression analysis, rank deficient singular value decomposition and ridge regression analysis were tested in [7]. Dou et al. presented a method based on recursive least-squares (RLSQ), and verified its ability [8]. In practice, besides maneuvers, the induced and the eddy-current interferences are associated to the magnitude of the geomagnetic field [9], which leads to the overlap between the frequency bands of the interference field and the geomagnetic field. Therefore, the trend in the geomagentic field remains in the filtered geomagentic signal, which brings a few errors into the results of compensation. Unfortunately, it is ignored by most of the current algorithms.

The paper is outlined as follows. In Sect. 2, the influence of the geomagnetic variation trend on estimating the coefficients of T-L model is studied, then an improved RLSQ method is proposed. In Sect. 3, the simulation results are exhibited and analysed. Lastly, the ability and the limitation of the proposed method are discussed in Sect. 4.

2 The Proposed Method

In this section, the influence of the trend in the filtered geomagnetic field is analysed at first. Thereout, an improved method based on RLSQ is proposed. For convenience, the proposed method is named as IRLS method.

2.1 The Theoretical Analysis

As mentioned above, the magnetic field measured by the airborne magnetometer can be separated into two parts depending on whether it is associated to the aircraft maneuvers. Since the main field in the part unrelated to the maneuvers is the geomagnetic field, the total field can be expressed as

$$H_T(t) = H_I(t) + H_E(t) \tag{1}$$

where H_T is the total field; H_I is the part related to maneuvers; H_E is the geomagnetic field; t means time. The T-L model explains H_I as

$$H_I(t) = \phi^T(t) \cdot \theta \tag{2}$$

where ϕ and θ are two column vectors with 18 elements. The vector ϕ is composed by direction cosines of the geomagnetic field. The vector θ consists of the coefficients of the T-L model.

The traditional methods suppose H_E is an invariable in the signal filtered by a band-pass filter bpf. It is expressed as

$$bpf(H_E(t)) \approx 0. \tag{3}$$

Hence, the filtered signal can be written as

$$bpf(H_T(t)) \approx bpf(H_I(t)) = bpf(\phi^T(t)) \cdot \theta. \tag{4}$$

Assuming that there are N sample points in a segment of the signal, (4) can be discretized, i.e., for each sample point,

$$y_m = \phi_m^T \cdot \theta + v_m \ (1 \le m \le N) \tag{5}$$

where y_m represents a sample point of $bpf(H_T)$, and v_m is the noise series. Then some parameter estimation methods are available, such as the recursive least-squares estimation method.

In practice, $bpf(H_E)$ could not be ignored, even though the magnitude of the filtered geomagnetic field is only a few nT. The reason is that it will lead to the nonstationarity of the noise series v_m, which is the unbiased condition of the least-squares estimation method. The primary cause of the phenomenon is that there is a trend in the variation of the geomagnetic magnitude. The trend remains in the filtered geomagnetic signal $bpf(H_E)$, because the pass-band of the filter and H_E overlap.

As in Fig. 1, the geomagnetic field and its filtered signal are exhibited. For the filtered geomagnetic field, we use the Dickey-Fuller test [12] to verify the existence of trends. The null-hypothesis will be accepted when $lag = 0$, or rejected when $lag = 1$. It indicates that the filtered geomagnetic field can be considered as a first-order difference stationary process.

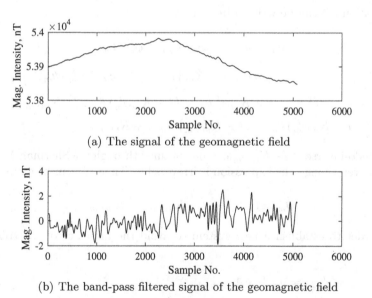

(a) The signal of the geomagnetic field

(b) The band-pass filtered signal of the geomagnetic field

Fig. 1. Exhibitions of the geomagnetic field signal and its band-pass filtered signal.

Theoretically, ϕ is only related to the variation of the geomagnetic direction caused by the aircraft maneuvers. The trend component is entirely attributed to the noise series v_m. In consideration of that the first-order differencing is a simple but useful detrend procedure [10], the noise series v_m can be represented as

$$v_{m+1} = \psi v_m + \zeta_m \qquad (6)$$

through the first-order differencing. Where ζ_m is a noise series which can be regarded as a stationary series.

2.2 The IRLS Method

The RLSQ updated equations are given by

$$\hat{\theta}_{m+1} = \hat{\theta}_m + K_{m+1}\left(y_{m+1} - \phi_{m+1}^T \cdot \hat{\theta}_m\right), \qquad (7)$$

$$K_{m+1} = P_m \cdot \phi \left(\lambda I + \phi_{m+1}^T \cdot P_m \cdot \phi_{m+1}\right)^{-1}, \qquad (8)$$

$$P_m = \frac{1}{\lambda}\left(I - K_m \cdot \phi_m^T\right)P_{m-1} \qquad (9)$$

where K_m is the gain matrix, and P_m is the covariance matrix. Their updated equations are (8) and (9), therein, I is an identity matrix, and λ is a forgetting factor.

Equation (7) can be written further as

$$\hat{\theta}_m = \hat{\theta}_{m-1} + K_m\left(y_m - \phi_m^T \cdot \hat{\theta}_{m-1}\right)$$
$$= \hat{\theta}_{m-1} + K_m\left(\phi_m^T \cdot \hat{\theta}_m + v_m - \phi_m^T \cdot \hat{\theta}_{m-1}\right)$$
$$= K_m\phi_m^T\hat{\theta}_m + \left(I - K_m\phi_m^T\right)\hat{\theta}_{m-1} + K_m v_m$$
$$\left(I - K_m\phi_m^T\right)\hat{\theta}_m = \left(I - K_m\phi_m^T\right)\hat{\theta}_{m-1} + K_m v_m. \qquad (10)$$

The pseudo-inverse of $I - K_m\phi_m^T$ can be obtained through the Sherman-Morrison formula even though the expression is irreversible. Then (7) can be expressed as

$$\hat{\theta}_m = \hat{\theta}_{m-1} + \left(I - K_m\phi_m^T\right)^{-1} K_m v_m. \qquad (11)$$

In order to establish a new system to avoid the influence of the trend, we define

$$y'_{m-1} = y_m - \psi A_m y_{m-1}$$
$$= \phi_m^T\hat{\theta}_m + v_m - \psi A_m\left(\phi_{m-1}^T\hat{\theta}_{m-1} + v_{m-1}\right)$$
$$= \phi_m^T\hat{\theta}_m - \psi A_m\phi_{m-1}^T\hat{\theta}_{m-1} + v_m - \psi A_m v_{m-1} \qquad (12)$$

where A_m is a short form for convenience. It is defined as

$$A_m = \phi_m^T\left(I - K_m\phi_m^T\right)^{-1} K_m + 1. \qquad (13)$$

Thus if (11) is substituted into (12), the new series y' can be transformed into

$$
\begin{aligned}
y'_{m-1} &= \phi_m^T \left[\hat{\theta}_{m-1} + \left(I - K_m \phi_m^T \right)^{-1} K_m v_m \right] - \psi A_m \phi_{m-1}^T \hat{\theta}_{m-1} + v_m - \psi A_m v_{m-1} \\
&= \phi_m^T \hat{\theta}_{m-1} - \psi A_m \phi_{m-1}^T \hat{\theta}_{m-1} + \phi_m^T \left(I - K_m \phi_m^T \right)^{-1} K_m v_m + v_m - \psi A_m v_{m-1} \\
&= \left(\phi_m^T - \psi A_m \phi_{m-1}^T \right) \hat{\theta}_{m-1} + A_m v_m - \psi A_m v_{m-1} \\
&= \left(\phi_m^T - \psi A_m \phi_{m-1}^T \right) \hat{\theta}_{m-1} + A_m \zeta_{m-1}.
\end{aligned}
\tag{14}
$$

Equation (14) is a new system which has the same set of coefficients with the original system (5). The expection of the new noise seires $A_m \zeta_{m-1}$ would be zero, because of the mutual independence of A_m and ζ_{m-1} and the stationarity of ζ_{m-1}. Performing the RLSQ process based on the new system could yield a more accurate result. The simulation results and analysis will be demonstrated in the next section.

3   Simulation and Analysis

The test environment is based on a digital simulation platform. For more details about the simulation platform, see [8]. The flight data is acquired through FlightGear Flight Simulator. And the magnetic data is calculated by World Magnetic Model 2015 (WMM2015), which is the only difference from the simulation platform in [8]. As an instance, Fig. 2 shows the total field and the three-axes components generated by the simulation platform.

In each test, there are two steps named as calibration and compensation respectively. The coefficients are calculated in the calibration procedure, then applied to remove the interference field in the compensation procedure. Hence, two sets of data should be prepared in a test. There are 9 sets of data with different quality. As Fig. 3, we illustrate the results in 9 subgraphs. Each of them exhibits the compensation indexes of all sets of data. The only difference in these subgraphs is that the coefficients are yielded by the corresponding set. The index mentioned above is FOM, short for figure of merit, which is a performance index of the compensation system proposed by Hood in 1967 [11]. Figure 3 indicates that the proposed method can promote the effectiveness of RLSQ for aeromagnetic compensation.

A test is analysed in details as an instance. Essentially, the coefficients are used to estimate the interference field produced by aircraft maneuvers. As (1), if the interference field is removed from the total field, the residual field should approximate the geomagnetic field. Hence, the performance of compensation methods can be evaluated through comparing the residual field to the geomagnetic field. Herein, the result of comparison between the RLSQ method and the IRLS method is exhibited in Fig. 4. We separate the three signals in the y-axis for clarity. It is obvious that the residual field of the IRLS method (below) is more similar to the geomagnetic field (center). Thereout, the ability of the proposed method has been verified.

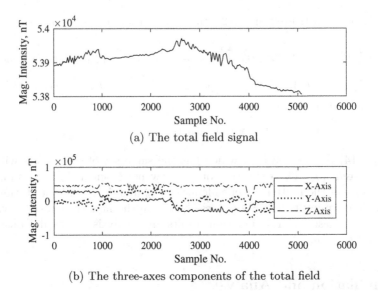

(a) The total field signal

(b) The three-axes components of the total field

Fig. 2. The total field and its three-axes components generated by the simulation platform.

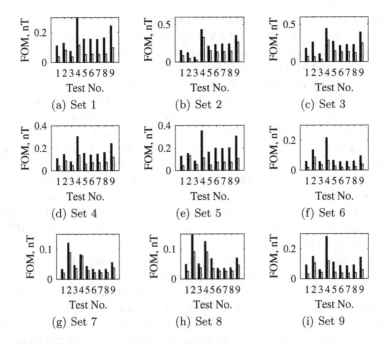

(a) Set 1

(b) Set 2

(c) Set 3

(d) Set 4

(e) Set 5

(f) Set 6

(g) Set 7

(h) Set 8

(i) Set 9

Fig. 3. Each of the subgraphs shows the compensation indexes of all sets of data. The coefficients are yielded by corresponding set. In each subfigure, the left and right bars represent the compensation results of the RLSQ method and the IRLS method respectively.

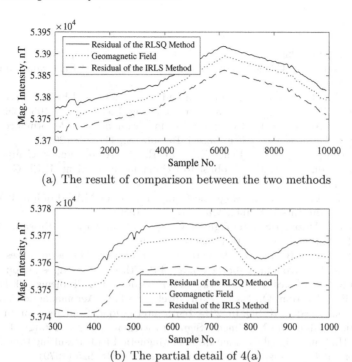

(a) The result of comparison between the two methods

(b) The partial detail of 4(a)

Fig. 4. The middle curve is the geomagnetic field. The upper and lower curves are the residuals of the RLSQ method and the IRLS method respectively.

4   Discussion

The nonstationarity of the geomagnetic field is studied in this paper. The overlap between the frequency domains of the interference field and the geomagnetic field leads to a problem that the band-pass filter can not remove the trend in the geomagnetic field. Based on this, an improved RLSQ method has been proposed. Although its performance has been verified by simulation results, there is still a problem, which is the assumption that the trend component is entirely attributed to the noise series v_m. Actually, the magnitude of H_E would change because of the aircraft maneuvers which lead to the varying position of the magnetometer. Therefore, there might also be trends in ϕ. This problem needs to be resolved in future.

Acknowledgements. This work is supported by the National Natural Science Foundation of China (Grant Number: 61471141, 61361166006, 61301099), and BasicResearchProject of Shenzhen, China (Grant Number: JCYJ20150513151706561).

References

1. Sheinker, A., Frumkis, L., Ginzburg, B., Salomonski, N., Kaplan, B.-Z.: Magnetic Anomaly Detection Using a Three-Axis Magnetometer. IEEE Trans. Magn. 45, 160–167 (2009)
2. Leliak, P.: Identification and Evaluation and of Magnetic and Field Sources and of Magnetic and Airborne. IRE Trans. Aerosp. Navig. Electron., 95–105 (1961)
3. Noriega, G.: Aeromagnetic Compensation in Gradiometry—Performance, Model Stability, and Robustness. IEEE Geosci. Remote. S., 117–121 (2015)
4. Dou, Z., Han, Q., Niu, X., Peng, X., Guo, H.: An Aeromagnetic Compensation Coefcient-Estimating Method Robust to Geomagnetic Gradient. IEEE Geosci. Remote. S., 611–615 (2016)
5. Tolles, W.E., Lawson, J.D.: Magnetic Compensation of MAD Equipped Aircraft. Airborne Instruments Lab. Inc., Mineola, N.Y. (1950)
6. Leach, B.W.: Automatic Aeromagnetic Compensation, National Research Council of Canada, National Aeronautical laboratory (1979)
7. Groom, R.W., Jia, R., Lo, B.: Magnetic Compensation of Magnetic Noises Related to Aircrafts Maneuvers in Airborne Survey. In: 17th EEGS Symposium on the Application of Geophysics to Engineering and Environmental Problems (2004)
8. Dou, Z., Ren, K., Han, Q., Niu, X.: An Novel Real-Time Aeromagnetic Compensation Method Based on RLSQ. In: 2014 Tenth International Conference on Intelligent Information Hiding and Multimedia Signal Processing, pp. 243–246 (2014)
9. Bickel, S.H.: Small Signal Compensation of Magnetic Fields Resulting from Aircraft Maneuvers. IEEE Trans. Aerosp. Electron. Syst. 15, 518–525 (1979)
10. Hamilton, J.D.: Time Series Analysis. Princeton University Press, Princeton (1994)
11. Hood, P.J.: Magnetic Surveying Instrumentation a Review of Recent Advances. In: Mining and Groundwater Geophysics (1967)
12. Dickey, D.A., Fuller, W.A.: Distribution of the estimators for autoregressive time series with a unit root. Journal of the American statistical association, 74, 427-431 (1979)

An Improved Adaptive Median Filter Algorithm and Its Application

Rui Ha[1,2,3] ,Pengyu Liu[1,2,3] and Kebin Jia[1,2,3]

1 Beijing Laboratory of Advanced Information Networks, Beijing, China
2 College of Electronic Information and Control Engineering, Beijing University of Technology,
Beijing, China
*liupengyu@bjut.edu.cn

Abstract. Compared with traditional median filter, the filter performance of adaptive median filter has been improved at the cost of high computation complexity. An improved adaptive median filter algorithm is proposed in this paper. First, the filter window size is determined according to the distance between the valid pixels and the center pixels in the proposed algorithm, which can avoid the waste of pixels repeated sort in window expand process. Second, the improved algorithm only to take the median value from valid pixels within the window, effectively weakening the interference with noise point, it will improve the quality of image. Compared with the original algorithm, this proposed algorithm reduced the complexity of algorithm and improved effective of noise reduction, PSNR value average increases 10dB. Finally, this paper applied the algorithm on noise reduction with depth image captured from Kinect.

Key words: Adaptive Median Filter; Valid Pixel; The Best Window Size; Depth Image Filtering

1 Introduction

It is inevitably doped into the noise in the process of image gathering, especially for depth image, such as sensitive elements of a photoelectric conversion sensitivity inhomogeneity, quantization noise, in the process of digital transmission there have different levels of noise, most impact on the visual effect. One of the most common is the salt and pepper noise. Salt and pepper noise in digital images shown as the distribution in the image, some pixels grayscale value are maximum or minimum and has no relationship with surrounding pixels. Median filter is always considered as the high efficiency and fast method to suppress salt and pepper noise, but when the noise concentration is large, median filtering effect will be sharp decline. So some scholars proposed the adaptive median filter algorithm [1], and solve the problem when high concentrations of salt and pepper noise. However, there will be a part of pixels repeated calculation in the process of window iteration, and it is sort for all pixels are make efficiency of algorithm greatly reduced. Aiming at the problem, some scholars put forward according to the relationship between concentration and noise filtering window to determine the best window size, such as literature [2] and [3].This approach makes frequency of iteration drop and complexity of

© Springer International Publishing AG 2017   179
J.-S. Pan et al. (eds.), *Advances in Intelligent Information Hiding*
and Multimedia Signal Processing, Smart Innovation, Systems and Technologies 64,
DOI 10.1007/978-3-319-50212-0_22

algorithm significantly reduced, but the filter effect has been decline. In terms of improved effective of adaptive median filtering, some scholars put forward to change the filter window shape[4][5], introduced weight coefficient in function[6][7][8],and further distinguish the noise points[9][10].Compared with the former methods, literature[11]put forward a method only sorting valid pixels, good control of the complexity of algorithm and reduction effect to noise is significant. But this method still have some pixels repeated sort problem, so, the complexity of the algorithm Still needs to be improved. This paper proposed a method through distance between the valid pixels and the center pixels to determine the filter window size, and only to take the median value from valid pixels within the window, these two aspects ensure the low complexity of Algorithm and effect of reduction the noise.

2   Adaptive Median Filter

2.1   Adaptive Median Filtering Principle

There is a hypothetical picture of gray and the digit is 8 bits, if pixels were judged to belong to the "salt and pepper", it grayscale value is 0 or 255. Using the probability density function said is:

$$P(x) \begin{cases} Pa, x = a \\ Pb, x = b \\ 0, other \end{cases} \qquad (1)$$

Among the formula, a=0 and b=255, these are the black and white noise points, Pa and Pb are probability corresponding with a and b. The basic principle of median filter is ordering points within the filtering window (window size generally is odd), and then using intermediate value of the sort sequence to replace the certain point of the window. The mathematical expression of median filter is as follows:

$$I(i, j) = Median(n(k)) \qquad (2)$$

Among the formula, K is the number of pixels in the window, n is a sort of gray value sequence, i is the pixel horizontal coordinate, and j is the pixel vertical coordinate.

For the traditional median filter, when the noise probability $P(x)$ is larger than 20%, filtering effect will sharply decline. This is because the traditional median filter is not discrimination noise points within the window, When the window invalid noise point accounted for more than 50%, the median point will replaced by a noise point, which makes the filter loss significance. Adaptive median filter establishment a judge condition to distinguish the replaced median point whether or not a noise point, effective to avoid the problem of median filter in the filter failure, as follows: First, set up a layer of an outer loop, find the maximum value within the windows set as Wmax ,minimum is Wmin and median is Wmed.Set A1=Wmed-Wmin in the A layer, A2=Wmed-Wmax, if A1>0 and A2<0 jump to inner loop, set as B, otherwise the window size in accordance with the

W+2 to continue expansion; In layer B, B1 = Wi. j-Wmin, B2= Wi. j-Wmax. If the B1>0 and B2<0 unchanged current pixel values, otherwise Wi, j = Wmed. Flow chart as shown in Figure 1.

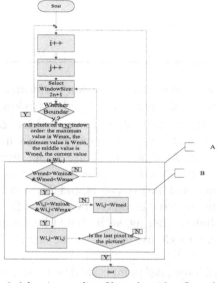

Fig. 1. Adaptive median filter algorithm flow chart

2.2   Algorithm complexity analysis

From the above steps of adaptive median filtering can be seen, a part of pixel points will be wasted when the window size expand. As shown in Figure 2, when the filter window from a square (3x3) to (5x5), the former will be included 3x3=9 pixels to repeat calculation. Similarly, if the window is enlarged to the maximum to (9x9), there are nine points sorted 3 times, 25 points repeated sorted 2 times, 49 points sorted 1 times, so 83 points in a single operation will repeat sorted, efficiency of the algorithm greatly reduced. Table1 shows that when the window size lower limit is 3x3, the upper limit is 11x11, in different window size, pixel points repeat sorted times in filtering.

3   Adaptive Median Filter

3.1   Adaptive Median Filtering Principle

This paper focuses on how to quickly determine the size of the window, how

Tab.1 Iterative use window repeat points in sorting number under different conditions

Window Interation	The Number of Repeated Points
3×3 to 5×5	9
3×3 to 7×7	34
3×3 to 9×9	83
3×3 to 11×11	164

to use valid pixels filtering. The algorithm performance are improved and the complexity of algorithm was reduced，Specific as follows:

In this paper, for the problem described in section 2.2 , this method can quickly determine the size of the filtering window by the distance between valid pixel and the center pixel. As shown in Figure 3, the distance between center point I and valid pixel point Y set as r. It is important to note that r value by subtracting the coordinates of the two points, and If let I to Y distance $r=|i-m|$ will make the window value occurred error, can be taken as EFGH, that the Y can not be included in the window filtering, so it should be compare the vertical and horizontal length difference, choose the larger one to make Y in MINO, and for H, $r=|i-z|$.

In this paper, First, if the pixel point is valid point, skip filtering operation. Compared with the traditional algorithm, it is not necessary to judge after sorted, and it can save a lot of unnecessary operation process. If the total pixel number of the image is D=M * N, the concentration of salt and pepper noise is 20%, number of (1-0.2)D pixels computational complexity would reduced from $O(n2)$ to $O(1)$[13]; Compared with the adaptive median filter, window size was taken as the most include valid pixel point, it will make the filtered pixel value more reasonable.

3.2 Algorithm Flow

In this paper, the improved algorithm is to determine the pixels in the maximum window size when selecting the window size, find the valid pixels and record it coordinate, calculate distance between the center pixel. Then through this distance to determine the best window size, and finally to sort the valid pixels within the window, Specific processes are as follows:

① Set the starting point as $I(i,j)$, if $I(i,j)$ is valid pixel it will skipped, Otherwise the window Wi,j size would initialization as the maximum value, it is set to M, the salt and pepper noise maximum gray level is a, the minimum value is b; Otherwise set window Wi,j size as the maximum value M. The salt and pepper noise maximum gray level is a, minimum is b;

② Traversing pixel points in M*M window except the center $I(i,j)$, If

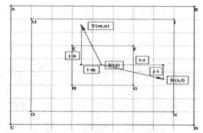

Fig.2 To determine distance of effective pixel

Tab.2 Complexity of different algorithms

Window Size	Complexity (comparison times)		
	Adapt Median Filter	Literature [11][12] Algorithm	Improved Algorithm
3×3	36	≤36	≤40
5×5	336	≤336	≤304
7×7	1521	≤1521	≤1180
9×9	4761	≤4761	≤3244
11×11	12021	≤12021	≤7264

$I(i \pm m, j \pm n)$ point is valid pixels ($I(i \pm m, j \pm n) < a$ and $I(i \pm m, j \pm n) > b$),mark its position (m, n), And calculate its distance from the center $I(i, j)$, denoted as the radius r, the mathematical expression is:

$$r = \begin{cases} |i - m|, |i - m| \geq |j - n| \\ |j - n|, |i - m| < |j - n| \end{cases} \qquad (3)$$

③Statistics different rk values by calculated, pick out the most repeated rk value, the window size is selected as $W_{i,j} = 2r_k + 1$;

④And then sort the valid pixel points in the window, set as $H[W_{i,j}][1]$, $I(i, j)$Value as median($H[W_{i,j}]$);

4 Analysis and Comparison of Experimental Results

In this paper, the peak signal to noise ratio (PSNR) is introduced as the evaluation of objective quality to image. The mathematical expression is as follows:

$$PSNR = 10 \times lg\{\frac{255^2 \times M \times N}{\sum_{i,j}(S(i,j) - S'(i,j))}\} \qquad (4)$$

MxN represents the image size, S(i,j) to represent the original reference imag
e, S'(i,j) to express the image after noise filtering.

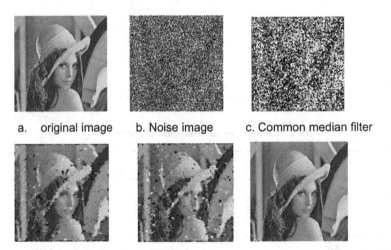

a.   original image   b. Noise image   c. Common median filter

d. adaptive median filtering   e. Literature [4] algorithm   f. Algorithm in this paper
Fig.3 Compare different algorithms of under 80% noise levels

5   Application Examples

Three dimensional reconstruction based on Kinect depth map applications increasingly widespread Kinect in the process of obtaining a depth map inevitably mixed lots of salt and pepper noise. As shown in Figure 8a, the noise in the depth map is very obvious, so we need to filter the collected depth map and then apply it to the actual system. For real time systems such as 3D reconstruction, the filtering process of the depth map is required as fast as possible.

In this paper, several filtering algorithms are used to reduce the noise of the depth map, In the Kinect commonly used C + + development environment VS (Visual Studio) were program run time statistics, Conclusions as shown in Table 4 and figure 7.

6   Conclusion

In this paper, an improved adaptive median filtering algorithm was proposed to solve the problem of low efficiency and high noise concentration for adaptive median filter. The proposed method firstly traverse to the noise as a "center", find the distance relationship between valid pixels and the "center", select the appropriate filtering window size based on this distance. Further, only operation the valid pixels in the window, The experiment results show that, the algorithm

Tab.3 Several kinds of algorithms under different noise levels PSNR value comparsion

Different Filtering Algorithms	Run Time Required
Common median filter	125ms
adaptive median filtering	481ms
Bilateral filtering	1680ms
Algorithm in this paper	198ms

Tab.4 Waste time of different algorithms reduction noise

Noise Concentration	Noise Figure	Medin Filter	Adaptive Median Filter	Literature [4]	Algorithm in This Paper
10%	15. 4dB	27. 9dB	29. 93dB	26. 5dB	38. 0dB
20%	12. 6dB	26. 4dB	28. 4dB	25. 2dB	35. 1dB
40%	9. 6dB	18. 2dB	24. 9dB	24. 7dB	34. 0dB
60%	7. 9dB	12. 3dB	22. 2dB	22. 0dB	33. 2dB
80%	6. 7dB	8. 4dB	17. 1dB	16. 8dB	28. 9dB

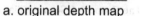

a. original depth map   b. common median filter   c. bilateral filtering

in this paper can not only reduce the complexity of adaptive median filtering, but also it can improve the filtering performance by the operation of the valid point ranking. At last, the algorithm is applied to the noise reduction of depth image captured from Kinect.

Acknowledgment

This paper is supported by the Project for the Key Project of Beijing Municipal Education Commission under Grant No. KZ201610005007, Beijing Postdoctoral Research Foundation under Grant No.2015ZZ-23, China Postdoctoral Research Foundation under Grant No.2015M580029, 2016T90022, ,the National Natural Science Foundation of China under Grant No.61672064. and Computational Intelligence and Intelligent System of Beijing Key Laboratory Research Foundation under Grant No.002000546615004

d. adaptive median filter   e. algorithm in this paper

Fig.4 Performance of different algorithms on depth image

References

1. SUN M,The basis of digital image processing and analysis[M].Publishing House of Electronics Industry,2013:64-73
2. PAN T,WU X B,ZHANG W W, et al. Application of Improved Adaptive Median Filter Algorithm in Image Denoising[J],Journal of Logistical Engineering Unversity,2015,31(5):93-95
3. XIAO L,HE K,ZHOU J L.Image Noise Removal On Improvement Adaptive Medium Filter[J],Laser Journal,2009,30(2):45-46
4. WANG X K,LI F.Improved Adaptive Median Filtering.Computer Engineering and Applications,2010,46(3):175-176
5. ZHANG X M,DANG L Q,XU J C.Fast Adaptive Image Median Filter Based On Crossing Windows[J], Computer Engineering and Applications,2007,43(27):37-39
6. ZHU Y H,ZHANG J C,JIANG T · Iterative Adaptive Median Filter for Impulse Noise Cancellation[J],SHANDONG SCIENCE, 2010,23(4):52-54
7. DENG X Q,XIONG Y,PENG H. Effective Adaptive Weighted Median Filter Algorithm[J],Computer Engineering and Applications,2009,45（35）:186-187
8. LONG Y,HAN L G,DENG W B,GONG X B, Adaptive Weighted Improved Window Median Filtering[J],Global Geology, 2013,32（2）:397-398
9. Guo Zheng-yang[C],Improved Adaptive Median Filters.2014 Tenth International Conference on Computional Intelligence and Security, The School of Science Tianjin University of Technology and Education, China.
10.HUANG B G,LU Z T,MA C M. Improved Adaptive Median Filtering Algorithm[J],Journal of Computer Applications, 2011,31(7):1836-1837
11.LI J,CHEN W J,ZOU G F. Imporovement of Median Filtering Algorithm based on Extremum With High Noise Density[J],Control and Automation, 2009,25(10):133-134
12.LI G,WEI J L.A Twice Adaptive Median Filtering Algorithm with Strong Salt and Pepper Noise Image[J],Microcomputer Applications,2011,32（4）:23-24
13.LI G,FAN R X.A New Median Filter Algorithmin Image Tracking Systems[J],Journal of Beijing Institute of Technology, 2002,22（3）:377-378
14.S. Samsad Beagum, Nisar Hundewale , M. Mohamed Sathik [C],Improved Adaptive Median Filters Using Nearest 4-Neighbors for Restoration of Images Corrupted with Fixed-Valued Impulse Noise. 2015 IEEE International Conference on Computational Intelligence and Computing Research ,India. Madurai :Vickram College of Engineering,2015.
15.Simon Perreault, Patrick Hébert.[J]. Median Filtering in Constant Time. IEEE Tansactions on Image Processing,2007,16(9):2389-2390

A Novel Approach to the Quadratic Residue Code

Ching-Fu Huang [1], Wan-Rong Cheng[2], and Chang-ching Yu[1]

Department of Information Engineering
[1]I-Shou University,Kaohsiung, Taiwan 84008, Republic of China
awic6039@yahoo.com.tw,fish995@yahoo.com.tw
Department of Biomedical Engineering
[2]Chung Yuan Christian University ,Taoyuan , Taiwan 32023, Republic of China
ssaamm5111@gmail.com

Abstract. In this paper presents a novel hash table algorithm for high-throughput decoding of the (n, k, d) Quadratic Residue (QR) code. The main ideas behind this decoding technique are based on one-to-one mapping between the syndromes "S1" and correctable error patterns. As compared with the binary lookup table method, the presented technique is faster than binary searching method for finding error pattern. In addition, it has the advantage that can be performed quickly with a masking operation as a choice hash table of size $M = 2^{\lfloor \log_2 \overline{M} \rfloor}$ for decoding QR codes. Moreover, the presented high speed of the decoding procedure has potential applications in modern communication systems and digital signal processing (DSP) systems.

Keywords: Quadratic Resid, Error Pattern; Syndrome ,DSP, Finite Field.

1 Introduction

The QR code is one of the best-known examples of binary cyclic codes [1], and it consists of a code rate greater than or equal to 1/2 with generally a large minimum distance, so that most of the known QR codes are best-known codes [2, 3]. Among them, non-algebraic techniques have been considered such as: t-designs [4], the permutation method [5]. In other methods, Reed [6] used algebraic decoding techniques to decode QR codes. Even further, Reed et al. [7, 8] used the Sylvester resultant to solve non-linear, multivariate equations obtained from the Newton identities. Recently, a new algebraic decoding method has been proposed by R. He et al. [9]to decode binary QR codes using irreducible generator polynomials. The key idea in [9] is based on evaluating the needed unknown syndromes [10] while utilizing an efficient inverse-free Berlekamp-Massey (BM) algorithm [11, 12]to determine the error-locator polynomial. However, these algebraic decoding methods recited above require a vast number of operations of addition and multiplication calculations over a finite field; thus, it would be difficult to implement in a Real-Time-System. As shown in [13, 14], by utilizing a

© Springer International Publishing AG 2017

J.-S. Pan et al. (eds.), *Advances in Intelligent Information Hiding and Multimedia Signal Processing*, Smart Innovation, Systems and Technologies 64,
DOI 10.1007/978-3-319-50212-0_23

one-to-one mapping between the syndromes and the error correctable patterns, the lookup table decoding (LTD) algorithm dramatically reduces the memory requirement in DSP systems. In this aspect, using the binary search can increase the performance of shift search algorithm for decoding QR codes. In this paper, improvement on decoding the binary systematic (n, k, d) QR codes with hash table method is proposed. The decoding of the (71, 36, 11) QR code using Scheme 2 can be approximately 63% faster than the LTD method [14]. These results would imply that using the proposed method would be faster than a binary search for decoding of QR codes.

2   Background of the binary QR code

A binary QR codeword (n, k, d) is defined algebraically as a multiple of its generator polynomial $g(x)$ over $GF(2^m)$. There are a total of n elements in finite field $GF(p^m)$, where p is a prime number and m is a positive integer. Let n be a prime number of the form $n = 8l \pm 1$ where l is a positive integer. Let k be the number of the form $k = \frac{n+1}{2}$. Let $t = \frac{d-1}{2}$ be the number of the errors that can be corrected in a block, where d is minimum distance of a code.

$$Q_n = \{j | j \equiv x^2 \bmod n \; for \; 1 \leq x \leq n-1\} \tag{1}$$

Let m be the smallest positive integer such that n divides $2^m - 1$ and let α be a generator of the multiplicative group of all nonzero elements in $GF(2^m)$. Then the element $\beta = \alpha^u$, where $u = \frac{2^m-1}{n}$, is a primitive n-th root of unity in $GF(2^m)$. A binary (n, k, d) QR code is a cyclic code with the generator polynomial $g(x)$ of the form,

$$g(x) = \prod_{i \in Q_n} (x - \beta^i) \tag{2}$$

In this paper, the systematic encoding method is utilized, in which m(x) is the information message polynomial with its associated vector $m = (m_0, m_1, ..., m_{k-1})$. The systematic encoding is obtained by $d(x) \equiv m(x)x^{k-1} \bmod g(x)$, the polynomial $c(x) = x^{k-1}m(x) - d(x)$. Let $d(x)$ be the remainder polynomial with its associated vector $d = (d_0, d_1, ..., d_{k-2})$ obtained from the division operation so that it can be expressed as the following vector: $c = (c_0, c_1, ..., c_{n-1}) = (c_d, c_m) = (d_0, d_1, ..., d_{k-2}, m_0, m_1, ..., m_{k-1})$. If the codeword c is transmitted through a noisy channel and the vector $r = c + e = (r_d, r_m) = (r_0, r_1, ..., r_{k-2}, r_{k-1}, r_k, ..., r_{n-1})$, where $r_d = (r_0, r_1, ..., r_{k-2})$, $r_m = (r_{k-1}, r_k, ..., r_{n-1})$ and the error vector is as follows: $e = (e_d, e_m)$, where $e_d = (e_0, e_1, ..., e_{k-2})$, $e_m = (e_{k-1}, e_k, ..., e_{n-1})$ then the polynomial $r(x)$ with its associated vector r is received. The set of known syndromes is obtained by evaluating polynomial $r(x)$ at the roots of the generator polynomial $g(x)$, i.e.,

$$S_i = r(\beta^i) = c(\beta^i) + e(\beta^i) = e(\beta^i) = e_0 + e_1(\beta^i) + \cdots + e_{n-1}(\beta^i)^{n-1}, i \in Q_n. \tag{3}$$

If, during the data transmission, v errors occur in the received vector r, then the error polynomial has v nonzero terms, namely $e(x) = x^{l1} + \cdots + x^{lv}$, where

$0 \leq l_1 < \cdots < l_v \leq n - 1$. And, for $i \in Q_n$, the syndrome S_i can be written as

$$S_i = \sum_{j=1}^{v} \beta^{il_j}, S_i \in GF(2^m) \, for \, 1 \leq j \leq t \tag{4}$$

Here l_j is called the error locators. Let $E = GF(2^m)$ be the finite field of the (n, k, d) QR codes and α be a primitive element of finite field and be a root of the primitive polynomial $p(x)$. The element $\beta = \alpha^{\frac{2^m-1}{n}}$ is n-th root of unity in E. The set Q_n of quadratic residues modulo n is the set of nonzero squares modulo n; that is,For example, let α be a primitive element of finite field $GF(2^{35})$ which can be constructed from the primitive polynomial $p(x) = x^{35} + x^2 + 1$. This implies that $P(\alpha) = \alpha^{35} + \alpha^2 + 1 = 0$, or equivalently $\alpha^{35} = \alpha^2 + 1$. The element $\beta = \alpha^{4839399} = 34120188749$ is 71-th root of unity in E.

3   Chossing hash function for decoding a QR code

One lists all β^j where $0 \leq j \leq 71$ could be utilized in computing syndrome the sum of different combination of error patterns and generate the lookup table for hashing search. For example, the decoding of the (71, 36, 11) QR code using the value of $\beta = \alpha^{48393997}$ in Table 1, one gets the syndrome by receiving the polynomial more quickly.

Table 1. Evaluating syndrome of the (71, 36, 11) QR code

$\beta^0 = 1$	$\beta^{14} = 4012678623$	$\beta^{28} = 1643213815$	$\beta^{42} = 26915919534$	$\beta^{56} = 4790277429$
$\beta^1 = 34120188749$	$\beta^{15} = 17830756444$	$\beta^{29} = 29035672107$	$\beta^{43} = 30701922701$	$\beta^{57} = 13230619667$
$\beta^2 = 1344641628$	$\beta^{16} = 22475928310$	$\beta^{30} = 31449461629$	$\beta^{44} = 13209079568$	$\beta^{58} = 7097200360$
$\beta^3 = 28503725773$	$\beta^{17} = 27396615072$	$\beta^{31} = 19675233275$	$\beta^{45} = 20650423370$	$\beta^{59} = 19346701503$
$\beta^4 = 5568205786$	$\beta^{18} = 33455564263$	$\beta^{32} = 33135943609$	$\beta^{46} = 23042326059$	$\beta^{60} = 1873263996$
$\beta^5 = 11660961459$	$\beta^{19} = 9515872963$	$\beta^{33} = 6789292169$	$\beta^{47} = 30825048026$	$\beta^{61} = 18199510701$
$\beta^6 = 3147231358$	$\beta^{20} = 11789239831$	$\beta^{34} = 23836582959$	$\beta^{48} = 6798557156$	$\beta^{62} = 9360604994$
$\beta^7 = 28829856380$	$\beta^{21} = 17177160930$	$\beta^{35} = 22710600763$	$\beta^{49} = 22782484442$	$\beta^{63} = 18276997635$
$\beta^8 = 2872005070$	$\beta^{22} = 16143965236$	$\beta^{36} = 21847856280$	$\beta^{50} = 25085509954$	$\beta^{64} = 17641071974$
$\beta^9 = 9746482619$	$\beta^{23} = 1991769057$	$\beta^{37} = 22277656296$	$\beta^{51} = 26161342046$	$\beta^{65} = 25980965732$
$\beta^{10} = 11513128231$	$\beta^{24} = 25926137209$	$\beta^{38} = 11931513381$	$\beta^{52} = 26157343362$	$\beta^{66} = 2628414179$
$\beta^{11} = 24703502821$	$\beta^{25} = 7751071256$	$\beta^{39} = 12059306574$	$\beta^{53} = 32614355110$	$\beta^{67} = 34120571586$
$\beta^{12} = 19311238654$	$\beta^{26} = 18133972355$	$\beta^{40} = 16812124981$	$\beta^{54} = 31602719787$	$\beta^{68} = 15505434202$
$\beta^{13} = 2602454769$	$\beta^{27} = 15761658209$	$\beta^{41} = 8347946267$	$\beta^{55} = 10121160842$	$\beta^{69} = 12522083912$
				$\beta^{70} = 11714203624$

Total number of the necessary error patterns is $N = \left|S_1^{(1)}\right| + \left|S_1^{(2)}\right| + \cdots + \left|S_1^{(t-1)}\right| = \sum_{i=0}^{t-1}\binom{n}{i} - \sum_{i=0}^{t-2}\binom{k-1}{i}$, where t is error correcting capability. In this paper, the number of decoding of (71, 36, 11) QR code required $N = 1024171$ error patterns. The value of syndrome set $S_1^{(v)}$, for $v = 1, 2..., t - 1$, is called a "key". The value of the key is generated by a received polynomial $r(x)$ as follows:

$$key = r(\beta) = \sum_{i=0}^{n-1} r_i\beta^i = \sum_{k=1}^{t-1}\beta^{u_k}, for\ 0 \le u_1 < u_2 < \cdots < u_{t-1} < n. \quad (5)$$

Here, r_i is given the coefficient of the polynomial $r(x)$ and u_k is given error location. The hash function is utilized in decoding QR codes as follows:

$$f(key) \equiv key\ mod\ M. \quad (6)$$

From (6), it is important to determine the value of M. To do this, two schemes for M in hash function are proposed in this section. Scheme 1 requires the usage of a hash table with size $\overline{M}$ where $\overline{M}$ has the maximum collisions less than binary search $\Theta(\log_2 N)$ time. On the other hand, Scheme 2 can use the same hash table size from Scheme 1 to force the value to the range of a power of two and that can be performed quickly with a masking operation, i.e., $M = 2^{\lfloor \log_2 \overline{M} \rfloor}$. The costs of these operations are related to the average length of a chain. According to (6), hash table size can help test for the number of maximum of collisions that occurs (P_m), which in turn will show that the worst case in chaining searches times would still be smaller than a binary search time, where $1 \le i < P_m$, as segment addressing for chaining entries in linked list as shown in Table 2.

Table 2. Number of collisions

Collision occurs i	Segment addressing for index point	Number of collisions A_i
$0 \sim p_m$	Segment[0]=M Segment[P_m-1]=$A_{p_m-1} \sim -$	$A_0 = A_{p_m}$

For example, the (71, 36, 11) QR code, the range is testing the hash table size from 93250 to 1024171, the number of maximum of the collisions is $P_{min}^{\overline{M}} = 9$ imes in hash table of size $\overline{M} = 717981$. So that hashing search times is less than the binary search $\lceil \log_2 971811 \rceil = 20$ times. Let the table size be $2^{\lfloor \log_2 (\overline{M}) \rfloor} = 524288$, that will speed up modulo operations in decoding QR code. There are two different numbers of collisions in distribution of syndrome by using different hash table of size $\overline{M} = 717981$ and $2^{\lfloor \log_2(\overline{M}) \rfloor} = 524288$ that is shown in Table 6.The number of syndromes has the same number of an entry to save error pattern in hash table. Formatting the Q-bits of the quotient syndrome, the P-bits of the error patterns, and the I-bits of the next entry index. An entry format is defined in Table 3.

Table 3. An entry format

Q-bits of Quotient syndrome	P-bits of Index-error pattern	I-bits of Next index
$x = \log_2 \frac{max\ S}{M}$	$2 \left\lceil pt = \left(\log_2 \sum_{i=1}^{t-1} \binom{k}{i} + 1 \right) \right\rceil$	$\lceil \log_2 max\ A \rceil$

Quotient syndrome of Q-bits is required $Q = \left\lceil x = \log_2 \frac{max\ S}{M} \right\rceil$ bits into entry format. The error pattern of P-bits is needed to set a codeword of length n. The next index of I bits for chaining entries in linked list, the maximum value of a set of the number of collisions $A = \{A_i\}_{i=1}^{p_m - 1}$ to denote $max\ A$, is required $\lceil \log_2 max\ A \rceil$ bits. If the value of a syndrome divided by $\overline{M}$ is called a quotient syndrome q, then it can reduce an entry size as shown in Table 6. However, an original value of syndrome would need to recover and that is shown by equation (7) when search key is in linked list.

$$S_q = q * \overline{M} + f(key). \tag{7}$$

For example, the maximum value of syndrome in the (71, 36, 11) QR code is $S_{max} = \beta^{60} + \beta^{51} + \beta^{43} + \beta^3 = 34359714914$, Quotient syndrome $\left\lceil \frac{S_{max}}{717981} \right\rceil$ is required $Q = \lceil \log_2 47856 \rceil = 16$ bits in formatting entry. The error pattern of P-bits setting with codeword of length n, it would require more memory size to store in hash table, i.e., n=71 over 64 bits. The FLTD method is only using k-bits to store the error pattern in portions of a message for the decoding QR codes. One can improve the error pattern size, which is the index of an error pattern in an entry.The P-bits of error patterns can has k bits of the message of even length k. Therefore, the number of the error pattern is $\sum_{i=1}^{t-1} \binom{k}{i} + 1$ and non-error pattern that involves variable p with $\left\lceil \log_2 \sum_{i=1}^{t-1} \binom{k}{i} + 1 \right\rceil$ bits to find an error pattern e_m in array $Err\_pattern\ []$,that is, $e_m = Err\_pattern\ [p]$. The index of error pattern needs $P = \left\lceil \log_2 \sum_{i=1}^{t-1} \binom{k}{i} + 1 \right\rceil$ bits for indexing error pattern as shown in Table 1. For example, the (71, 36, 11) QR code only establishes $\binom{36}{1} + \binom{36}{2} + \binom{36}{3} + \binom{36}{4} + 1 = 66712$ that needs 17-bits for indexing error pattern less than error pattern of length 36-bits.

If error pattern table are divided into two blocks, than the errors table size is only establishe $\binom{18}{1} + \binom{18}{2} + \binom{18}{3} + \binom{18}{4} + 1 = 4048$ which needs 12-bits. So the index of addressing needs 24bis for addressing errors pattern in an entry. However, this method enables to increase decoding time by indexing for finding error pattern. Therefore, error pattern table are divided into more blocks that can be saved to use memory size, but it will increase more size of index for addressing error patterns in an entry. It clear knows that memory size is required $\left(\sum_{i=0}^{t-1} \binom{n}{i} - \sum_{i=0}^{t-2} \binom{k-1}{i}\right) \times entry\ size\ bits$ for making a hash table. The errors pattern table is divided into different blocks for indexing errors patterns for requires as shown in Table 4.

Table 4. Different blocks for indexing error patterns

P bites of errorpattern	Number of Error patterns	Index size	$\left(\sum_{i=0}^{4} \binom{71}{i} - \sum_{i=0}^{3} \binom{35}{i}\right)$ $= 1024171 \times entry\ size\ bits$
P=36	66712	17	53256892
P=18	4048	24	60426089
P=9	256	32	68619457

Finally, a hash function can be used to make a hash table; that is, called an array $Hash\_table[\,]$ for decoding QR codes.

4   Using hash table for finding error pattern procedure

A hashing search is using a hash function f(key), where $key = \sum_{i=0}^{n-1} r_i \cdot \beta^i$, to find index of error pattern for decoding (n, k, d) OR codes up to t-1 errors. The finding index of the error pattern procedure is called **Find_err_index**(key) as follows:

(1) Let $i = 0, Index = f(key) = (key\ mod\ M) = r, Next\_index = Index$.
(2) If $Hash\_table[Next\_index] = -1, Index = -1$, then go to step 8.
(3) $\frac{S_q}{M} = q + Index$, if r=index and $q \neq q_{i+1}$ then $De\_table[Next\_index] = q_{i+1}$, where the value of q is obtained by the value of Q portion.
(4) If $key = S_q$, then the value of index p error pattern is obtained the value of P portion of $De\_table[Next\_index], Index = p$, go to step 8.
(5) If the value of the I portion of $Hash\_table[Next\_index]$equal -1, then $Index = -1$, go to step 8.
(6) $Next\_index \leftarrow$ The value of the I portion of $Hash\_table[Next\_index]$.
(7) $Next\_index \leftarrow Next\_index + Segment[i], i = i + 1$, go to step 3.
(8) Return $Index$.

5   Simmulation results

The number of collisions in the distribution of (71, 36, 11) QR codes are given in Table 5. The number of collisions to store in array *Segment* [ ], when the value of syndrome collisions has occurred, is used to point search chaining in a linked list.

Table 5. Number of collisions in the distribution of (71, 36, 11) QR code

Number of collision		0	1	2	3	4	5	6	7	8	9
M of QR-	717981		545218	299834	124179	40715	11111	2552	477	83	2	
code	$2^{19} = 524288$	452757	307074	162751	68827	23786	6807	1729	363	67	10

According to the proposed algorithm, for evaluation decoding (71, 36, 11) QR codes. The computational time of these QR codes over 1,000,000 times of codeword computations is listed in Table 6. The hash table of size is M = 524288, the memory needs 9.13Mbytes and 5.31 Mbytes for decoding the (71, 36, 11) QR code. The decoding of the (71, 36, 11) QR code using Scheme 2, as discussed above, can be approximately 63% faster than LTD method. Scheme 1 can save more memory than the LTD method approximately 42%.

Table 6. Complexities of decoding of (71, 36, 11) QR codes

Method	Memory	Search time
Binary Search(LTD)	9.19MBytes	4.15 second
Hash Search (M=mini-collision)	9.13MBytes	1.98 second
Hash Search ($M= 2^{19}$)	5.31MBytes	1.53 second

6   Conclusion

One presents a logarithmic hash search function as a hash table size to decode the binary systematic (71,36, 11) QR codes. In Table 6 as shown, the hash table could use a novel approach Scheme 2 for decoding of QR codes. As a result, Scheme 2 is actually faster than Scheme 1 in all instances, but Scheme 2 reduces more memory in certain occasions. Scheme 2, forces a value of hash table of size $\overline{M}$ to the range of a power of two, and it can be performed quickly with a masking operation as a choice hash table of size $M = 2^{\lfloor \log_2(\overline{M}) \rfloor}$; that is, suitable for decoding of QR codes as shown in Table 6. If memory consumption in embedded systems is an acceptable range, using the proposed method is readily adaptable for speeding up the decoding of QR codes. Thus, Scheme 2, the decoding of the (71, 36, 11) QR code, dramatically, not only reduces the decoding time with approximately 63% but also reducing memory with approximately 42% in comparison to the LTD method. Therefore, the HT algorithm is more suitable in real time system.

References

1. Kasami, T.: 'A decoding procedure for multiple-error-correcting cyclic codes', *IEEE Trans. Inf. Theory*, 1964, IT-10, pp. 134138
2. Chen, X., Reed, I. S., Truong, T. K.: 'A performance comparison of the binary quadratic residue codes with the 1/2-rate convolutional codes ', *IEEE Trans. Inf. Theory*, 1994 (40), pp. 126-136
3. MacWilliams, F. J., Sloan, N. J. A.: 'The theory of error-correcting code' (Amsterdam, the Netherlands: North-Holland, 1977)
4. Assmus, E. F. Jr., Mattson, H. F. Jr.: 'New 5-designs', *J. Comb. Theory*, 1969, (6), pp. 22-151
5. Wolfmann, J.: 'A permutation decoding of the (24, 12, 8) Golay code', *IEEE Trans. Inf. Theory*, 1983, 5, IT-29, pp. 748750
6. Reed, I. S., Yin, X., Truong, T. K., Holmesd, J. K.: 'Decoding the (24, 12, 8) Golay code', Proc. *IEE Inst. Elec. Eng.*, 1990, 137, (3), pp. 202206
7. Reed, I. S., Yin, X., Truong, T. K.: 'Algebraic decoding of the (32, 16, 8) quadratic residue code', *IEEE Trans. Inf. Theory*, 1990, 36, (4), pp. 876880
8. Reed, I. S., Truong, T. K., Chen, X., and Yin, X.: 'The algebraic decoding of the (41, 21, 9) quadratic residue code', *IEEE Trans. Inf. Theory*, 1992, 38, (3), pp. 974985
9. He, R., Reed, I. S., Truong, T. K., and Chen, X.: 'Decoding the (47, 24, 11) quadratic residue code', *IEEE Trans. Inf. Theory*, 2001, 3, 47, (3), pp. 11811186
10. Chen,Y. H., Truong, T. K., Fellow IEEE, Chang, Y, C., Lee D., Chen, S. H.: 'Algebraic Decoding of Quadratic Residue Codes Using Berlekamp-Massey Algorithm', *J. Inf. Sci. Eng.*, 2007, 23, (1), pp.127-145
11. Massey, J. L.: 'Shift-register synthesis and BCH decoding', *IEEE Trans. Inf. Theory*, 1969, IT-15, (1), pp. 122127
12. Youzhi, X.: 'Implementation of Berlekamp-Massey algorithm without inversion', Proc. *IEE Commun. Speech Vis.*, 1991, 138, (3), pp. 138140
13. Chen, Y. H., Truong, T. K., Huang, C. H., Chien, C. H.: 'A Lookup Table Decoding of Systematic (47, 24, 11) Quadratic Residue Code', *Inf. Sci.*, 2009, 179, (14), pp. 2470-2477
14. Chen, Y. H., Chien, C. H., Huang, C. H., Truong, T. K., Jing, M. H.: 'Efficient Decoding of Systematic (23, 12, 7) and (41, 21, 9) Quadratic Residue Codes', *J. Inf. Sci. Eng.*, 2010, 26, (5), pp. 1831-1843

A High Performance Compression Method Using SOC and LAS

Yu-Hsaun Chiu, Min-Rui Hou, and Yung-Chen Chou*

Department of Computer Sciences and Information Engineering, Asia University,
Taichuang 41354, Taiwan (R.O.C.)
E-mail: {a925d37, g81lsui, yungchen}@gmail.com

Abstract. With delivery and save in internet, processing and using of digital images is important issue. Due to the size of files will highly influenced efficiency in delivery and save, which can save part of data for use data compression. Digital image transfer is universal and high demand in internet, considering efficiency of delivery, image compression is important and necessity. In this paper, the propose of scheme are used Search Order Coding and Locally Adaptive Scheme based on VQ compression to reduce the amount of data of images. The proposed method achieves the goal of used lossless compression to improve compression rate in index table of VQ compression.

Keywords: Image Compression, Vector Quantization, Search Order Coding, Locally Adaptive Scheme

1 Introduction

In recent years, digital image becomes more refined and colorful, like high-quality photo. Accordingly, the size of image and the data storage are much larger than ever. Therefore, image compression be used to reduce the size of files in delivery and save. Many excellent researchers create several schemes about image compression, such as VQ, BTC. There are two most important factors for the compression scheme design namely compression rate and visual quality of decompressed image.

Image compression methods can be classified into the Lossy compression and Lossless compression. The Lossless compression is focus on visual quality preserving of decompressed image. However, the compression rate of Lossless compression is hard to transcend Lossy compression method. In some critical application, the Lossless compression is required, such as medical image and military image. Lossy compression also known as distortion compression [1, 2].That aims to sacrifice part of data in image for reduce data in transmission and storage and improve efficiency when image are compressed. Compared with Lossless image compression, most of Lossy compression is better for compression effect, it can reduce more information of load in transmission and storage. But for the

---

* Corresponding author

quality of image, the Lossy compression unable to recover integrated data after decompression. Also it still has much difference from human's eyes, which hard to distinguish from the recover image and original image.

In this paper, we are inspired from VQ, SOC, and LAS. Compress index table of VQ can be further improved, we try to use SOC and LAS to compression index table. VQ compression is one method of lossy compression. The block size, which used in original VQ, is usually 4 × 4 pixels, and the resulting bit rate reported in the literature is in the range of 0.5-0.6 bpp (bits per pixel) for grayscale images. In order to improve compression performance and visual quality further, according to the VQ concept extend many deformed method such as SMVQ [7], and TSVQ[8].

2   Related Works

2.1   Vector Quantization

Vector Quantization is a Lossy compression which was proposed by Linde et al. in 1980 [3]. In VQ compression requires a codebook CB composed by Nk-dimensional codewords. For more easily to describe, $CB = \{CW_i | i = 0, 1, \ldots, N-1\}$ and $CW_i = \{x_j | j = 0, 1, \ldots, k-1\}$, where k is denoted size of block. The codebook CB can be generated by applying LBG algorithm [6]. First, the input image I is divided into the non-overlapping block B_i and represented as $I = \{B_i | i = 0, 1, \ldots, h/H \times w/W - 1\}$ where h and w are used to represent as the height and width of block B_i. Also, H and W are the height and width of input image. Second, for a block B_i , which can be found a most similar codeword cw_p from codebook, and the most similar means that the codeword has smallest Euclidean distance between B_i and cw_p. Finally, the compression code is composed by all indexes values in binary format.

2.2   Search Order Coding

In natural image, the local area has similar pixel value distribution. Also, in index table, the index that values in a local area has similar value distribute. According to this property, Hsieh and Tsai proposed Search Order Coding in 1996 [4] to compress VQ index table again. The main idea of SOC is using a state code book (SCB) to remember some of past neighbor indexes for the further compression. The SCB construction is to collect the index values stay on the search order. Here, the indexes collection of SCB construction will not make the same index value appears twice. The State Code Book can be defined as $SCB = \{idx_i | i = 0, 1, \ldots, n\}$. The $idx_{i,j}$ is the current index need to be processed. The $idx_{i,j}$ will encoded by using less bits when SCB contains the same value with $idx_{i,j}$. When $idx_{i,j}$ can be encoded by SOC, then the position of $idx_{i,j}$ in SCB is used to encode. For preventing the ambiguous in decoding phase the indicator for each index is required. The indicator bit can be used to indicate the index is corresponding to VQ index or position in SCB. For instance, bit '0' indicates the index encoded by VQ and bit '1' indicates that the index encoded by SOC. For this point of view, high frequency of SOC case will get better compression rate performance.

2.3 Locally Adaptive Scheme

Locally Adaptive Scheme be applied VQ index table is proposed by Chang et al., in 1997 [5]. In index table for generated by natural images, the index value in same neighborhood is high percentage identical and similar, with that situation which can used SOC coding. Unfortunately, if same of index value did not appear in neighborhood, that unable to using SOC coding then further waste bits to save. In order to solve the problem of above, LAS coding be applying in index table. For LAS, that need produce HIL(History Index List) to record index value was appear. In encode of $idx_{i,j}$, that will be coding by location information of HIL.

3 Proposed Method

3.1 Compression Phase

Because the target of the proposed compression method is VQ index table, the index table is generated by VQ compression and it will further be compressed by SOC which cooperated with LAS. To simplify the description, the index table was denoted as $IT = \{idx_{i,j} | i = 1, 2, \ldots, N_{bh}; j = 1, 2, \ldots, N_{bw}\}$, where $idx_{i,j}$ represented the index value in IT which located in (i, j), N_{bh} and N_{bw} was denoted as the height and the width of IT, respectively. Because $idx_{1,1}$ has no previous index information, $idx_{1,1}$ must be encoded by VQ (i.e., using bits to encode the index). For the remaining indexes, the proposed compression will adopt SOC and LAS to encode the indexes as many as possible. In order to distinguish encoding cases of SOC and LAS in decompression phase, the indicator is required. The examples of descriptions that encoded by SOC and LAS are as follow:

Case 1 : $idx_{i,j}$ encode by SOC (i.e., Indicator $\parallel SCB_k$)
Case 2 : $idx_{i,j}$ encode by LAS (i.e., Indicator $\parallel$ preceding code $\parallel HIL_k$)

The k represents the location information of $idx_{i,j}$ on SCB and HIL. In the case of our proposed method which using SOC coding, the searching range (N_{SCB}) are 4 neighboring index values $\{idx_{i,j-1}, idx_{i-1,j-1}, idx_{i-1,j}, idx_{i-1,j+1}\}$ from IT. After that, the minimum and maximum values can be obtained by $idx_{\min} = \min\{idx_{i,j-1}, idx_{i-1,j-1}, idx_{i-1,j}, idx_{i-1,j+1}\}$ and $idx_{\max} = \max\{idx_{i,j-1}, idx_{i-1,j-1}, idx_{i-1,j}, idx_{i-1,j+1}\}$. The range of the SCB construction is between $idx_{\min}$ and $idx_{\max}$ which mean the range of the SCB construction is from minimum to maximum of VQ code book and it was denoted as $SCB = \{idx_{\min}, idx_{\min+1}, \ldots, idx_{\max}\}$. Eventually, the encoding of $idx_{i,j}$ will be represented as the location information of SCB.

For example, Fig. 1 shows the IT which was sized into 4×4 by the codebook that included 256 code words. The $idx_{1,1}$ will be encoded as 11000100. For the remaining indexes, because the $idx_{3,2}$ satisfied the conditions of SOC, it will start to search four index values in neighborhood of $idx_{3,2}$ and the neighborhood index values are 211, 200, 193, 197 respectively. After that, choosing the minimum and maximum values 193 and 211 to generate $SCB=\{193, 194, 195, 196, 197, 198, 199, 200, 201, 202, 203, 204, 205, 206, 207, 208, 209, 210, 211\}$. For the encoding of $idx_{3,2}$, there are 19 values in SCB and the minimum requirement of bits is 5. Therefore, it will use 5 bits to record the location information of SCB (State Code Book). After adding the indicator, the $idx_{3,2}$ will be encoded as 101101_2.

196	195	195	194
200	193	196	196
211	206	195	200
190	211	190	190

Fig. 1. An example for compression phase

When the SOC method is invalid, we will apply the LAS coding method. The LAS coding used N_{SCB} to calculate the average value idx_a and then applied Jumping Strategy to create the history index list HIL. The HIL can be denoted as $\{idx_a, idx_{a+1}, idx_{a-1}, idx_{a+2}, idx_{a-2}, idx_{a+3}, \ldots\}$. Before the coding step of $idx_{i,j}$, the HIL needs to remove all the values that are same as the values in SCB. It is necessary for LAS strategy to add the preceding code between indicator and locating information of $idx_{i,j}$. The preceding code is denoted as "0" and the number of bits is $\lceil \log_2 k \rceil$.

For example, the $idx_{2,3}$ is not satisfy the conditions of SOC. Therefore, the value of $idx_{2,3}$ will be represented as the current value and the idx_a value of $idx_{2,3}$ will be used to produce the HIL. The idx_a can be denote as $idx_a = \text{avg}\{193, 195, 195, 194\}$, and then Jumping Strategy (Fig. 2) will be applied to generate $HIL = \{194, 195, 193, 196, 192, 197, 191, 198, 190, 199, 189, 200, \ldots, 0\}$. As it mentioned before, because the value of $idx_{2,3}$ is not correspond to the value in SCB, the HIL needs to delete all the values of $SCB = \{193, 195, 194\}$ to avoid the waste of bits. After deleting all the values of SCB, the HIL will be denoted as $\{196, 192, 197, 191, 198, 190, 199, 189, 200, \ldots, 0\}$. Next, the LAS will be applied to encode the $idx_{2,3}$ which recorded the location information of HIL in compressed code string. Therefore, in our proposed method, we added the preceding code to identify how many bits were used to record the location information of HIL for the decompression step. As the example shows, after adding the indicator and preceding code, the $idx_{2,3}$ will be coded as 00010. But there will be an exception for LAS strategy, if the location information in HIL of $idx_{i,j}$ is '0', the $idx_{i,j}$ will be encoded as 010.

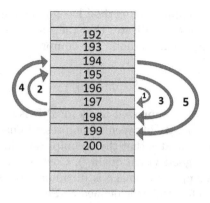

Fig. 2. Example of jumping strategy

3.2 Decompression Phase

In decompress situation, depends on identification indicator of compression code string from extract the compression code string of compressed idx, if compression code string indicator is 1 that would be judged for using SOC coding; if compression code string indicator is 0, that be judged for using LAS coding.

Procedure: Image Decompression

Input: Compression code C
Output: Index table IT
Step 1: Take 8 bits from C to restore index of first value in IT.
Step 2: Take indicator bits from C.
Step 3: If indicator $==$ '1' then go to **Step 4**, else go to **Step 6**.
Step 4: Generate SCB by using minimum and maximum of 4 values.
Step 5: Take compression code from C to form the position and take the value in SCB to restore the index value in IT. Go to **Step 7**.
Step 6: Use Jumping Strategy to create HIL and take compression code from C to form the position value in HIL. Then, using the index value in HIL to restore the index value in IT.
Step 7: Repeat **Step 2** to **Step 7** until all of compression code have been decompressed.
Step 8: Output index table IT.

4 Experiment Results

The compression rate is the most important factor for performance evaluation. Also, Peak-Signal-to-Noise-Ratio (PSNR) is adopted for evaluating the visual quality of decompressed image. The compression rate is using total bits of compressed code to compare with total bits of original image. A small value of

CR indicates that the compression method has good compression performance. Contrary, a large value of CR indicates that the compression method has worse performance of a compression method. The compression rate is defined as Eq. 1.

$$CR = \frac{||I'||}{||I||} \tag{1}$$

For visual quality of decompressed image, PSNR is an objectively measurement to compare the similarity between decompressed image and original image. A large value of PSNR indicates that the decompressed image is most similar to the original image (i.e., good visual quality). Contrary, a small value of PSNR indicates that the decompressed image is dissimilar to its original image (i.e., worse visual quality). The $PSNR$ is defined as Eq. 2-3.

$$PSNR = 10 \times \log_{10} \frac{255^2}{MSE} dB \tag{2}$$

$$MSE = \frac{1}{H \times W} \sum_{i=1}^{H} \sum_{j=1}^{W} (p_{i,j} - p'_{i,j})^2 \tag{3}$$

Where H and W denote to the height and width of the image. The $p_{i,j}$ and $p'_{i,j}$ denote the pixels of the original image and decompression image located on (i, j), respectively. Nine commonly used gray-scale images sized 512×512 are shown in Fig. 3. The codebook training is adopted the LBG algorithm and using nine gray scale images to be the training images. In the experiment of proposed method, the codebook size is 128, 256, 512, and 1024, and the dimension is 4×4 for experiment.

Figures Fig. 4-5 shown the compression rate comparison results in different code book size. As we see, the proposed method has better compression rate in different codebook size. We found that the proposed method has better outcome in smooth content image than complexity content images. Such as in Baboon image has no significant improvement in compression rate.

From the experimental results, the smooth content image provide high chance to find the index in the front part of HIL in the proposed method LAS situation. That is the proposed method is more suitable for the smooth content image.

5   Conclusions

The proposed method using the SOC and the LAS coding to decrease the bits requirement of storage of index table. It means the proposed method is not only the same image quality of traditional VQ, but also have better compression rate performance than traditional VQ, LAS, and SOC. From experimental results, the proposed method has better compression rate performance in any codebook size. To consider the computation cost of compression phase and the visual quality of the decompressed image, we suggest that codebook size set to 256 is a suitable choice. To consider the complexity of image content, the proposed method has

(a) Alan   (b) Baboon   (c) Boat

(d) Lena   (e) Jet(F16)   (f) Pepper

(g) Tiffany   (h) Toys   (i) Zelda

Fig. 3. The test images sized 512×512

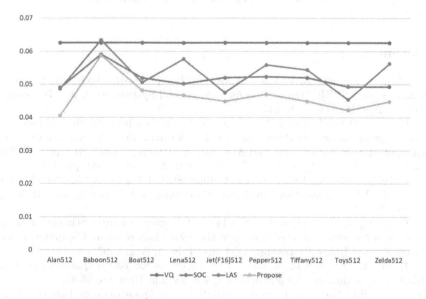

Fig. 4. The comparison results of compression rate ($N = 256$)

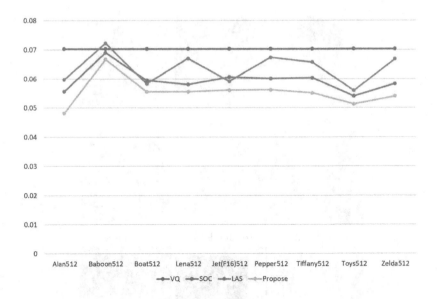

Fig. 5. The comparison results of compression rate ($N = 512$)

better compression rate improvement with smooth content image. The reason is that the smooth image contains more similar indexes distribution in local area. The proposed method more suitable for smooth content image for doing the image compression.

References

1. S. Chen, Z. He, and B. L. Luk, "A Generic Postprocessing Technique for Image Compression," IEEE Transactions on Circuits and Systems for Video Technology, Vol. 11, Apr. 2001, pp. 546-533.
2. J. J. Shen and H. C. Huang, "An Adaptive Image Compression Method Based on Vector Quantization," in Proceedings of the 1st International Conference Pervasive Computing Signal Processing and Applications, Harbin, China, 2010, pp. 377-381.
3. R. M. Gray, "Vector Quantization", IEEE ASSP Magazine, 1984, pp. 4-29.
4. C. H. Hsieh and J. C. Tsai, "Lossless Compression of VQ Index with Search-Order Coding," IEEE Transactions On Image Processing, Vol. 5, No. 1, Nov. 1996, pp. 1579-1582.
5. C. C. Chang, C.H. Sung, and T. S. Chen, "A Locally Adaptive Scheme for Image Index Compression," in Proceedings of the 1997 Conference on Computer Vision, Graphics, and Image Processing, Taichung, Taiwan, Aug. 1997, pp. 93-99.
6. Y. Linde, A. Buzo and R. M. Gray, "An Algorithm for Vector Quantizer Design," IEEE Transactions on Communications, Vol. 28, Jan 1980, pp. 84-95.
7. T. Kim, "Side Match and Overlap Match Vector Quantizers for Images," IEEE Transactions on Image Processing, Vol. 1, Apr 1992, pp.170-185.
8. C. C. Chang, F.J. Shiue, and T. S. Chen, "Tree Structured Vector Quantization with Dynamic Path Search," in Proceedings of International Workshop on Multimedia Network Systems(MMNS), Aizu, Japan, 1999, pp.536-541

The Reduction of VQ Index Table Size by Matching Side Pixels

Ya-Feng Di [1], Zhi-Hui Wang [1], Chin-Feng Lee[2, *], Chin-Chen Chang[3]

[1] School of software, Dalian University of Technology
Dalian, China 116620
dyf.dlut@gmail.com, wangzhihui1017@gmail.com

[2] Department of Information Management, Chaoyang University of Technology
Taichung, Taiwan 41349
*Whom correspondence: lcf@cyut.edu.tw

[3] Department of Information Engineering and Computer Science, Feng Chia University
Taichung, Taiwan
e-mail: alan3c@gmail.com

Abstract. The vector quantization (VQ) technology is applied to compress an image based on a local optimal codebook, and as a result an index table will be generated. In this paper, we propose a novel matching side pixels method to reduce the index table for enhancing VQ compression rate. We utilize the high correlation between neighboring indices, the upper and the left of the current index, to find the side pixels, and then reformulate the index. Under the help of these side pixels, we can predict the adjacent elements of the current index and then partition the codewords into several groups for using fewer bits to represent the original index. Experimental results reveal that our proposed scheme can further reduce the VQ index table size. Compared with the classic and state-of-the-art methods, the results reveal that the proposed scheme can also achieve better performance.

Keywords: vector quantization; matching side pixels; image compression

1   Introduction

With the rapid development of information technology, more and more people utilize the Internet to communicate and exchange multimedia information, such as images, audios, and videos. Among them images occupy a large proportion, which brings storage issues that need urgent solution. In order to reduce the cost of storage and increase the transmission speed, the images need to be compressed before transmitting [1][2][3]. Vector quantization [4], as a typical and effective compression technique is widely used in field of image compression [5][6]. VQ produces an index table by quantizing a given image in block-wise manner. The indices can stand for the whole image, in which the size of the former is much smaller than the latter, so VQ

© Springer International Publishing AG 2017
J.-S. Pan et al. (eds.), *Advances in Intelligent Information Hiding
and Multimedia Signal Processing*, Smart Innovation, Systems and Technologies 64,
DOI 10.1007/978-3-319-50212-0_25

can achieve a good compression rate. While the codebook design is a quite important work before compressing the image, it is usually trained with the well-known Linde-Buzo-Gray (LBC) algorithm [4].

However, people find that the index table can still be reduced to some degree, so some techniques which can further compress the index table are proposed to improve the compression effect. The famous compressing index table methods includes search-order-coding (SOC) and improved search-order-coding (ISOC) which can be applied to further losslessly compress VQ index table. SOC was proposed by Hsieh and Tsai in 1996 [7] and the algorithm encodes each index one by one in a raster scan order to find the same index with the current index along a predefined search path. After finding the same index, replace the original index with the search order code which is shorter than the original one. In 1999, Hu and Chang proposed an improved SOC method [8] to further compress an index table. Before doing the search-order-coding operation, sort the codewords in the codebook according their mean values in the descending order. By doing so, the index value can be more approximate with the neighboring indices. Besides the compression methods, a data hiding scheme based on search-order-coding and state-codebook mapping was proposed by Lin *et al* in 2015 [9]. Lin *et al* first employ the SOC method to reduce index table and then use the sate-codebook mapping method to deal with the indices which cannot be proposed by SOC. Lin *et al*'s method has good performance on reducing the index table; however, this method is a bit complicated and is not easy to implement. In this paper, we propose a novel method for reducing a VQ index table by matching side pixels. By exploiting the high correlation between the neighboring indices, we utilize the side pixels of the neighboring indices to predict and find the correct index. The experimental results demonstrate that our proposed scheme has a better performance in compression rate compared with SOC, ISOC and Lin *et al*'s method.

The reminder of this paper is organized as follows. Section 2 gives a brief introduction of the conventional VQ technique. The proposed scheme is described in Section 3 in detail. Experimental results are demonstrated in Section 4 and the conclusion part follows.

2   Related Work

In this section, a brief introduction of vector quantization is described and an example is also provided for better understanding.

Vector quantization (VQ) is proposed in 1980 by Linde *et al.* [4]. Due to its simple and cost-effective advantage, VQ is used in numerous applications, such as image and video compression. A complete VQ system consists of three parts, codebook design, compression and decompression. Designing a proper codebook has significant effect on the compression results. Generally, the popular Linde-Buzo-Gray (LBG) algorithm is utilized to train a representative codebook with several images. In the compression procedure, the image first needs be partitioned into a series non-overlapping blocks, and the size of each block is $l \times l$. Then each block is encoded with the index of the best matched codeword in the codebook. The codebook has C codewords, and each codeword C_i is a $l \times l$ dimensional vector, where $i = 0, 1, 2, \ldots, C\text{-}1$. For the current block assumed as X, calculate the Euclidean distance between X and each codeword C_i with following equation (1)

$$Dis(X, C_i) = \|X - C_i\| = \sqrt{\sum_{j=1}^{l \times l} (x_j - c_{ij})^2} ,\tag{1}$$

where x_j is the jth pixel of the current block, c_{ij} represents the jth element of the ith codeword C_i, $j = 1, 2, \ldots, l \times l$.

The codeword which has the minimal Euclidean distance with the current block X, is the best matched codeword. Then utilize the index of the best matched codeword to stand for the current block. After obtaining the indices of all blocks in the image, an index table is generated and need be stored. When decompressing the image, the receiver can refer the indices in the index table and the codewords in the codebook using a simple table look-up method to reconstruct the image. There is one thing to point that, VQ compression is a lossy algorithm.

Figure 1 shows a 512×512 image *Lena* and each block size is 4×4. A codebook with 256 codewords generated by the LBG algorithm is illustrated in Figure 2. Simultaneously, after the VQ compression phase, we can get an index table which has 512×512/ (4×4) = 16384 indices. To measure the compression efficiency, bit rate is the most commonly used metrics. The bit rate is 16384×8/ (512×512) = 0.5 bpp in the example.

index	codeword
0	(22, 35, 22, ..., 24)
1	(57, 53, 51, ..., 55)
2	(75, 52, 63, ..., 58)
3	(81, 79, 81, ..., 107)
4	(117, 92, 70, ..., 94)
5	(86, 73, 92, ..., 106)
6	(103, 109, 116, ..., 105)
253	(46, 55, 77, ..., 45)
254	(167, 152, 122, ..., 164)
255	(87, 142, 153, ..., 104)

Fig. 1. *Lena image*   **Fig. 2.** An example of codebook

3   Proposed Scheme

Driven by the motivation of reducing the index table, in this section, we propose a novel method to reduce the VQ index table by matching the side pixels.

3.1   Side pixels matching

In the index table, except the indices encoded using the conventional VQ algorithm at the first row and first column, each of the residual indices has its upper and left adjacent index neighbors which can be utilized to compress further by the proposed matching side pixels. Figure 3 demonstrates some part of the generated index table in the case of block size 4×4. The variables t_1, t_2, ..., t_{16} are the specific elements of the current index T. For a current index T, which belongs to the residual indices, side pixels refers to the elements $\{t_1, t_2, t_3, t_4, t_2, t_9, t_{13}\}$ at the first row and column elements in the current index and their adjacent elements in the neighboring blocks, which are the shaded pixels u_{13}, u_{14}, u_{15}, u_{16} in the upper index U and l_4, l_8, l_{12}, l_{16} in the left index L. We design a method of side pixel matching which utilizes $\{u_{13}, u_{14}, u_{15}, u_{16}, l_4, l_8, l_{12}, l_{16}\}$ to predict the adjacent elements $\{ t_1, t_2, t_3, t_4, t_2, t_9, t_{13}\}$ in the current index to realize the index table reducing.

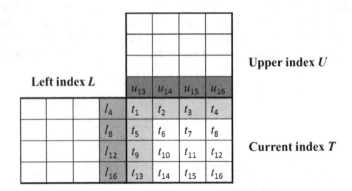

Fig. 3. Side pixels of the index T

3.2   Reduce the index table

The fewer bits to represent the indices, the higher compression rate can be achieved. First we utilize the side pixels to predict the adjacent elements of the current index. Assume the block size is $l \times l$, there are $2l-1$ adjacent elements in the current index. The first element in the corner of the current index is forecasted by calculating the average of its upper and left element. The other adjacent elements in the current index are directly predicted by equaling their neighboring elements. Take the index table in Figure 3 as an example, in the current index T, the value t_1 can be forecasted by solving the equation $t_1 = (l_4 + u_{13})/2$. The other values of the current index are directly predicted by $t_2 = u_{14}$, $t_3 = u_{15}$, $t_4 = u_{16}$, $t_5 = l_8$, $t_9 = l_{12}$, $t_{13} = l_{16}$. Then calculate the Euclidean distance d_T between the predicted elements of the current index T and the corresponding elements of the codewords using following Equation (2)

$$d_T = \sqrt{\sum_{j=1}^{2l-1} (p_j - c_{ij})^2} , \tag{2}$$

where p_j is the jth predicted element of the current index, and c_{ij} is the corresponding element of the ith codeword C_i, $j = 1, 2, 3, 4, 5, 9, 13$.

After obtaining the distances, a sorting operation is utilized on them. Specifically, by sorting the distances in the ascending order between the seven values with the corresponding elements of the codewords in the codebook, we can get a distance list. Then we divide the sorted distances list into four non-overlapping groups, G_0, G_1, G_2 and G_3 as follows,

$$\begin{cases} G_0 = \{0, 1, 2, \ldots, n-1\} \\ G_1 = \{n, n+1, \ldots, 2n-1\} \\ G_2 = \{2n, 2n+1, \ldots, 4n-1\} \\ G_3 = \{4n, 4n+1, \ldots, N-1\} \end{cases} ,$$

where n stands for the number in the first group and N means the codewords number in the codebook. So we can use two bits stand for the correct index is in which group, Specifically, "00" represents G_0, "01" represents G_1, "10" and "11" stands for G_2 and G_3 respectively. As can be seen from the probability, if the correct index lies in the first three groups, we can use fewer bits represent the index instead of original 8 bits. Though the unfortunate situation still has the possibility to occur, experimental results reveal that the index lies in the first three groups occupy the majority. An example is given to illustrate the index table reducing procedure and prove the compression effect.

In the example, we set n equals 8 and the current index T is 7. And the codebook has 256 codewords. Then the Euclidean distance is calculated between the seven side pixels with the elements in the corresponding positions of the codewords. After sorting the codewords, the method partitions the sorted codewords into four groups. Figure 4 shows the grouping situation. If the correct CW_7 is in G_0, the correct index 7 can be represented by "00" adding three bits, that is five bits; if CW_7 is in G_1, the correct index code can be represented by adding two bits "01" to the head of the index binary representation such that CW_7 can be presented as five bits; if CW_7 is in G_2, the correct index 7 can be represented by left-padding "10" to the four-bit representation of CW_7 to form a six-bit index code; unfortunately if CW_7 falls in G_3, the correct index 7 can be represented by ten bits due to two leading bits "11" should be added to its original eight-bit representation so the length of index code is ten. Figure 5 reveals the distribution of correct index lying in groups when n equals 8. From the figure, we can find that the probability of the correct index lies in the first three groups is far greater than that in the last group, which proves that our proposed method can reduce the index table significantly. Moreover, we assign the variable n with different values and compare the different compression rates to obtain the best performance outcome.

G_0		G_1		G_2		G_3	
0	CW_5	8	CW_{18}	16	CW_{20}	32	CW_{95}
1	CW_{17}	9	CW_{35}	17	CW_{155}	33	CW_{28}
2	CW_{78}	10	CW_{15}	18	CW_{234}	34	CW_{198}
⋮	⋮	⋮	⋮	⋮	⋮	⋮	⋮
7	CW_{23}	15	CW_{179}	31	CW_{254}	255	CW_{223}

Fig. 4. An example of grouping result

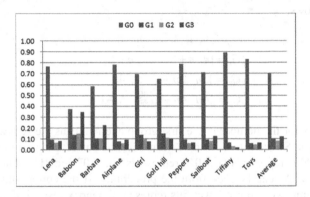

Fig. 5. The distributions of index grouping when $n = 8$

4   Experimental Results

In this section, experimental results are demonstrated to verify our proposed scheme with satisfactory performance in terms of reducing the index table after VQ operation. In our experiments, ten typical gray images sized 512×512 are used as the test images. Each image is partitioned into 4×4 blocks without overlapping and a codebook consisting 256 codewords is also prepared preliminarily. To obtain best experimental results, the variable n is assigned with 16, 8 and 4, respectively.

Table 1 illustrates the performance of the test images with various n values. We can find that all of the results are better than the original VQ compression bit rate which is 0.5 bpp. When n equals 4, the proposed scheme can get the smallest bit rate which means the compression rate is the most satisfactory.

Table 1. Comparison of bit rate (bpp) when n takes different values

	Lena	Baboon	Barbara	Airplane	Girl	Gold hill	Peppers	Sailboat	Tiffany	Toys	Average
$n=16$	0.3897	0.4402	0.4194	0.3923	0.3869	0.3900	0.3874	0.3977	0.3792	0.3877	0.3971
$n=8$	0.3440	0.4304	0.3899	0.3461	0.3443	0.3523	0.3384	0.3574	0.3225	0.3378	0.3563
$n=4$	0.3122	0.4429	0.3773	0.3121	0.3247	0.3386	0.3122	0.3122	0.2743	0.2974	0.3316

The conventional VQ compression technique cannot gain satisfactory compression rate, so some improved compression algorithms such as SOC and ISOC based on VQ have been also proposed to get lower bit rate. We also make comparisons with these two typical improved algorithms. Table 2 demonstrates the bit rate of these two techniques as well as our proposed scheme in which *Lena* is taken for example.

Table 2. Comparison of bit rate (bpp) with SOC and ISOC

Methods	m	BR
	$m=2$	0.357
SOC	$m=4$	0.341
	$m=8$	0.359
	$m=2$	0.329
ISOC	$m=4$	0.322
	$m=8$	0.344
Ours	0.3122	

Table 3. Comparison of bit rate (bpp) with Lin *et al*'s method

	Lena	Baboon	Airplane	Girl	Peppers	Sailboat	Tiffany	Average
$n=4$	0.3122	0.4429	0.3121	0.3247	0.3122	0.3122	0.2743	0.3316
Lin *et al* [8]	0.3165	0.4526	0.3190	0.3662	0.3284	0.3883	0.2976	0.3527

5   Conclusions

In this paper, a novel VQ index table representation by matching side pixels is proposed. By exploiting the correlation between the side pixels of the neighboring indices, we can use fewer bits to represent the original index. Experiment results present that the proposed scheme can achieve better compression performance compared with the classic compression index table methods and the state-of-the-art methods such as SOC and ISOC methods.

In the future, we plan to explore the proposed index table reduction to information hiding. The proposed information hiding scheme can hide a huge amount of information in the index map of an image and allows complete reconstruction of the indexes of the image. We would like to propose an information embedding scheme to hide a huge amount of information in the reduction of an index map and allows complete reconstruction of the indexes of the image.

References

1. C. Shi, J. Zhang and Y. Zhang: Content-based onboard compression for remote sensing images. Neurocomputing, Vol. 191, pp. 330-340 (2016)
2. H. S. Li, Q. Zhu, M. C. Li, and H. Ian: Multidimensional color image storage, retrieval, and compression based on quantum amplitudes and phases. Information Sciences, Vol. 273, pp. 212-232 (2014)
3. L. Zhang, L. Zhang, D. Tao, X. Huang and B. Du: Compression of hyperspectral remote sensing images by tensor approach. Neurocomputing, Vol. 147, No. 1, pp. 358–363 (2015)
4. Y. Linde, A. Buzo, and R.M. Gray: An algorithm for vector quantization design. IEEE Transactions on communications. Vol. 28, No. 1, pp. 84-95 (1980)
5. M. Lakshmi, J. Senthilkumar, and Y. Suresh: Visually lossless compression for Bayer color filter array using optimized Vector Quantization. Applied Soft Computing, Vol. 46, pp. 1030-1042.
6. Y.K. Chan, H.F. Wang, and C.F. Lee,: "A refined VQ-Based image compression method," Fundamenta Informaticae, Vol. 61, No. 3-4, pp. 213-221 (2004)
7. C. H. Hsieh and J. C. Tsai: Lossless compression of VQ index with search-order coding. IEEE Transactions on Image Processing, Vol. 5, No. 11, pp. 1579-1582 (1996)
8. Y. C. Hu and C. C. Chang: Low complexity index-compressed vector quantization for image compression. IEEE Transactions on Consumer Electronics, Vol. 45, No. 1, pp. 1225-1233 (1999)
9. C. C. Lin, X. L. Liu and S. M. Yuan: Reversible data hiding for VQ-compressed images based on search-order coding and state-codebook mapping. Information Sciences, Vol. 293, pp. 314-326 (2015)

Ultrasonic Ranging-based Vehicle Collision Avoidance System

S.H. Meng[1,1], A.C. Huang[2], T.J. Huang[3], Z.M. Cai[1], Q.Z. Ye[1] and F.M. Zou[1]

[1] Fujian University of Technology, School of Information Science and Engineering, Fuzhou, Fujian 350118, China
[2] Department of Electrical Engineering, National Sun Yat-Sen University, Kaohsiung, Taiwan
[3] Guo-Guang Laboratory School, National Sun Yat-sen University, Kaohsiung, Taiwan.
E-mail address: menghui@fjut.edu.cn

Abstract. With the increase in automobile traffic on a global scale, a corresponding increase in vehicle accident rates has occurred. To improve driver safety, finding an automated solution that is cost-effective and will reduce the global vehicle accident rate is necessary. This paper proposes a vehicle collision-avoidance system based on ultrasonic technology. We present a design for a device that will detect vehicle distance and provide the driver with timely warning to avoid vehicle collisions. The design includes both hardware and software components. The hardware consists of a minimal system board that supports a single-chip microcomputer (SCM), a power supply module, a liquid crystal display (LCD) 1602 module, and an ultrasonic module. The software algorithms that implement the vehicle collision function are written in the 'C' language using the Keil compiler to generate the microcomputer machine code. The code is debugged using a serial port module.

Keywords: Auto ranging; Ultrasound; Single-Chip Microcomputer (SCM); liquid-crystal display (LCD)

1   Introduction

With the global proliferation of automobiles, automotive safety has gained substantial attention. In response, several types of collision-avoidance sensors have been implemented [1-3]. Marketing surveys suggest that microwave-radar technology is the most popular implementation. There are several categories of sensors that deploy this technology: (a) "short distance", (b) "medium distance", and (c) "long distance". Each category has different characteristics and requirements. The long-distance collision-avoidance radar requires a high degree of sensitivity to its environment. In addition to a requirement for maximal miniaturization and large production costs, this type of radar requires advanced technology and a large number of development and deployment resources. There are additional important considerations in the design of collision-avoidance radar. These include:
   1.Radio band selection;

© Springer International Publishing AG 2017   211
J.-S. Pan et al. (eds.), *Advances in Intelligent Information Hiding*
and Multimedia Signal Processing, Smart Innovation, Systems and Technologies 64,
DOI 10.1007/978-3-319-50212-0_26

2.Mutual interference between multiple radars;

3.Signal processing requirements;

4.Weather condition impact on signal;

5.Influence of cloud density on satellite reception, particularly in the presence of typhoons;

6.Consumption of gasoline;

7.Vulnerability to rain fade;

8.Unsuitability of the lane-recognition system to operate for sustained periods of time;

All of the issues presented above represent restrictions on the development of any collision-avoidance system. To address these issues we propose an ultrasonic collision-avoidance system that will enhance the accuracy of object positioning, and thus reduce the accident casualty rate by 90%.

Ultrasound refers to the sound wave or vibration whose frequency exceeds the maximum human hearing threshold 20 kHz. The advantages of ultrasound over audible sound include better directivity, better object penetration, and ease of focusing. As a result, ultrasound waves are characterized by less absorption and less wave deformation when contacting objects. Significant benefits include lower cost, simpler structure, and easier implementation [4-5].

In this paper, we describe the distance detection method that is based on ultrasonic technology. This algorithm can greatly improve the reliability and speed of detecting the distance to an object. The flowchart of the algorithm is as follows:

1. The Microcontroller drives the ultrasonic module to transmit ultrasonic signals in real time;

2. The Microprocessor LCD1602 displays information;

3. The Microcontroller drives the ultrasonic collection module to compute the distance between two vehicles;

4. The Microcontroller uses the computed distance between the two vehicles to compute a safe distance to maintain from the detected vehicle based on a threshold comparison, and estimates the collision danger based on the real-time situation.

2   Hardware Design and Main Control Chip

A smart chip is essential for an implementation of an intelligent vehicle distance detection system. The selection of the main control chip is the core of the entire system design. Considering instruction set, cost and size factors, a STC89C52 chip is selected as the main control chip that will provide the processor to support the required software. The minimum system board of the chip is shown in Figure 1. The STC89C52 chip, produced by STCmicro Technology, is an 8-bit controller with high performance and low power consumption. It employs an 8 KB high-performance online-programmable flash memory, making it very convenient for user development. In addition, although the STC89C52 uses the classical MCS-51 core, the manufacturer has made substantial improvements in the original chip. As a result, with more advanced functionality, the STC89C52 chip provides effective, convenient, and efficient development choices for developers of embedded software. The main features of the STC89C52 [6] chip are as follows:

1. It has more function than the 8051 SCM. The original 8051 SCM supports only 12 clock cycles per machine cycle, but the modified STC89C52 supports both 6 clock cycles per machine cycle and 12 clock cycles per machine cycle. Furthermore, its instruction set is fully compatible with the 8051 SCM and is very convenient to use.

2. It has a wider range of working voltage. The working voltage of the 5 V SCM is 5.5 V-3.3 V, and the working voltage of the 3 V SCM is 3.8 V-2.0 V. Low working voltage that making it suitable for long-term use in practical applications.

3. With a frequency range of only 0-40 MHz, its working performance is equivalent to a regular 8051 SCM with a frequency range of 0-80 MHz.

4. The chip is integrated with a 512B RAM.

5. It has a total of 32 general IO ports. It is noted that a pull-up resistor is not required if the P0 port is used as a bus extension, but an IO port must be added if the P0 port is used as the only IO port.

6. It has a total of three 16-bit timer counters, making it more versatile in practical applications.

7. It supports online programming. Unlike other SCMs, it does not require a dedicated programmer. Instead, it can download programs through the serial port in a rapid and convenient fashion.

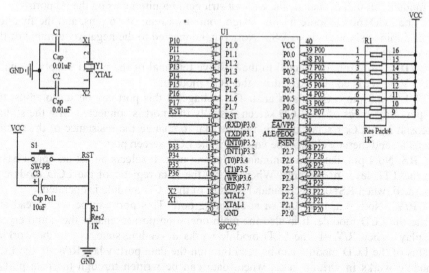

Fig. 1. Circuit of minimum system board of SCM STC89C52.

As shown in Figure 1, the minimum system board of the STC89C52 contains the crystal oscillation circuit and the reset circuit. The crystal oscillation circuit is primarily composed of two capacitors and a crystal oscillator. This oscillation circuit generates clock signals at a frequency of 12 MHz to trigger the normal operation of the SCM. The reset circuit is composed of an independent button and capacitors.

A reset voltage signal is generated when the reset circuit of button is pressed. Then the output is shifted from high level to low level, and the SCM is reset at the falling edge, thereby triggering the program to restart.

2.1 LCD Module

With the emergence of various new semiconductor materials, the liquid crystal display devices market is extremely competitive. Among them, LCD and OLED devices are very popular among consumers and developers due to their excellent performance. For this design, there is a single requirement from the LCD module; it need only display the prompt information. For this reason, the inexpensive LCD1602 module is selected as the display device for the entire system.

The LCD1602 module has the following features:

1. Able to display 16*2 characters. This fully meets our normal display needs;
2. Adjustable working voltage within a certain range based on need. This ensures that it will not fail to display characters or display garbled characters when the voltage decreases.
3. Low power consumption. The LCD1602 module requires only a 2 mA driving current to read and write the LCD screen, greatly saving energy and increasing the service life of other system components.
4. Simple driving circuit. The SCM can easily set the internal register of the LCD module through data analysis, without stringent requirements on the IO ports.

The LCD1602 module in this design contains a total of 16 pins, and the function of each pin is shown below: **VSS**: No.1 pin, connected to the negative terminal of the power supply;

VDD: No.2 pin, connected to the positive terminal of the power supply. Usually the 5 V power is connected to drive the LCD module;

VL: No.3 pin, LCD bias signal. The voltage of this port can be set to adjust the display brightness of the LCD screen. Usually this port is connected with the sliding rheostat by VCC. Users can slide the rheostat to change the resistance of the sliding rheostat, and thereby change the voltage of the LCD screen port.

RS: No.4 pin, data and command selecting port. It selects an action object based on the TTL level of the port. When RS=1, the data register of the LCD module is operated; when RS=0, the command register of the LCD module is operated.

R/W: No.5 pin, read and write selecting port. This port can be set to read and write the LCD module. It is also the main operation port to enable the liquid crystal display. When R/W=1, the LCD module works in reading state, where the working status of the LCD module can be read through the data port; when R/W=0, the LCD module works in writing state, where data can be written through the data port to specify the characters displayed on the LCD screen.

E: No.6 pin, enabling port. This port can be set to enable or disable the LCD module. When E=1, the LCD module is working properly and ready for character display; when E=0, the LCD module is closed and the LCD screen will not be subject to any control;

D0-D7: No.7-14 pins, 8-bit data ports. Data can be written to data ports through any one of SCM ports P0, P1, P2, and P3.

BLA: No.15 pin, the positive pole of backlight. Usually it is connected to the positive pole of a 5 V voltage to supply power for backlight;

BLK: No.16 pin, the negative pole of backlight. Usually it is connected to the negative pole of a 5 V voltage to supply power for backlight.

2.2   Ultrasonic Module

The ultrasonic emitting part allows the ultrasonic emission transducer TCT40-16T to send out square wave pulse signals of about 40 kHz. There are two commonly used methods to generate 40 kHz square wave pulse signals: use of hardware such as a 555 oscillator or use of software such as SCM software programmed output. Our developed system deploys the latter. Since the output power of the SCM port is not sufficient to support our design, programmed output of square wave pulse signals at about 40 kHz are generated through SCM P1.0 port in two parts. One part is a push-pull circuit composed of 74HC04 hex inverters amplifies the power of the signals to ensure a sufficiently long emission distance and meet the requirements for distance measurement; the other part is transmitted to the ultrasonic emission transducer TCT4016T and eventually emitted into air in the form of sound waves. The circuit of the emitting part is R31 and R32 are the pull-up resistors at the output end. They can not only improve the ability of driving the inverter 74HC04 to output at a high level, but they also enhance the damping effect of the ultrasonic transducer to shorten the time of free oscillation.

The ultrasonic receiving part allows the ultrasonic receiving transducer TCT40-16R to effectively receive the reflected wave (echo). The reflected wave is then converted into an electrical signal, which will be amplified, filtered, and shaped by an integrated SONY CX20106 chip. This creates a negative pulse sent to P3.2 (INT0) pin of the SCM to generate an interruption.

3   'C' pseudo code and Initialization of Ultrasonic Module

```
Trig=0;
EA=1;
TMOD=0x10;
while(1)
{
EA=0;
Trig=1;
delay_20us();
Trig=0;
while(Echo==0);
succeed_flag=0;
EA=1;
EX0=1;
TH1=0;
TL1=0;
```

```
TF1=0;
TR1=1;
delay(20);
TR1=0;
EX0=0;
```

The initialization code is entered and provides the input to the ultrasonic module
with a 20 us pulse trigger so that the ultrasonic module can emit ultrasonic signals.
When the value of the detection indicator "succeed flag" is 1, the reflected ultrasonic
signals are received. Otherwise, no reflected ultrasonic signals are received, indicating
that there are no vehicles in front of the detector carrier and it is safe.

3.1   Distance Calculation Display

This design uses the ultrasonic module to detect the distance between two vehicles,
and displays the results on the LCD screen. The specific implementation program is
as follows [7]:

```
if(succeed_flag==1)
    {
        time=timeH*256+timeL;
        distance=time*0.0172;
    }
if(succeed_flag==0)
{
        distance=0;
}
        display(distance);
```

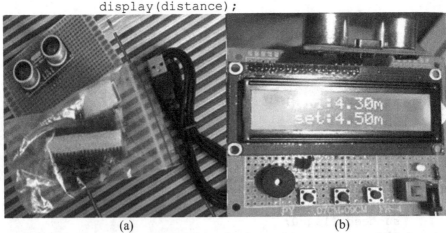

(a)   (b)

Figure 2. The ultrasonic module (a) of the circuit material components, and (b) the actual
component circuit of the sensing module.

When reflected ultrasonic signals are received, the calculated distance is stored in the "distance" variable, and then displayed in the "display()" function (It is quotes the ultrasonic ranging C program of the 51 SCM) [7].

The traffic control department specifies that when the vehicle speed is 60 km/h, the safe distance between vehicles should be 60 meters or longer; when the vehicle speed is 80 km/h, the safe vehicle distance should be 80 meters or longer, and so on. In common, ultrasonic ranging can be well used less than 10 meters. So this system is suggested to be used in URBAN areas, where vehicles run slowly. Moreover, if it is used to detect the distance of left or right side of a car, less 5-meter detection will be enough. In the short-distance simulations performed for this design, the safe distance is set to 5 m, and the onboard ranging system will sound the alarm if there is any other vehicle within this distance. Figure 2 shows the actual circuit of the sensing components.

There is no ultrasonic sensor in the car that driving distance 60m will getting bump in 3.6seconds when the vehicle speed is 60km/hr. Adding ultrasonic sensors with a 4.5 meter distance sensing to the car. It has more 0.27 seconds reaction time to avoid collision than a car without ultrasonic sensors. If a car adding ultrasonic sensors such with 45 meter distance sensing. It provides the driver more 2.7 seconds reaction time with timely warning to avoid vehicle collision than a car without ultrasonic sensor, shows the Figure 3.

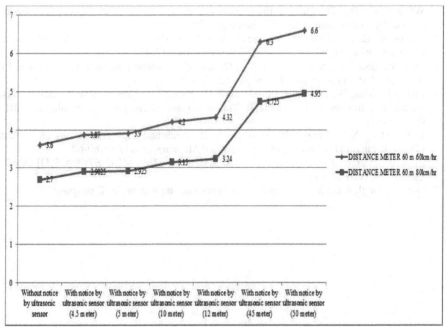

Figure 3.   Reaction time data.

Conclusion

In common, ultrasonic ranging can be well used less than 10 meters. So this system is suggested to be used in URBAN areas, where vehicles run slowly. Moreover, if it is used to detect the distance of left or right side of a car, less 5-meter detection will be enough. This proposal presents a design for a vehicle safety alarming system to prevent danger caused by poor vision (e.g., due to weather conditions and geographical positions) or lack of driver concentration, with the intent that the design and setup of the intelligent vehicle collision avoidance system/sound wave multiple-capture system can improve the computed accuracy of vehicle positioning and reduce the accident rate and prevent collisions. The hardware deployed in the design includes the STC89C52 development board, a power supply module, the LCD1602 module, and the ultrasonic module. The STC89C52 development board sends out ultrasonic signals that will be reflected back upon encountering an obstacle. Then the actual distance between the two vehicles is calculated according to the time difference, which can be displayed on the LCD screen.

References

1. Wang C.R.: The Design and Implementation of a Vehicle Active Safety Pre-warning System[D]. Dalian University of Technology, (2013).
2. Jiang-wei D.U.: Research on Safe Vehicle Distance Based on Vehicle-Vehicle Communication [D].Wuhan University of Technology, (2012).
3. Shen Y., Gao X.R., Sun Z.Y., Li J.L.: Design of ultrasonic ranging instrument based on MCU[J]. Modern Electronics Technique, (2012), 07:126-129.
4. Zhang J., Wang W., Jun-feng H.E., Xiang F.U., Zheng T.J.: New positioning method based on double ultrasonic wave module[J]. Transducer and Microsystem Technologies, (2012), 09:22-24+31.
5. Li G., Meng X.J., Wang X.H., Wang Z.Q., Research and application status on domestic ultrasonic ranging [J]. Science of Surviving and Mapping, (2011), 04:60-62.
6. Zhou P.: Design of temperature detection system based on MCU STC89C52[J]. Modern Electronics Technique (2012), 22:10-13.
7. Using 51 Single-Chip Microcomputer Ultrasonic ranging distance of C program.

Part III
Hardware Design and Assisting Systems

Development of an Easy Japanese Writing Support System with Text-to-Speech Function

Takeshi Nagano, Hafiyan Prafianto, Takashi Nose, and Akinori Ito

Graduate School of Engineering, Tohoku University,
6-6-05, Aramaki Aza Aoba, Aoba-ku, Sendai city, Japan
nagano@spcom.ecei.tohoku.ac.jp
hafiyan@spcom.ecei.tohoku.ac.jp
tnose@m.tohoku.ac.jp
aito@spcom.ecei.tohoku.ac.jp

Abstract. Many foreigners visit and stay in Japan. Natural disasters such as earthquake, flood and volcano often occur in Japan. When a disaster occurs, the authority needs to give announcement to people including non-Japanese. Easy Japanese(EJ) is focused on conveying information to non-Japanese. EJ is a kind of Japanese designed to be easily understood by non-Japanese. We implemented an Easy Japanese writing support software "YANSIS" by Java. YANSIS runs any platform where Java runs. Under a disastrous condition, not only text information but also speech announcement is required. Thus we implemented a text-to-speech(TTS) function to YANSIS. To integrate the text-to-speech function with YANSIS, we implemented the Japanese TTS system Open JTalk with Java. In this paper, we describe our software YANSIS, and compare the quality of the synthesized speech by Open JTalk and our implemented TTS software.

Keywords: Easy Japanese, YANSIS, text-to-speech, speech synthesis, HMM, Java

1 Introduction

Nowadays, many foreigners visit Japan because of sightseeing or business. Many foreigners also live in Japan because of study, work or marriage. In the places in Japan where many foreigners visit, signs are written not only in Japanese but also in other languages such as English, Chinese and Korean.

In Japan, we have many natural disasters such as earthquake, flood, volcano, and tsunami. When a disaster occurs, the authority needs to give announcement such as evacuation route, shelter and so on precisely. To tell non-Japanese the information precisely, multilingual announcement is needed. Under the disastrous condition, however, the multilingual announcement is impossible because of limitation of time and human resources.

Easy Japanese(EJ) was proposed to convey important information under the emergency such as a big disaster[1]. EJ is composed of words and phrases

© Springer International Publishing AG 2017
J.-S. Pan et al. (eds.), *Advances in Intelligent Information Hiding
and Multimedia Signal Processing*, Smart Innovation, Systems and Technologies 64,
DOI 10.1007/978-3-319-50212-0_27

which can be understood by non-Japanese easily. Nowadays, EJ has been widely used in the public documents, websites of local authorities or broadcast[2][3]. Ordinary Japanese people needs some training to write sentences in EJ, because Japanese people don't know which words and phrases are easy or difficult for non-Japanese. To help writing sentences in EJ, we developed an Easy Japanese writing support software[4][5]. We named this software YANSIS, which stands for "YAsasii Nihongo sakusei SIen System"(Easy Japanese composition support system) in Japanese. YANSIS was written in Java so that it runs on any platform where Java runs. When a disaster occurs, the network might be shut down because the network lines and systems will be damaged. Thus YANSIS runs as a standalone software without network connection. Not only text information but also voice announcement is required under such a situation. We can use the Text-to-speech(TTS) function via HTML5 with network connection, but we aim the TTS system built into YANSIS so that the system can be used without network connection. Thus, we decided to implement TTS function to YANSIS.

We can use several open-source TTS software, including Festival(C++)[6], Flite(C)[7], MaryTTS(Java)[8] for English or other languages. Galatea Talk(C)[9], Open JTalk(C)[10] are also open-source TTS software for Japanese. (A symbol in parentheses of the previous list denotes the programming language for implementation). In Japanese TTS, intonation and accent sandhi must be estimated for synthesizing high quality Japanese speech. We selected Open JTalk as our base software because Open JTalk has been maintained well and quality of synthesized speech is high. Then, we rewrote Open JTalk in Java so that the TTS function could be integrated with other parts of YANSIS. We named our software Gyutan.Gyutan was named after a famous food at Sendai city at which our university is located. In this paper, we describe the function of YANSIS, structure of Gyutan and comparison quality between Open JTalk and Gyutan.

2 Structure of YANSIS

Fig. 1 shows the structure of YANSIS. The functions of YANSIS are described in the following subsections.

2.1 Morphological analyzer module

The input text by the user is analyzed using a morphological analyzer to split the string into words, estimate pronunciation of the words, and estimate the prosodic features such as accent and accent phrase boundary. We exploited the morphological analyzer Sen[11], which is a Java clone of the morphological analyzer MeCab[12].

2.2 Japanese level analyzer module

The Japanese level analyzer module analyzes the difficulty level of a word in the text. The difficulty level of a word is determined according to Japanese Language Proficiency Test(JLPT) level of the word.

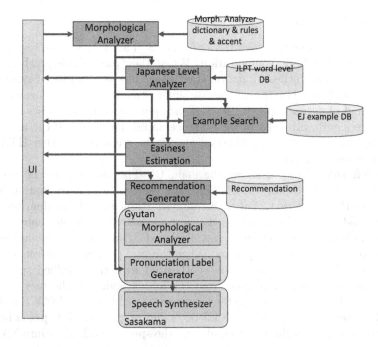

Fig. 1. Structure of YANSIS

2.3   Example search module

Example sentence of EJ for disaster mitigation is prepared in database in YAN-SIS. The example search module searches sentences from the database that include a word in the input sentence.

2.4   Recommendation generator module

The recommendation generator module shows the results of the analysis of the input text. This module points out the difficult words and phrases in the text. This module also advises how the sentence is difficult (such as the sentence to be long) and how to make the text easier.

2.5   Easiness estimation module

The easiness estimation module shows the easiness score of a sentence. The easiness score is estimated using the feature of the sentence such as ratio of Chinese characters, length of the sentence and so on[13].

2.6   Speech synthesis module

We implemented a speech synthesis module. This module is composed of Japanese text analyzer Gyutan and speech synthesizer Sasakama.In section 3, we describe this module in detail.

2.7   User Interface

Fig. 2 shows the User Interface(UI) of YANSIS. The right side of Fig. 2 shows UI for speech synthesis configuration. When the "use speech synthesis" box is checked, speech synthesis function is enabled. The speech rate of synthesized speech can be changed by the speech rate slider or by directly inputting value into the text field of current speech rate. The adequate speech rate varies according to sentence length, difficulty of the sentence and so on[14][15]. If the user wants to change the voice of the synthesized speech, user can select other HTS voice which user wants to use.

The left side of Fig. 2 shows the main UI. The user inputs the text into the text area at the top of the UI. Next, the user pushes the evaluation button to evaluate the input text. Then, the analysis result and the evaluation points are displayed. When the user uses the example search, example sentences containing the keyword are retrieved and displayed. If speech synthesis is enabled, the speech synthesis button appears. When the user pushes the speech synthesis button, speech of the input text is synthesized with the specified speed rate, and synthesized speech is played from the audio device. If the user wants to save the synthesized speech, the user can save the speech as a WAV file. As explained, a user can easily compose EJ texts and prepare the EJ speech because of integration of EJ evaluation module and the speech synthesizer into YANSIS.

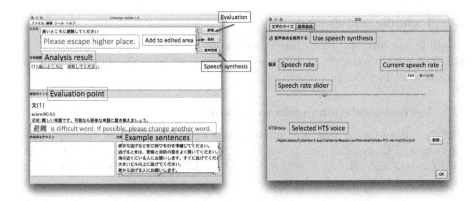

Fig. 2. UI of YANSIS

3   Structure of Gyutan

Gyutan is a Japanese text-to-speech synthesizer module composed of the morphological analyzer module, pronunciation label generator module and speech synthesizer module.

3.1 Morphological analyzer module

The text analysis is indispensable for a high-quality TTS. MeCab is used as the morphological analyzer in Open JTalk. In Gyutan, Sen is used as a morphological analyzer that can be used in common with YANSIS. When synthesizing a Japanese speech, the accent is an important information. To determine the accent precisely, the accent analysis is needed. The accent analysis is based on the lexicon. Open JTalk uses NAIST-jdic[16] as a dictionary, which includes accent information. The original NAIST-jdic is to be used with MeCab, which is not compatible with a dictionary for Sen. Thus we modified NAIST-jdic so that it works with Sen. We used NAIST-jdic 0.4.3 as dictionary of Gyutan. In the later experiment, we used the dictionary that comes with Open JTalk 1.08 as the dictionary of Open JTalk. Number of entries of the dictionaries are 456169 and 483156 for Gyutan and Open JTalk respectively.

3.2 Pronunciation label generator module

We need pronunciation labels for synthesizing speech, which include information of phoneme and intonation. The pronunciation labels are generated based on the result of morphological analysis. The labels should be generated considering the accent sandhi rules and other rules of pronunciation change by combining words. Fig. 3 shows an example of accent sandhi, where the vertical positions of the circles indicate relative pitch. The left side of Fig. 3 shows the pattern of the pitch accent of the word, *Sendai* and *eki*(station). The right side of Fig. 3 shows the intonation of the two words pronounced as a compound word, *Sendai-eki*. As shown, the intonation of the compound word is not a simple concatenation of the individual accents. Gyutan processes the accent sandhi and pronunciation

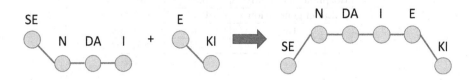

Fig. 3. Example of accent sandhi rule

change based on the rules used in Open JTalk 1.08.

3.3 Speech synthesizer module

This module synthesizes Japanese speech using pronunciation label based on the HMM speech synthesis[17]. Open JTalk uses hts_engine API[18] as a speech synthesis library. We implemented the speech synthesis engine in Java so that all the systems were integrated using Java. We rewrote hts_engine API written in C using Java. This speech synthesis library written in Java is called Sasakama, which is also a name of food in Sendai. Sasakama can synthesize Japanese speech using an acoustic model in HTS Voice format as same as Open JTalk. Sasakama can play the synthesized speech via an audio device. Sasakama is based on hts_engine API 1.10.

4   Experiment

4.1   Speed of the speech synthesis

Gyutan is supposed to be slower than Open JTalk because of difference of the programming language of implementation. In this subsection, we examined the synthesis speed of each software. In the experiment, we measured the times of following processes: (1)morphological analysis, (2)pronunciation label generation, (3)speech synthesis and (4)total. The computer used in this experiment had Intel Xeon L5640 processor at 2.27GHz clock, with 32GB memory. In the experiment, we used 503 sentences from ATR Japanese Speech Database was used. Average times spent by the processes were measured. The acoustic model was a female model developed by our laboratory.The average character length of synthesized sentences was 34.

Table 1 shows the average process time of a sentence. These times include I/O time. From Table 1 it is found that Gyutan was 3 times slower than Open JTalk. The morpheme analysis results were different between Open JTalk and Gyutan

Table 1. Comparison of average calculation time(ms)

	Open JTalk	Gyutan
Morphological analysis	0.34	30
Label generation	0.62	104
Synthesis	565	770
Total	631	1851

because of the difference of dictionaries. Since the generated pronunciation labels depends on the result of the morphological analysis, the labels of the two systems may differ each other. Thus we evaluated the synthesized speech signals generated by the both systems by the objective evaluation and the subjective evaluation. In the objective evaluation, we compared the mel cepstrum and log F0 of the utterances used as training data to those of the synthesized utterances with the same contents. Two experiments were conducted: in the first experiment, the speech signals were synthesized from the texts; in the second one, the signals were synthesized using the same pronunciation labels. The evaluation results are shown in Table 2. In Table 2, hts_engine and Sasakama were used for synthesis from the pronunciation labels, while Gyutan with Sasakama and Open

JTalk with hts_engine API were used for synthesis from texts. These results show that the two systems gave almost same quality of synthesized speech. In sub-

Table 2. Distance between the synthetic speech and the real speech

	Synthesis from labels		Synthesis from texts	
	hts_engine API	Sasakama	Open JTalk	Gyutan
Mel cepstrum distance[dB]	5.60	5.60	6.84	6.85
Log F0 RMSE[cent]	89.32	89.67	194.05	195.95

jective evaluation, we evaluated naturalness of the synthesized speech by Mean Opinion Score(MOS). For the evaluation, ten sentences were selected from ATR Japanese speech database B set randomly The evaluators were 8 adults aged from 22 to 24. The acoustic model used in evaluation was the same as that used in the objective evaluation. All sentences used in this evaluation were included in the training set of the acoustic model. The evaluation result is shown Table 3, which shows that the naturalness of the speech synthesized by both systems were exactly same. We also evaluated pronunciation error rate of each software using 503 sentences. Table 4 shows the pronunciation error rate. In Table 4, a word in parentheses denote morphological analyzer. The error rate of Gyutan with Sen was slightly higher than Open JTalk with MeCab, but the absolute error rate was still very small.

Table 3. Naturalness of synthesized voice

	MOS
Open JTalk	3.0
Gyutan	3.0

Table 4. Error rate of pronunciation of morpheme

	error rate(%)
Open JTalk(MeCab)	0.4
Gyutan(Sen)	0.9

5   Conclusion

In this paper, we described the text-to-speech function of Easy Japanese writing support system, YANSIS. We rewrote the Japanese speech synthesis software Open JTalk by Java, which is named Gyutan. A user can easily synthesize the

text written in EJ because the speech synthesis function was integrated into
YANSIS.

Acknowledgments. We appreciate Prof. Kazuyuki Sato of Hirosaki University
and other members of Yasasii Nihongo Workshop. We also appreciate HTS working group that publish and maintain HMM based speech synthesis software for
free. Part of this work is supported by JSPS Kakenhi Grant-in-Aid for Scientific
Research (B) Grant Number 26284069.

References

1. "Easy Japanese",http://human.cc.hirosaki-u.ac.jp/kokugo/EJ1a.htm
   (accessed 2016-07-04)
2. M. Moku and K. Yamamoto.: Investigation of Paraphrase of Easy Japanese in Official Documents. Proc. The 17th Yearly Meeting of Association for Natural Language
   Processing, 2011.
3. NHK: "NHK News Web Easy". http://www3.nhk.or.jp/news/easy (accessed 2016-
   07-04).
4. Takeshi Nagano and Akinori Ito: YANSIS:An "Easy Japanese" writing support
   system. Proceedings of 8th International Conference ICT for Language Learning,
   pp. 1–5, 2015.
5. YANSIS, http://www.spcom.ecei.tohoku.ac.jp/~aito/yansis/ (accessed 2016-
   07-04)
6. Festival,http://www.cstr.ed.ac.uk/projects/festival/ (accessed 2016-07-04)
7. Flite,http://www.festvox.org/flite/ (accessed 2016-07-04)
8. MaryTTS,http://mary.dfki.de/ (accessed 2016-07-04)
9. Galatea Talk,https://osdn.jp/projects/galateatalk/ (accessed 2016-07-04)
10. Open JTalk,http://open-jtalk.sourceforge.net/ (accessed 2016-07-04)
11. Sen,https://java.net/projects/sen (accessed 2016-07-04)
12. MeCab,http://taku910.github.io/mecab/ (accessed 2016-07-04)
13. M. Zhang, A. Ito and K. Sato: Automatic Assessment of Easiness of Japanese for
    Writing Aid of "Easy Japanese". Proc. Int. Conf. on Audio, Language and Image
    Processing, Shanghai, pp. 303-307, 2012.
14. Hirosaki University Sociolinguistics Laboratory:Verification result of broadcast
    reading speed that is easy to be understood by foreigners in the event of disaster.
    http://human.cc.hirosaki-u.ac.jp/kokugo/onnseikennsyoukekka.html (in Japanese).
15. H. Prafianto, T. Nose, Y. Chiba, A. Ito and K. Sato: A study on the effect of speech
    rate on perception of spoken easy Japanese using speech synthesis. Proc. Int. Conf.
    on Audio, Language and Image Processing, pp. 476–479, 2014.
16. NAIST Japanese Dictionary,https://osdn.jp/projects/naist-jdic/ (accessed
    2016-07-04)
17. H. Zen, K. Tokuda and A. W. Black: Statistical parametric speech synthesis.
    Speech Communication, 51(11), pp. 1039–1064, 2009.
18. hts_engine API,http://hts-engine.sourceforge.net/ (accessed 2016-07-04)

Design of a Hardware Efficient Antenna-Configurable Multiple-input Multiple-output Detector

Syu-Siang Long, Chin-Kuo Jao, Shih-Kun Lin, Chang-Hsuan Hsu, Wei-Cheng Wang, and Muh-Tian Shiue

Department of Electrical Engineering,
National Central University, Zhongli, Taiwan 32001
945401026@cc.ncu.edu.tw, mtshiue@ee.ncu.edu.tw

Abstract. In this paper, a hardware efficient multiple-input multiple-output (MIMO) detector which flexibly supports multiple antenna configurations ($2 \times 2, 4 \times 4$ and 8×8)) and data modulations (QPSK, 16-QAM and 64-QAM) is presented. A breadth-first decoding algorithm known as distributed K-best (DKB) is applied in this design. The DKB greatly reduces required visiting nodes and excludes sorting operation at each layer to save computing complexity. The proposed antenna configurable pipelined multi-stage architecture is composed of elementary function blocks of DKB. Each DKB blocks requires only K clock cycles to find the best K candidates, where K denotes the number of calculated survival paths. The proposed MIMO detector has been implemented by a 90-nm CMOS technology with the core area of 1.014 mm^2. The average power consumption is 17.76 mW when operates at the maximum frequency of 74 MHz and 1 V power supply voltage.

Keywords: Multiple-input multiple-output (MIMO) detection, antenna configurable, distributed K-best (DKB), shift multiplier.

1 Introduction

In the last decade, *spatial multiplexing* in multiple-input multiple-output (MIMO) technologies is the most researched MIMO topic to improve bandwidth efficiency. It allows transmitting and receiving several independent data streams simultaneously in the same space [1]. Let us consider the spatial multiplexing MIMO system with N_t transmitting antennas and N_r receiving antennas. The real-valued baseband equivalent model can be written as:

$$\mathbf{Y} = \mathbf{H}\mathbf{s} + \mathbf{n}, \tag{1}$$

where $\mathbf{s} = [\Re\{s_1\, s_2 \cdots s_{N_t}\}\, \Im\{s_1\, s_2 \cdots s_{N_t}\}]^T$ is the transmitted symbol vector. For each element s_i is independently chosen from M-QAM constellation set Ω and transmitted by i^{th} antenna, where $\Re\{\cdot\}$, $\Im\{\cdot\}$ and $[\cdot]^T$ imply real and imaginary part and matrix transpose, respectively; $\mathbf{Y} = [\Re\{y_1\, y_2 \cdots y_{N_r}\}\, \Im\{y_1\, y_2 \cdots y_{N_r}\}]^T$

© Springer International Publishing AG 2017
J.-S. Pan et al. (eds.), *Advances in Intelligent Information Hiding*
and Multimedia Signal Processing, Smart Innovation, Systems and Technologies 64,
DOI 10.1007/978-3-319-50212-0_28

is the received symbol vector by N_r antennas; $\mathbf{H}$ denotes the real-valued channel matrix of size $2N_r \times 2N_t$, whose entries are identically independent distributed (i.i.d) zero-mean Gaussian random variables; and $\mathbf{n} = [n_1\, n_2 \cdots n_{2N_r}]^T$ is a white Gaussian noise vector with variance $\frac{\sigma^2}{2}$. The optimum solution for (1) is called maximum-likelihood (ML) decoding, which finds out the most possible symbol vector $\hat{\mathbf{s}}$ according to the following minimum-distance criterion:

$$\hat{\mathbf{s}} = \arg\min_{\mathbf{s}\in\Omega^{2N_t}} ||\mathbf{Y} - \mathbf{Hs}||^2. \tag{2}$$

The ML decoder exhaustively computes $\sqrt{M}^{2N_t}$ possible symbol vectors to determine one $\hat{\mathbf{s}}$ vector. The computing complexity grows exponentially with the antenna number and constellation size and is not feasible for realization. Consequently, two kinds of simplified recursive decoding algorithms known as *depth-first* type and *breadth-first* type [2]-[5] have been developed for hardware implementation.

To transforming ML decoding method into tree-branch searching, thanks to QR decomposition, i.e., $\mathbf{H} = \mathbf{QR}$, where $\mathbf{Q}$ and $\mathbf{R}$ are unitary matrix and upper triangle matrix, respectively. By applying $\mathbf{Q}^H$ at both sides of (1), results in $\tilde{\mathbf{Y}} = \mathbf{Q}^H\mathbf{Y} = \mathbf{Rs} + \mathbf{Q}^H\mathbf{n} = \mathbf{Rs} + \mathbf{w}$. Since $\mathbf{R}$ is an upper triangular matrix, (2) can be rewritten in recursive form as follows:

$$\hat{\mathbf{s}} = \arg\min_{\mathbf{s}\in\Omega^{2N_t}} \sum_{i=1}^{2N_r} \underbrace{\left| \tilde{y}_i - \sum_{j=i}^{2N_t} R_{ij}s_j \right|^2}_{i^{th}\, layer\ PED:\ e_i(s^{(i)})}. \tag{3}$$

The tree-searching procedure starts from bottom layer-$2N_t$ to top layer and calculates accumulated *partial Euclidean distance* (PED): $T_i(s^{(i)}) = T_{i+1}(s^{(i+1)}) + e_i(s^{(i)})$ at i^{th} layer to find the best candidate which owns minimum accumulated PED in (3).

In this paper, the proposed antenna-configurable MIMO detector can flexibly support 8×8, 4×4 and 2×2 and data modulations, such as QPSK, 16-QAM and 64-QAM, is presented. In our design, we adopt DKB algorithm [6] as elementary functional blocks to achieve multiple antenna configurations. The computing complexity is independent to the applied constellation size. In the following sections, the hardware architecture and circuit design will be detailedly depicted.

2   Hardware Architecture

The conventional K-best scheme concurrently expands and sorts the children nodes from all parent nodes and preserve K best candidates with minimum PEDs, while DKB scheme sequentially listed best K candidates one by one within K clock cycles. Owning to this feature, the DKB avoids the sorting circuit and only requires a minimum finder circuit to select the child node with minimum PED in i^{th} layer. The number of visited nodes at each DKB layer is $2K - 1$,

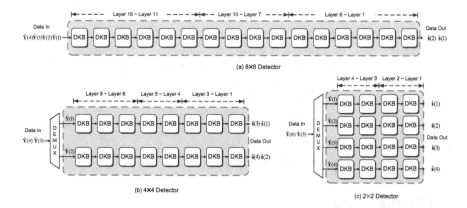

Fig. 1. Block diagram of the proposed configurable MIMO detector: (a) 8×8 detector, (b) 4×4 detector, and (c) 2×2 detector.

which is independent of the constellation size M. Therefore, the total number of visited nodes for DKB is $\min(K, \sqrt{M}) + (2N_t - 1)(2K - 1)$. Compared to the conventional K-best algorithm, the DKB scheme is an efficient method and is feasible to compose the proposed 2×2, 4×4, and 8×8 antenna configurable MIMO detector.

2.1 Antenna-configurable Architecture

To support multiple antenna configurations, the proposed pipelined architecture is built by elementary functional blocks which can flexibly reconstruct 2×2, 4×4, and 8×8 antenna configurations shown in Fig.1. Based on the aforementioned algorithm, the elementary functional block "DKB" is the key circuit in our design.

The 8×8 MIMO detector structure has the deepest tree searching layers and the largest pipelined stages should be firstly determined. With the elementary building block, the proposed 8×8 detector architecture is depicted in Fig. 1(a). The main advantage of this reconstructible building-block architecture is that these building blocks of 8×8 detector can reconstruct the two streams of 4×4 and four streams of 2×2 MIMO detectors as shown in Fig. 1(b) and Fig. 1(c) respectively. Therefore, the hardware utilization achieves maximum for various antenna configurations. The overall throughput and BER performance can be maintained through all antenna configurations. In the other hand, it is possible to construct other antenna configurations in the proposed design, such as 3×3, 5×5, 6×6 and 7×7. However, the hardware utility and throughput decreases most to 62.5% at non power of two configurations.

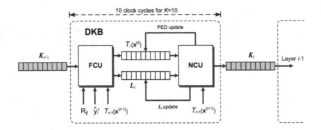

Fig. 2. Block diagram of the DKB building block.

3 Circuit Design

The key building block "DKB" will be described in this section. The ith layer DKB circuit diagram is depicted in Fig. 2. It consists of a *first child unit* (FCU), a *next child unit* (NCU), and registers for data buffering. According to the K parent nodes fed from the $i + 1^{th}$ layer, the FCU sequentially finds out the corresponding K first child (FCs) and stores them into the contender register. Their corresponding accumulated PEDs are also calculated and stored into the PED register. When all the K FCs have been found by the FCU, then the NCU begins comparing the PEDs of FCs and selecting the corresponding FC with minimum PED into the K-best candidate register K_i. In the meanwhile, the NCU enumerates the best sibling node next to the selected FC as the next child (NC). Then the contents of the selected FC in the contender register L_i and PED register $T_i'(\mathbf{x}^{(i)})$ are replaced by this NC. As long as the registers in each DKB stage are full filled with data, the FCU and NCU can determine one FC and one NC in every clock cycle, respectively. Therefore, for K-size candidate list in each DKB stage, only K clock cycles are needed to complete the K-best list and deliver to the next stage. The circuit operation principles of FCU and NCU in the DKB block will be depicted in the following sub-sections.

3.1 Shift Multiplier (SM)

The PED computation costs major hardware resources in the DKB procedure. From (3), it can be seen that the multiplications of R_{ij} and s_j costs most parts of computing complexity in the PED calculation. In order to avoid the multiplication complexity, we propose a *shift multiplier* (SM) which reduce the area and critical path delay of the multiplication significantly. Since $s_j \in \Omega = \{-\sqrt{M} + 1, ..., -1, 1, ..., \sqrt{M} - 1\}$ is an odd number in the real-valued axis, we uses a new signal set $\Omega' = \{-\sqrt{M}, ..., -2, 0, ..., \sqrt{M} - 2\}$ instead of Ω. The new signal set Ω' is represented in Fig. 3(a). In the case of 16-QAM, the signal values in the signal set are power of two, i.e., $\Omega' = \{-4, -2, 0, 2\}$. Therefore, R_{ij} multiplied by the shifted signal set can easily be implemented by simple bit-shifting and addition. The received signal vector $\hat{\mathbf{Y}}$ can be represented by

$$\hat{\mathbf{Y}} = \mathbf{R}(\mathbf{s} - \mathbf{e}) + \mathbf{R}\mathbf{e} + \mathbf{w}, \tag{4}$$

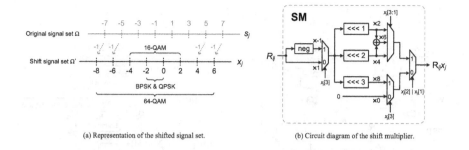

(a) Representation of the shifted signal set.   (b) Circuit diagram of the shift multiplier.

Fig. 3. (a) Representation of the shifted axis. (b) Hardware implementation of the shift multiplier.

where $\mathbf{e} = [1\,1\,...\,1]^T$ is a $2N_t \times 1$ all one vector. Let $\mathbf{x} = \mathbf{s} - \mathbf{e}$ be the new shifted signal vector in set Ω'. Then (4) can be rewritten as:

$$\hat{\mathbf{Y}}' = \hat{\mathbf{Y}} - \mathbf{Re} = \mathbf{Rx} + \mathbf{w}, \tag{5}$$

which is identical to (2) except the new shifted signal $\mathbf{x}$ and the new received vector $\hat{\mathbf{Y}}'$.

3.2   First Child Unit (FCU)

In Eq. (3), the expected FC of a K_{i+1} parent node will has the minimum distance increment $|e_i(\mathbf{x}^{(i)})|^2$, i.e.,

$$x_{i,\mathrm{FC}} = \underset{x_i \in \Omega'}{\arg\min} |e_i(\mathbf{x}^{(i)})|^2 = \underset{x_i \in \Omega'}{\arg\min} |\,\hat{y}_i' - \underbrace{\sum_{j=i+1}^{2N_t} R_{ij}x_j}_{\text{center } C_i(\mathbf{x}^{(i)})} - R_{ii}x_i|^2, \tag{6}$$

where $C_i(\mathbf{x}^{(i)})$ is referred to as *center*. Since the new signal set Ω' is defined and can be represented with simple binary values. Then the *center* is calculated by simply bit-shifting and adding the input value R_{ij} to perform R_{ij} multiplied by $x_j \in \Omega'$. In upper left corner of Fig. 4(a), the *center* is calculated by shift multipliers labelled "SM". Then the FC can be found by quantizing $C_i(\mathbf{x}^{(i)})/R_{ii}$ to the nearest constellation signal in Ω':

$$x_{i,\mathrm{FC}} = Q\left[\frac{C_i(\mathbf{x}^{(i)})}{R_{ii}}\right], \tag{7}$$

where $Q[\cdot]$ denotes the quantizing operator, which rounds the operand to the nearest integer value in Ω'. In order to avoid the divider operation in (7), our design directly compares the center value $C_i(\mathbf{x}^{(i)})$ with $\pm R_{ii}$, $\pm 3R_{ii}$, $\pm 5R_{ii}$, and $-7R_{ii}$ as shown in the right half of Fig. 4(a).

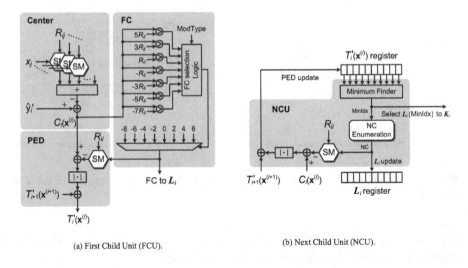

(a) First Child Unit (FCU).   (b) Next Child Unit (NCU).

Fig. 4. (a)Circuit diagram of the first child unit (FCU). (b)Circuit diagram of the Next child unit (NCU).

With these comparison results, the FC can be selected and stored in the contender list L_i register and waiting to be chosen into the K-best candidates. According to (3), letting $T_i(\mathbf{x}^{(i)}) = |T_i'(\mathbf{x}^{(i)})|^2$, it can be seen that the PED is defined as a squared ℓ^2-norm, i.e.,

$$|T_i'(\mathbf{x}^{(i)})|^2 = |T_{i+1}'(\mathbf{x}^{(i+1)})|^2 + |e_i(\mathbf{x}^{(i)})|^2. \tag{8}$$

This PED calculation of $|e_i(\mathbf{x}^{(i)})|^2$ requires a large bit-size multiplication, which cost plenty of computing complexity. A simplified ℓ^1-norm approximation is presented in [3], [7] to reduce the circuit complexity and critical path delay. The ℓ^1-norm approximation is given by

$$T_i'(\mathbf{x}^{(i)}) \approx |T_{i+1}'(\mathbf{x}^{(i+1)})| + |e_i(\mathbf{x}^{(i)})|. \tag{9}$$

Comparing with the squared ℓ^2-norm, the BER performance loss of ℓ^1-norm applied in K-best algorithm is negligible [3]. Therefore, our design adopts the ℓ^1-norm approximation to save computing complexity and power consumption. In bottom left corner of Fig. 4(a), the determined FC is multiplied by R_{ii} using an SM and subtracted by *center* $C_i(\mathbf{x}^{(i)})$ to get $e_i(\mathbf{x}^{(i)})$. Then, the PED is accumulated according to the ℓ^1-norm approximation. In the meanwhile, the accumulated PED is stored into the PED register for later processing by the NCU circuit.

3.3 Next Child Unit (NCU)

In the previous subsection, the FCU finds an FC and compute the corresponding PED in one clock cycle. After K clock cycles, all the K FCs and PEDs are ready to proceed. Then the NCU takes over the data and finds the minimum PED index (MinIdx) by the minimum finder circuit shown in Fig. 4(b). According to this index, the first K-best candidate denoted as $\boldsymbol{L}_i\{\text{MinIdx}\}$ is determined and selected into the $\boldsymbol{K}_i$ register for passing to the next layer. Once an FC is stored into the K-best list, the NC enumeration block enumerates the best sibling node of $\boldsymbol{L}_i\{\text{MinIdx}\}$ as the NC to replace the register position $\boldsymbol{L}_i\{\text{MinIdx}\}$. We adopt an 1D SE-enumerated strategy to flexibly support various modulation schemes.

Eq. (6) shows that the center $C_i(\mathbf{x}^{(i)})$ needed in NCU is independent of x_i. Hence, the center values that have been already computed in the FCU can be reused to calculate the updated PEDs of the enumerated NCs as shown in Fig. 4(b). It should be noted that the best K candidates are decided one by one and from small to large PED in this DKB architecture. Therefore, the sorting circuit for the conventional K-best algorithm is absent in our design.

4   Hardware Comparison

The proposed MIMO detector is implemented in 90-nm CMOS technology. The core area is 1.014mm^2 and the power consumption is 17.76mW when operates at 74MHz clock frequency and 1V power supply voltage. The throughput of this design is given by $\Phi = (N_t \log_2 M)f_{\text{clk}}/K$, where f_{clk} is the clock frequency. Since our design supports two K-value configurations,i.e., $K = 5, 10$, it can provide the maximum throughput up to 710Mbps. Table 1 lists the performance comparison between this work and other MIMO detectors. To eliminate the process factor, the normalized power is formulated as [10]:

$$P_{\text{norm}} = \text{power} \times \left(\frac{1.0}{V_{\text{dd}}}\right) \times \left(\frac{0.09}{\text{Tech}}\right) \times \left(\frac{1}{\text{Throughput}}\right). \qquad (10)$$

5   Conclusion

In this paper, we have presented the flexible antenna configurable MIMO detector. From the algorithmic aspect, the DKB scheme provides efficient decoding procedure. In the hardware point of view, shift-multiplier with new signal set Ω' and ℓ^1-norm approximation greatly reduce the overall computing complexity. Furthermore, various modulation schemes from QPSK to 64-QAM are also supported in our design. From Table 1, our design consumes less power than other literatures, thanks to DKB scheme and low complexity hardware design.

Acknowledgments. The authors would like to thank Chip Implementation Center of National Applied Research Laboratories (CIC/NARL) for EDA tools service. This work was supported by Ministry of Science and Technology, Taiwan, under Grant MOST 104-2220-E-008-004.

Table 1. Hardware comparison of the proposed MIMO detectors

Reference design	JSAC' 06 [2] IC1	JSAC' 06 [2] IC2	TVLSI' 07 [5]	JSSC' 05 [8]	ISSCC' 09 [9]	This work	
Antenna	4×4	4×4	4×4	4×4	4×4	8×8, 4×4, 2×2	
Modulation	16-QAM	16-QAM	64-QAM	16-QAM	64-QAM	QPSK 16-QAM 64-QAM	
Algorithm	K-best	K-best	K-best	Depth-first	DKB	DKB	
Model (Real/Complex)	Real	Real	Complex	Complex	Real	Real	
Output type	Hard	Soft	Soft	Hard	Hard	Hard	
Technology	0.35 μm	0.13 μm	0.13 μm	0.25 μm	0.13 μm	0.09 μm	
Gate count	91K	97K	280K	117K	114K	362K	
Area (mm^2)	5.76	0.56	2.38	N/A	2.31	1.014	
Max. clock rate	100 MHz	200 MHz	270 MHz	51 MHz	282 MHz	74 MHz	
Max. throughput (Mbps)	53.5	107	8.57	73.5	675	$K=5$ 710	$K=10$ 355
Power (mW)	626	N/A	94	360	135	17.76	17.76
Normalized power (mW/Mbit)	0.384	N/A	5.273	0.282	0.082	0.025	0.049

References

1. S. M. Alamouti, "A simple transmit diversity technique for wireless communications," *IEEE J. Sel. Areas Commun.*, vol. 16, no. 8, pp. 14511458, Oct. 1998.
2. Z. Guo and P. Nilsson, "Algorithm and implementation of the K-best sphere decoding for MIMO detection," *IEEE J. Sel. Areas Commun.*, vol.24, no.3, pp.491-503, 2006.
3. M. Wenk, M. Zellweger, A. Burg, N. Felber and W. Fichtner, "K-best MIMO detection VLSI architectures achieving up to 424 Mbps," *Proc. of ISCAS*, pp.1151-154, 2006.
4. M. Shabany and P.G. Gulak, "Scalable VLSI architecture for K-best lattice decoders," *Proc. of ISCAS*, pp.940-943, 2008.
5. S.Chen, T. Zhang and Y. Xin, "Relaxed K-best MIMO signal detector design and VLSI implementation," *IEEE Trans. on Very Large Scale Integr. (VLSI) system*, vol.15, no.3, pp.328-337, Mar. 2007.
6. M. Shabany, K. Su and P.G. Gulak, "A pipelined scalable high-throughput implementation of a near-ML K-best complex lattice decoders," *Proc. of ICASSP*, pp.3173-3176, 2008.
7. C. Studer, A. Burg, and H. Bolcskei, "Soft-output sphere decoding: Algorithms and VLSI implementation," *IEEE J. Sel. Areas Commun.*, vol. 26, no. 2, pp. 290–300, 2008.
8. A. Burg, M. Borgmann, M. Wenk, M. Zellweger, W. Fichtner and H. Bolcskei, "VLSI implementation of MIMO detection using the sphere decoding algorithm," *IEEE J. Solid-State Circuits*, vol.40, pp.1566-1577, 2005.
9. M. Shabany and P.G. Gulak, "A 0.13um CMOS 655Mb/s 4×4 64-QAM K-best MIMO detector," *Proc. of IEEE Int. Solid-State Circuits Conf.(ISSCC)*, pp.256-257, 2009.
10. C.-H. Liao, T.-P. Wang and T.-D. Chiueh, "A 74.8mW soft-output detector IC for 8×8 spatial-multiplexing MIMO communications," *IEEE J. Solid-State Circuits*, vol.45, no.2, pp.411-421, 2010.

Design of Dipole Slot Antenna for 700/900/1800 MHz Applications

Yuh-Yih Lu[1*], Chih-Wei Fan[1], and Hsiang-Cheh Huang[2]

[1]Department of Electrical Engineering, Minghsin University of Science and Technology,
Hsinchu 304, Taiwan, R.O.C.
yylu@must.edu.tw
[2]Department of Electrical Engineering, National University of Kaohsiung, Kaohsiung 811,
Taiwan, R.O.C.
hch.nuk@gmail.com

Abstract. A coplanar symmetric dipole slot antenna is proposed for dual band wireless communication applications. The rectangular slot with T tuning stub is etched on the metallic layer of a single sided printed circuit board to form the coplanar dipole slot antenna. The dimensions of T stub are changed to design and fabricate the antenna which can be operated at 700/900/1800 MHz successfully. We use IE3D software to design this dipole slot antenna and choose the better parameters to manufacture the proposed antenna. The influences of dimension parameters of the proposed antenna on impedance bandwidth are described. The proposed antenna with the volume of 154mm×41mm×1.6mm has been fabricated and this antenna can be used in LTE 700/900/1800 MHz frequency bands.

Keywords: T tuning stub, dipole slot antenna, LTE

1 Introduction

Compact size, lower cost and easy fabrication are important factors to design antenna that can be used in wireless commercial products. Coplanar antennas possess these attractive features. Hence, many studies about planar antennas had been proposed and widely used in Ultra Wide Band (UWB), Radio Frequency Identification (RFID), and Wireless Local Area Network (WLAN) systems [1-9]. Bandwidth enhancement in antenna design can fit the need for increasing the information transfer. This requirement can be achieved using the slot antenna techniques with the tuning stub [10]. Today, wireless communication devices relating to the field of the internet of things are developed rapidly. These information devices should be capable of large bandwidth and operating at multiple frequency bands. Therefore, researches have been reported for multi-band planar antenna [11-13].

© Springer International Publishing AG 2017
J.-S. Pan et al. (eds.), *Advances in Intelligent Information Hiding and Multimedia Signal Processing*, Smart Innovation, Systems and Technologies 64,
DOI 10.1007/978-3-319-50212-0_29

In this study, a simple uniplane dipole slot antenna with T tuning stub is proposed. The return loss and impedance bandwidth are obtained from IE3D simulations. The T tuning stub of the proposed slot antenna controls the operating frequency bands of dipole slot antenna. The coplanar dipole slot antenna designed with the T tuning stub excites the resonant frequency that can be used for LTE 700/900/1800 MHz applications. The suitable geometric parameters of the T tuning stub are chosen to fabricate the proposed antenna. Therefore, a simple coplanar symmetric dipole slot antenna with the size of 154mm×41mm×1.6mm is presented in this paper. The proposed antenna can be built on a single sided printed circuit board. The single metal layer structure is suitable for mass production and reduces the manufacturing cost.

2   Antenna Design

The proposed coplanar dipole slot antenna with T tuning stub structure is printed on a single metallic layer of FR4 dielectric substrate which has permittivity of 4.4 and thickness of 1.6mm. The configuration of this proposed antenna is depicted in Fig.1. In this figure, the symmetric rectangular slot with T tuning stub is etched on the metallic layer to create the operating frequency bands. Points A and B are the feeding points of the coplanar dipole slot antenna. We adjust the T stub length parameters L3 and L6 to observe the variations with respect to the impedance bandwidth of the proposed antennas. The dimension parameters of the proposed antenna shown in Fig.1 are listed below: W1=76.5mm, W2=39mm, W3=73.5mm, W4=3mm, W5=2mm, G=1mm, L1=63mm, L2=30mm, L4=18.5mm, L5=4mm, L7=8mm, L8=10.5mm, D1=5mm, D2=11mm and D3=4mm. The 50 ohm coaxial connector was adopted for testing.

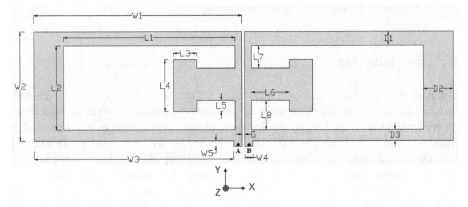

Fig. 1. Geometry of the proposed coplanar dipole slot antenna.

3   The Simulations

We adopted various dimension parameters L3 and L6 shown in Fig.1 of the coplanar symmetric dipole slot antenna to observe the characteristics of the proposed antenna. The numerical simulation and analysis for the proposed antennas are performed using IE3D simulation software. The simulated curves of return loss against frequency for varying the T stub parameter L3 of the proposed antenna with L6=14mm are shown in Fig.2. From this figure, two obvious operating frequency bands are observed and the lower resonant frequency is nearly unchanged but the higher resonant frequency is slightly shifted to higher frequency with increasing the value of L3. At the lower frequency band, the impedance bandwidth is nearly unchanged. But it is influenced at the higher frequency band. The T stub with L3=8.5mm exhibits the better characteristics of the proposed antenna. The simulated curves of return loss against frequency for varying another T stub parameter L6 of the proposed antenna with L3=8.5mm is shown in Fig.3.

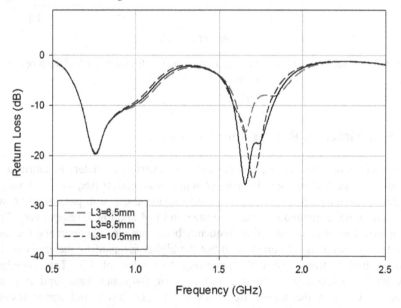

Fig. 2. Simulated curves of return loss against frequency for varying L3 of the proposed antenna with L6=14mm.

From Fig. 3, two obvious operating frequency bands is also observed. The lower and upper resonant frequencies are changed with varying the value of L6 shown in figure 3. At lower frequency band, the impedance bandwidth decreases with increasing the value of L6. While the impedance bandwidth increases and then decreases with increasing the value of L6 at higher frequency band. Therefore, carefully adjusting the T stub parameter L6 is very important to design the antenna that can be used in LTE700, LTE900 and LTE1800 bands.

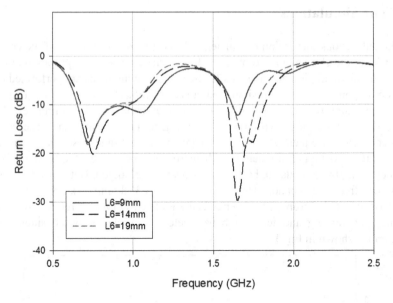

Fig. 3. Simulated curves of return loss against frequency for varying L6 of the proposed antenna with L3=8.5mm.

4   Experimental Results and Discussion

From the simulation results, we use the same geometric parameters to fabricate the proposed antenna. The measured curves of return loss against frequency for varying the T stub parameter L3 with L6=14mm and varying the T stub parameter L6 with L3=8.5mm of the fabricated antenna are shown in Fig.4 and Fig.5, respectively. From these figures, two obvious operating frequency bands are also obtained and the lower resonant frequency is nearly unchanged but the higher resonant frequency is slightly shifted to higher frequency with increasing the value of L3. The impedance bandwidth is also nearly unchanged at the lower frequency band and it is also influenced by L3 at the higher frequency band. The lower and upper resonant frequencies are slightly changed with varying the value of L6. At lower frequency band, the impedance bandwidth decreases with increasing the value of L6. While the impedance bandwidth increases and then decreases with increasing the value of L6 at higher frequency band. The trends of the measured results relate well with the simulated results. To reach the operating frequencies covering LTE700, LTE900 and LTE1800 bands, the better performance antenna with L3=8.5mm and L6=14mm is used to study the characteristics of the proposed antenna. The photography of this fabricated antenna is shown in Fig.6. The curves of return loss against frequency of the simulated and fabricated antenna are illustrated in Fig.7. The simulated and measured results of the proposed antenna are listed in Table 1.

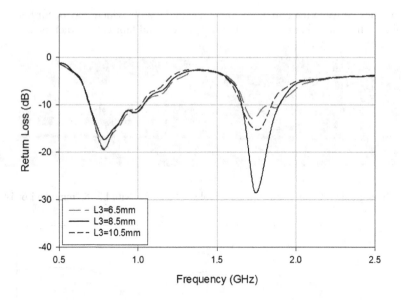

Fig. 4. Measured return loss of the fabricated antenna with L6=14mm.

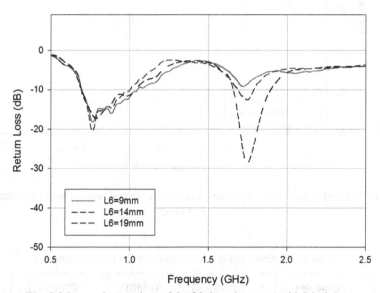

Fig. 5. Measured return loss of the fabricated antenna with L3=8.5mm.

The measured impedance bandwidths of the fabricated antenna for return loss less than -10dB at lower and upper frequency band are 343MHz and 247MHz, respectively. The measured return loss (RL) and impedance bandwidth (BW) of the fabricated antenna show better performance than that in simulation condition. There are discrepancies between the computed and measured results which may occur because of the effect of the coaxial connector soldering process and fabrication tolerance.

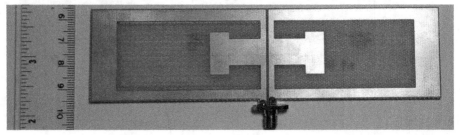

Fig. 6. Photography of fabricated coplanar dipole slot antenna with L3=8.5mm and L6=14mm.

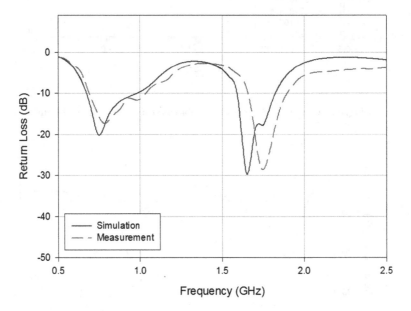

Fig. 7. Simulated and measured return loss of the proposed antenna with L3=8.5mm and L6=14mm.

Table 1. Simulated and measured results of the proposed antenna.

Condition	f (MHz)	RL (dB)	BW (MHz)
	700	-14.91	305
Simulation	900	-11.57	
	1800	-18.48	246
	700	-11.88	343
Measurement	900	-13.01	
	1800	-25.31	247

The measured radiation patterns of the fabricated antenna with L3=8.5mm and L6=14mm at 700/900/1800 MHz are shown in Fig.8. The measured peak gains for testing frequencies at x-z and y-z plane of the fabricated antenna are listed in Table 2. From Fig.8, it can be observed that the radiation patterns are almost omnidirectional in the y-z plane. The omnidirectional antenna radiation pattern indicates that the fabricated antenna is good for mobile devices.

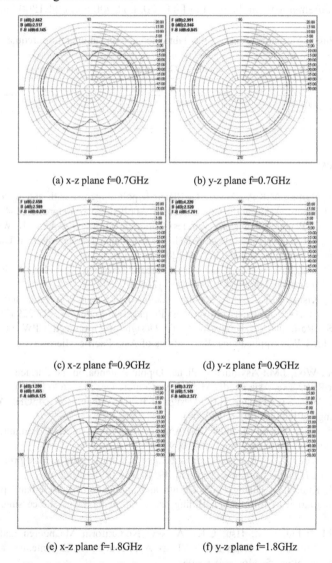

(a) x-z plane f=0.7GHz   (b) y-z plane f=0.7GHz

(c) x-z plane f=0.9GHz   (d) y-z plane f=0.9GHz

(e) x-z plane f=1.8GHz   (f) y-z plane f=1.8GHz

Fig. 8. Measured radiation patterns of the fabricated antenna.

5   Conclusions

In this study, the fabricated coplanar dipole slot antenna with tuning T stub exhibits simple structure and dual band characteristics. The size of T stub is designed and the frequency bands covering 692-1035MHz and 1653-1900MHz are obtained. Therefore, carefully choose the designed parameter of the T stub would implement the suitable antenna that can be used in LTE700, LTE900 and LTE1800 bands.

Table 2. Measured results of the fabricated antenna at operating frequency.

f (MHz)	x-z plane Peak Gain (dBi)	y-z plane Peak Gain (dBi)
700	2.66	2.99
900	2.66	4.22
1800	1.59	3.73

References

1. Ayop, O., Rahim, M.K.A., Masri, T.: Planar Dipole Antenna with and without Circular Parasitic Element. Asia-Pacific Conference on Applied Electromagnetics, 1-4 (2007).
2. Wong, K. L.: Planar Antennas for Wireless Communications. Hoboken, NJ: Wiley (2003).
3. Dubrovka, F.F., Vasylenko, D.O.: A Bell-Shaped Planar Dipole Antenna. Ultrawideband and Ultrashort Impulse Signals, 82 – 84 (2006).
4. Zhang,J.P., Xu, Y.S., Wang, W.D.: Ultra-wideband Microstrip-fed Planar Elliptical Dipole Antenna. Electronics Letters, 42 , 144 – 145 (2006).
5. Chair, R., Kishk, A.A., Lee, K.F.: Ultrawide-band Coplanar Waveguide-fed Rectangular Slot Antenna. IEEE Antennas and Wireless Propagation Letters, 3, 227-229 (2004).
6. Gupta, S., Ramesh, M., Kalghatgi, A.T.: Design of Optimized CPW Fed Monopole Antenna for UWB Applications. Proceedings of Asia-Pacific Microwave Conference, 12, (2005).
7. Lee, H.R., Woo, J.M.: Asymmetric Planar Dipole Antenna on the Surface of Conducting Plane for RFID Tag. Asia Pacific Microwave Conference, 633 – 636 (2009).
8. .Lu,Y.Y., Wei, S.C., Huang, H.C.: Design of RFID Antenna for 2.45GHz Applications. Proceedings of ICICIC, 601-604 (2009).
9. Khaleghi, A.: Dualband Meander Line Antenna for Wireless LAN Communication. IEEE Trans. Antennas Propagat., 55, 1004-1009 (2007).
10. Sze, J.Y., Wong, K.L.: Bandwidth enhancement of a microstrip-line-fed printed wide-slot antenna. IEEE Trans. Antennas Propagat., 49, 1020-1024 (2001).
11. Lu, Y.Y, Kuo, J.Y., Huang H.C.: Design and application of triple-band planar dipole antennas. Journal of Information Hiding and Multimedia Signal Processing, 6, 792-805 (2015).
12. Wu, C.M., Chiu, C.N., Hsu, C.K.: A New Nonuniform Meandered and Fork-Type Grounded Antenna for Triple-Band WLAN Applications. IEEE Antennas and Wireless Propagation Letters, 5, 346 – 348 (2006).
13. Xu, P., Yan, Z.H., Wang, C.: Multi-band Modified Fork-shaped Monopole Antenna with Dual L-shaped Parasitic Plane. Electronics Letters, 47, 364 – 365 (2011).

Modular Design of a General Purpose Controller for Hybrid Electric Power Conversion System

Yu-Cian Syu[1], Jian-Feng Tsai[1], Lieh-Kuang Chiang[2], Wen-Ching Ko[2]

[1] Department of Electrical Engineering National Formosa University, Huwei, Yunlin, Taiwan
{40025222,jeff.tsai}@nfu.edu.tw
[2] Industrial Technology Research Institute, Tainan, Taiwan
{kuang,wcko}@itri.org.tw

Abstract. A modular design procedure of the controller for hybrid electric power conversion system (HEPCS) with CAN bus protocol is proposed in this paper. By CAN bus protocol, users can easily set individual modules and change parameters with pre-defined data structures. Three basic modules are implemented in a HEPCS to demonstrate this concept. The experimental result shows the feasibility and advantage of this design procedure.

Keywords: Modular design, hybrid electric power conversion system (HEPCS), CAN bus.

1 Introduction

With growing impact of the global climate change and threat of poor air quality, electric motors are gradually adopted in vehicles to fully or partially replace the internal combustion engines. Various types of electrified vehicles, such as EV (Electric Vehicle), PEV (Plug-in Electric Vehicle), and FCEV (Fuel Cell Electric Vehicle), are commercially viable to achieve environmental sustainability. For EV, one of the most concerning problems is the cruising ability. Normally, it takes several hours to completely charge an EV and makes EV not be competitive with conventional ICE (Internal Combustion Engine) vehicles. FCEV provides an alternative solution that can significantly shorten refueling time to several minutes.

Different to ICE, most FCs would not immediately work well after the fuel being feeding into, such that the procedure for starting up FCs is unavoidable. For example, when using HT-PEM (High Temperature Proton Exchange Membrane) FCs, it needs to heat up FC stacks to be above operation temperature. Even for LT-PEM PCs, starting in a frozen environment is a critical issue, and thus a pre-heating mechanism is still needed. An auxiliary electric energy storage unit is able to deal with this issue, and it is used as the main energy source during this interval [1]. Second, the dynamic response of FCs to the instantaneous power change is typically slower than ICE. For these reasons, an electric battery is usually adopted, and it means that at least two energy sources are in the FCEV. Based on this topology, a simple example of hybrid

© Springer International Publishing AG 2017
J.-S. Pan et al. (eds.), *Advances in Intelligent Information Hiding and Multimedia Signal Processing*, Smart Innovation, Systems and Technologies 64,
DOI 10.1007/978-3-319-50212-0_30

electric power conversion system (HEPCS) is given, and an energy management system (EMS) which dominates the energy conversion strategies is therefore important.

Practically, specifications of all energy sources in a HEPCS are well defined according to the application. Afterwards, the EMS coordinates them to satisfy specific load requirements [2, 3]. This development procedure is dedicated to one particular application. Thus, for other applications, even with only slight deviations, the design parameters would not be proper and the same procedure should be repeated again [4]. In order to accelerate and simplify the development, a modular design of a general purpose controller for HEPCS is proposed in this paper. With CAN bus protocol [5], data structures for modules used in HEPCS are defined to generalize physical measurement, design parameters, and system information.

The remainder of this paper is organized as follows. The essentials of modular design are explained in Section II. Next, an example with three modules is demonstrated in Section III. A practical design with some experimental results is then presented in Section IV to verify this concept. Finally, conclusions are given in Section V.

2   Essentials of Modular Design

A HECPS consists of energy sources, power converters, energy storage units, power regulators, etc.. Each of them can be generally defined as a module with a general purpose controller of individual function. Physical measurements by sensors are transferred into modules via analog signal, such as voltage and current, and information among these modules is interchanged by either analog signal or specific communication protocol. While the number of modules increases, the communication protocol is obviously preferred, and CAN bus protocol is therefore suggested [5].

2.1   Definition of Module and General Purpose Controller

The fundamental architecture for all modules in HEPCS is presented in Fig. 1. All types of signals into the hardware are transmitted to others through a general purpose controller. Modules in HEPCS communicate with each other via CAN bus protocol. These signals are further defined in an n-array data structure, which includes module name, parameter, function, instruction, message, and so on, as shown in Fig. 2.

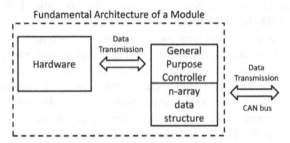

Fig. 1. Fundamental architecture for all modules in HEPCS

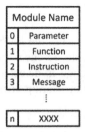

Fig. 2. N-array data structure for modules

2.2 Module Integration

Based on the aforementioned concept, each module is defined as a CAN node and connected to CAN network. Three modules: charger module, power relay module, and EMS module, generally included in a HECPS are given here as a demonstration. Besides, 4-array data structure is used for simplification. Thus, the result of the whole system is shown as Fig. 3. For other modules being inserted into HECPS, they should follow the same hardware architecture and 4-array data structure.

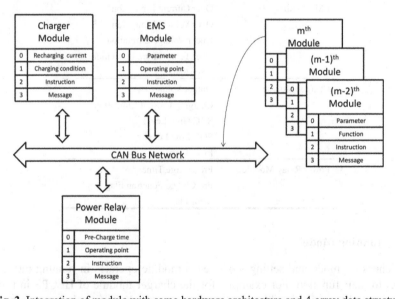

Fig. 3. Integration of module with same hardware architecture and 4-array data structure

3   Operation Modes of HEPCS

According to the definition in section II, the essentials for modular design are clearly explained. The next is to define the operation modes for HEPCS. Normally, it includes checking mode, setting mode, running mode, and error mode, determined by EMS module

3.1   Checking Mode

Initially, when the HEPCS is starting up, the EMS module would listen to the CAN network. All message received would be check according to the pre-load data. If a non-standard or unauthorized message is detected, startup error is recognized, and then it would enter error mode with the startup-error flag being set.

3.2   Setting Mode

After HEPCS passing the initial checking mode, the EMS module sends information-requiring message to CAN network. Module-characteristic messages from other modules are responded. The EMS module would process these data to decide the parameters for setting itself or other modules, as shown in Table 1. Then, it enters setting mode according to the flow chart in Fig. 4 (a). In this mode, indicating LEDs in all modules keep blinking until every module is ready.

Table 1. Parameter list for setting mode

Module Name	Parameter Array
EMS Module	Over Current Limitation
	Over Voltage Limitation
	Under Voltage Limitation
	Over Temperature Limitation
	Error Flag
Charger Module	Output Voltage
	Charging Current Limitation
	SOC High Level
	SOC Low Level
	Error Flag
Power Relay Module	Pre-Charge Time
	Pre-Charge Function Flag
	Error Flag

3.3   Running Mode

After checking mode and setting mode, each module operates in running mode with respect to their function. For example, for the charger module of HECPS in running mode, it would continue polling SOC and charge the battery or being standby. According to parameters of "SOC High Level" and "SOC Low Level", it is explained in Fig. 4 (a).

3.4   Error Mode

This mode is triggered when any abnormal condition is observed in one module and "Error Flag" is set. The EMS module will continuously inquire the error status and suspend whole system. Until "Error Flag" is removed, it returns to checking mode and then operates in running mode.

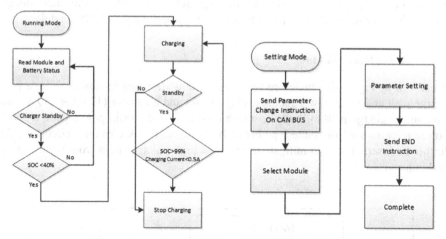

Fig. 4. Flow chart. (a) Setting mode. (b) Running mode

4   Hardware Implementation and Experimental Results

A practical design of HEPCS is implemented as Fig. 5. There are two energy sources included: a 48V fuel cell system provided by ITRI (Industrial Technology Research Institute of Taiwan) and a lithium-iron-phosphate (LFP) battery (52.6V, 87.5Ah) [6]. The others are EMS module, charger module, and power relay module. Obviously, all modules and energy sources in HEPCS are connected to CAN bus network and have their own 4-array data structure as described previously. The hardware implementation is detailed as follows.

4.1   EMS Module

The EMS module is designed with a dsPIC33EV family MCU [7] to provide CAN function and handle power management. As shown in Fig. 6, at the front panel, there exist a LCD (for sequentially displaying current, voltage, output power and other electrical measurements), LED indicator (for indicating immediate warring or failure status), a switcher (for changing the operation to manual control) and a key switch for further integration on the golf cart.

4.2   Charger Module

The charger module consists of a DC/DC converter and dsPIC33EV family MCU, as shown in Fig. 7 (a). It is designed to operate according to EMS module or stand-alone. With CAN bus protocol, it would receive the information from two energy sources or EMS Module. In the standalone condition, it would automatically charge the LPF battery. Depending on the SOC, voltage and current, there are two charging modes for this module: constant voltage mode and constant current mode. At low SOC, it operates in constant current mode. Instead, for high SOC, it operates in constant voltage mode. The threshold can be assigned by LPF battery or mainly by EMS module.

4.3   Power Relay Module

As shown in Fig. 7 (b), there are two power paths designed in this module, a pre-charge path and main power path. A relay (30 A) and a resistor (120Ohm, 80W) are used in pre-charge path to limit the inrush current. For the main power path, a 500A power relay is used to connect power in both directions. A clamping circuit would further be inserted here to mitigate unexpected high voltage stress from regenerating braking.

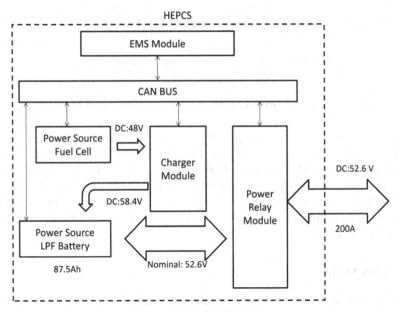

Fig. 5. Example of HEPCS

Fig. 6. The hardware implementation of EMS module

(a)   (b)

Fig. 7. The hardware implementation. (a) Charger module. (b) Power relay module

4.4   Experimental Results

The experiments for HCPS are temporarily conducted on test bench. All modules are connected to the same CAN network and communicate with each other in the same data array. It shows that the HCPS follows the startup procedure from checking mode, setting mode and then into running mode. Below a certain small charging current 0.5A defined by EMS module, the charger module would stop for power saving. As shown in Fig. 8, once the charging current decreases below 0.5A, the function indictor is pulled low to disable the charging function and the output voltage is cut-off. Besides, the output power relay is assumed to being connected to a 300uF capacitive load. With the resistance (120Ohm) of pre-charge resistor passed to the EMS module, a rising time about 0.2s could be calculated to determine the maximum power good time. The voltage rising behavior is depicted in Fig. 9.

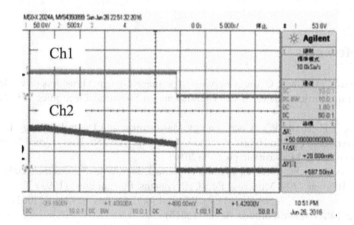

Fig. 8. Experimental results, CH1: Charge Voltage (50V/div), CH2: Charge Current (0.5A/div)

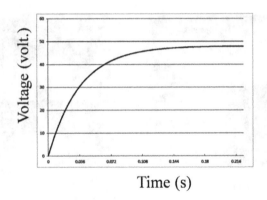

Time (s)

Fig. 9. Pre-charge time and output voltage

5 Conclusion

In this paper, modular design procedure of a general purpose controller for HEPCS is presented. A prototype of HEPCS with two energy sources and three basic modules is constructed to demonstrate this concept. Experimental results verify the feasibility and effectiveness. It is worth mentioning that this HEPCS will be further integrated into a golf cart. More on-road data would be obtained for advanced study.

Acknowledgment

This research is sponsored by Industrial Technology Research Institute of Taiwan, under the contract 105-AZ-13. The author would like to express their appreciation to Dr. Huan-Chan Ting (Institute of Nuclear Energy Research) for his valuable suggestions.

References

1. K. T. Chau, C. C. Chan, C. Liu: Overview of permanent-magnet brushless drives for electric and hybrid electric vehicles. IEEE Transactions on Industrial Electronics 55, 2246–2257 (2008)
2. C.H. Wu, Y.H. Hung, P.Y. Chen: Energy Management and Control Strategy Design of a Generation Set in Plug-in Hybrid Electric Vehicles. In: The 25th International Electric Vehicle Symposium and Exposition, pp. 5–9 (2010)
3. J. Larminie, A. Dicks: Fuel Cell Systems Explained. John Wiley& Sons Inc. (2000)
4. L.J. M.J. Blomen, M.N. Mugerwa: Fuel Cell Systems. Plenum Press (1993)
5. H. Boterenbrood: CANopen High-Level Protocol for CAN-bus. NIKHEF (2003)
6. Datasheet LPF Battery. Formosa Battery (2015)
7. User's Guide dsPIC33EV 5V CAN-LIN Starter Kit. Microchip (2014)

A Protective Device for a Motor Vehicle Battery

Rong-Ching Wu [1], Yen-Ming Tseng [2,1], En-Chih Chang[3], and Jia-Yi
Chen [4]

[1],3,4 Department of Electrical Engineering, I-Shou University,

Kaohsiung, Taiwan, R.O.C.

rcwu@isu.edu.tw

[2] Collegue of Information Science and Engineering,

Fujian University of Technology, China

swk1200@gmail.com

ABSTRACT. Electronic products have been developed motor vehicle
using. When these external devices in a car, they will consume power
continuously even after the car engine is turned off. In order to avoid
and prevent these problems, this paper has developed a device to protect
the car battery; This device provides three functions:
(1) When the engine is started, the power will immediately be provided
to the external devices.
(2) Warning for the battery power status.
(3) The stand-by current is very low, and it will not consume much
power.
This device has been applied in the actual vehicle. The device could
effectively resolve external device power consumption power
continuously after the vehicle engine was turned off, and it could
remind the driver about battery life. When the capacity of the battery
deteriorated to obviously less than the set voltage value of the device, it
will alert the driver. This device owns advantages in easy set up and
simple operation, and it can effectively prevent excessive power
consumption.

Keywords: vehicle, external device, battery

1. Introduction

Currently, many electronic products can be set up in the car to use. Whether
electronic device is connected to the battery directly, or connected with the cigarette
lighter socket to supply power. Those electronic products power is from the car
battery.

The car brands in Taiwan included both Japanese and European. Japanese cars are
designed in order to prevent power being used with unnecessary external devices [1].

© Springer International Publishing AG 2017
J.-S. Pan et al. (eds.), *Advances in Intelligent Information Hiding
and Multimedia Signal Processing*, Smart Innovation, Systems and Technologies 64,
DOI 10.1007/978-3-319-50212-0_31

When the engine is turned off, the power supply system turns off, so there isn't need to protect the battery as the external device will not waste power. Because of the design of European cars, however, consider the driver being still inside the car. When the car engine is turned off, it will not turn off the power supply automatically. In this case, if the driver does not remove or disable these external devices from running, it will consume power continuously, and it is therefore possible to run out of battery power so that the engine does not start normally; if this problem occurs and is encountered in an emergency situation [2], this would be a troubling situation for the driver.

Although the car battery can provide stable power for electronic products, the basic requirement is to start up the car successfully. Currently, some vehicles in the market cannot detect the power state of the engine, and this will lead to vehicles running out of battery power easily. This issue will directly or indirectly affect the battery power status and the subsequent launch of the starting program of vehicles. There is no suitable power supply provided for the normal car, so this study endeavors to develop a suitable power supply apparatus for vehicular use that protects the battery from running out of energy by use of a stand-by device. This work is to develop an energy-saving concept, for early detection of vehicle power problems, in order to achieve life relevance.

It is a very high degree that vehicle driving power is demand for the motor car. The external cigarette lighter socket extension cord for external devices has more recently become a general utility [3]. Then, this device's effect can exert the effect by the cigarette lighter socket. It can remind problems of the power [4].

This work solves the problem when the ignition is turned off but the power continually provides external devices. This equipment able to automatically cut the power supply after ext ignition; after the engine is started normally, than the power provides to external devices. This work not only can it effectively solve this situation by providing many additional functions, but it is also capable of being a machine having a variety of uses of the device.

2. Circuit Design

The complete circuit is displayed in Figure 1. When the input voltage of car battery exceeds 14V, it may make a surge voltage, and may damage the equipment circuit [5]. Therefore, there is a varistor to be a protective device [6]. If the voltage exceeds the rated voltage, the resistance of varistor will rapidly decline to the short-circuit state. The surge voltage will be imported to the rheostat to be dissipated by heat. The surge voltage will be absorb and the input voltage will be stable [7]. The series diode is sat for prevent reverse-connected [8]. The voltage detection chip is always used above 6V. If the voltage is smaller than 6V, it will be become an open circuit. There is a series of a Schottky diode sat behind the varistor. This diode makes the input voltage of car battery from 14V step-down to 6V, in order to achieve the detection range of the voltage detective chip. If the input voltage of car battery is smaller than 14V, this voltage detection device chip will make this circuit open-circuit. On the other hand, if the input voltage is higher than 14V, this voltage detection device chip will make this device short-circuit.

The voltage detection chip is also used to decide whether the car battery voltage when fully charged is more insufficient than the original full-charge voltage in the

ACC. If the full voltage of the battery is less than 11V, then the buzzer will make a noise for about 2-3 seconds to alert the driver that the car battery storage is obviously insufficient, so it is time to check whether the battery should be replaced [9]. In this case, the car can also still be started, but the voltage detection chip can help avoid the situation where a car battery is completely without power leading to the car being unable to be started. This device is used to supply power to the external device. The voltage detective chip is a series 2N7002 transistor. It reduces the voltage back to 12V; therefore, this apparatus can smoothly supply power to external devices. If the user wants to manually stop any external device, there is a convenient switch on this chip detective device, and it can stop the car battery supplying power to external devices. The driver does not need to stop them one by one. For preventing the reverse surge voltage when the switch is turned off, A series diode and a capacitor are sat to absorb this voltage.

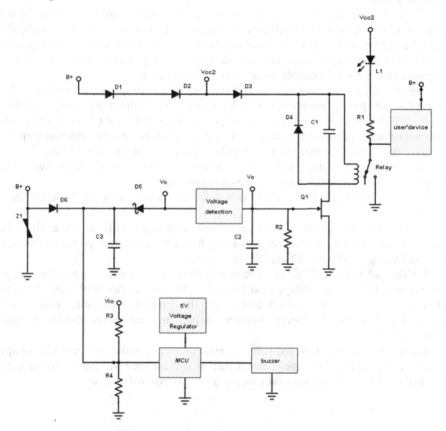

Figure 1, Circuit

3. Circuit operation

Function one is the engine is launched and power supplying: When the key is turned to ACC, the input voltage will begin to be detected via the voltage detection chip. The voltage detection chip keeps operating. When the key is turned to ON, the engine is launched, and the voltage will rise until 14V. The current will flow through the coil. So, the magnetic switching operates to become a short-circuit switch, and it will begin to supply power to external devices.

Function two is battery life detection: When the key is turned to ACC, the input voltage will begin to be detected via the voltage detection chip. The voltage detection chip keeps operating. If the voltage is lower than 11V, the buzzer will begin to operate to make a noise. The voltage regulator can regulate the voltage into 5V for the buzzer to operate. The range of function two is from 9V to 11V. The driver should consider whether change their battery at this range. When the voltage lowers than 9V, the engine isn't launched. The driver has to change their battery.

4. Test plan

Figure 2 is the test process of the device and the consequent expectation. In the original design of the normal vehicles (European), when the engine is turned off, the cigarette lighter socket will not cease supplying power. In this case, if the cigarette lighter socket connects with external devices, these products will continuously consume power, so it is possible to run out of battery power.

This device can effectively avoid the above situation occurring, and can also detect whether the car engine has been launched and is supplying power to external devices. The battery will not run out of power because of the power drain of the external devices. The works' features mainly depend on the voltage detection chip. Its function is to detect whether the battery voltage achieves the set value or not.

The car battery voltage switching is to use the power supply to be switching voltage. It simulates switch which is launch situations.

The key sub-divided into three sub-stalls, and every stall all have different functions:

(a).The stalls turn to off: the device will automatically cut off the power from the battery to save power; it can prevent battery to supply additional power to external devices leading to which can't start the car engine.

(b).The stalls turn to ACC: the voltage detection device wafer to detect the voltage has reached 11V. If the voltage is smaller than 11V, the device will make noise to alarm for driver about 2-3 seconds and remind the driver that the battery charge isn't as good as before. The battery becomes aging, and the driver should consider replacing battery.

(c).the stalls turn to ON: the voltage detection device wafer to detect the voltage has reached 14V. If the voltage is higher than 14V, the device will supply power to an external device; or the device won't supply power to external devices.

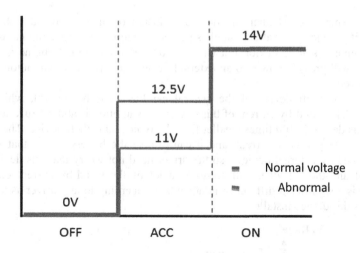

Figure 2, Test Results

5. Experimental results and Comparison

The key sub-divided into three sub-stalls, and every stall has different functions:

The stall are turned to off: the device will automatically cut off the power from the battery to save power; it can prevent the battery from supplying additional power to external devices leading to the car engine not starting.

The stalls are turned to ACC: the voltage detection device chip detects the voltage has reached 11V. If the voltage is smaller than 11V, the device will make a noise for about 2-3 seconds to notify the driver and remind the driver that the battery charge is not as good as before. The battery is aging, and the driver should consider replacing it; its warning range is shown in Figure 3.

The stalls are turned to ON: the voltage detection device chip detects the voltage has reached 14V. If the voltage is higher than 14V, the device will supply power to an external device; or the device will not supply power to external devices.

All parts of the plan have stepped instructions as Figure 4. This paper compares results of the device is used or not. When the device is not used, the battery supplies power to the external device continuously whether the engine is started or not. When the device is used, after the car engine is started, the battery will provide power to external devices. Otherwise, it will automatically cut off power; the battery does not provide power to external devices.

If the voltage is normal in the ACC, this device does not supply power to external devices. It will avoid supplying power unnecessarily. However, if the voltage of the battery is lower than 11V, this device will make a noise. This noise can remind the driver to consider whether the battery should be replaced soon.

This device will supply power to external devices in the ON gear, and the red light of the device will turn on. But if the car battery power is lower than 14 volts, the red light of the device will turn off and it will not supply power to an external device to alert the driver that the battery has insufficient power.

In comparing whether the device is installed, or not, consider this. Without installing device, the battery supplies power to external devices continuously whether the engine is started or not, but with this device, after the car engine is started; the battery will provide power to an external device. Otherwise, it will automatically cut off power.

The anti-theft device of the general car has stand-by current, which is about 1.5mA. The stand-by current of this device consumption is about 13uA, and consider that this device is 100 times smaller than the typical anti-theft device. The system has a very low quiescent current, and it can protect the battery power that is normally consumed by external devices, so the driver need not worry that this device, like the typical anti-theft device, will consume a lot of the stand-by current continuously. There is a significant difference in advantages accruing to any driver as to whether a car has this device installed or not.

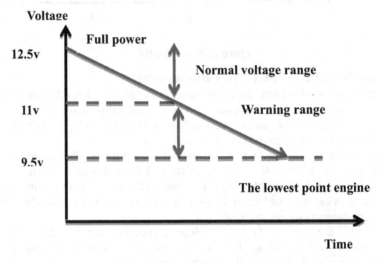

Figure 3, Battery life detection range

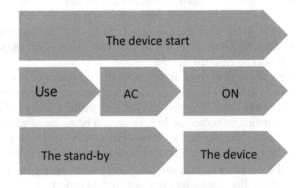

Figure 4, Provides electricity

6. Conclusion

Currently, there are some products that can display the voltage value of the power flowing through the cigarette lighter socket of a motor vehicle, but most drivers do not understand the meaning of the voltage values. This new developed device can alert the driver as to when the voltage is too low. It can help the normal drivers; especially, women drivers can understand the condition of their car.

The circuit of the work is designed simply, and it is cost-effective, as complexity of the circuit design has been kept to a minimum; the device is not only useful in function, but is also convenient to install. The price is low, and the aesthetics are very pleasing.

This working device has three functions: the battery supplies power when the engine is started; the condition of the battery power warning is indicated; and its stand-by current is very low, therefore not consuming a lot of power. Now, the device has been successfully installed, tested and used for battery protection. It can effectively solve the above problems. The authors believe that this will be a reference or modification for study in the future.

REFERENCES

[1]   Xuliang Ming Huang Wangen, car school car electricity articles III, Second Edition, Tai Keda Books Ltd., 2003.

[2]   David Zhang, how to diagnose and troubleshooting car electrical fault, Sanmin Internet bookstore, 2011.

[3]   Yu Yan, Research and implementation of the optimal charging algorithms for lithium batteries, National Taiwan University of Science and Technology, 2008.

[4]   Zhang, the electric car battery research residual estimates, the National Changhua University master's thesis, 2001.

[5]   Mohan / Undeland / Robbins, power electronics: converters applications and design 3rd Edition ,, John Wiley and Sons Ltd, 2002.

[6]   Li, Qian-Jin , Power System and method for inhibiting the switching surge test of , National Cheng Kung University master's thesis , 1980.

[7]   Fu-Tsai Cheng , Hazard Filtering Technique for Low-Power Circuit Design , Da-Yeh University master's thesis , 2001

[8]   Owen hung, Europe Jiajun, industrial electronics, Quan Hua Book Co., 2013.

[9]   Zhang Yongchang, lead acid battery residual lifetime prediction and analysis, the National Changhua University master's thesis, 2001.

[10] Shiren Lin, On Discharging Modeling and Capacity Estimationof Li-Ion Battery, National Chiao Tung University master's thesis, 2007.

Feeder Losses Sensitivity Analysis by Real or Reactive Power Node Injection of Distributed Generators

Yen-Ming Tseng[1], Rong-Ching Wu[2,1], Jeng-Shyang Pan[3] ,Rui-Zhe Huang[4]

1, 3, 4: Colleague of Information Science and Engineering, Fujian University of Technology, China.
swk1200@gmail.com

2: Department of Electrical Engineering, I-Shou University, Kaohsiung, Taiwan
rcwu@isu.edu.tw

Abstract. Feeder distance longer of transport power not only affects the power flow parameters also increase feeders losses. In this paper, apply the distributed power generation device on the feeder node injected real and reactive power to reduce the current and losses for short and middle term distances power delivery system to be a good solution . However, this method must be injected with real or reactive power to specify node to decide the feeder losses sensitivity analysis to find the optimal node of feeder and minima feeder's losses. Input feeder node load and line parameters by PSS/E power system simulation software were calculated feeder losses, firstly. Then inject real or reactive power of distributed generator (DG) to calculate the feeder loss reduced sensitivity, secondly and finally assessment the feeder losses reduction level.

Key words : Feeders Losses, Distributed Generator, Losses Sensitivity.

1. Introduction

Meet each user safety and reliability of electricity supply quality will be power provider want to pursue the goal [1]. In power system, the distribution systems is connected to the user which direct impact on power quality and challenge of power supply, mainly the varying is from the loadings changing[2]. A feeder load characteristic is composed by commercial, residential, agricultural and industrial demands. Feeder load characteristics is composed by commercial, residential, agricultural and industrial demands, so that different users depending on the amount of electricity the proportion to distinguish commercial-oriented, residential-oriented, agricultural-oriented and industrial-oriented feeders. However, different load characteristics feeder by the rapid growth of electricity load make power transmission among the feeders is complicated such as the problems of power delivery congestion of distribution system, transformer overload, power quality reducing and losses increased of feeders . Etc. Therefore, experts and scholars actively to solve the issues related to its distribution system has the following manner using the theory of fuzzy sets [3, 4], neural networks [5], the expert system approach [6], the rule base [7]

© Springer International Publishing AG 2017
J.-S. Pan et al. (eds.), *Advances in Intelligent Information Hiding
and Multimedia Signal Processing*, Smart Innovation, Systems and Technologies 64,
DOI 10.1007/978-3-319-50212-0_32

genetic algorithms [8] and Patricia network algorithms [9], etc., are to be explored for the loss of the distribution system, the purpose is to enhance the quality of the power distribution system and reducing feeder loss. Perform the steps of this study are as follows:

Step1: Select the 11.4kV overhead feeders of Chao-Liao substation with load characteristic of residential-oriented and agricultural-oriented.

Step2: Feeder node sets and collects load and line parameters of feeders.

Step3: Feed into the feeder parameters to PSS / E software and solving the load flow under its original state.

Step4: Real or reactive power of decentralized generators added to the PSS / E software to solve the load flow.

Step5: Calculated loss reduces the sensitivity obtained optimum set point.

Step6: Assessments the reduction rate of Feeder loss.

2. Feeder load characteristics analysis and node set

2.1. Load characteristics

Sampling every 15 minute form feeder output terminal of Chao-Liao substation obtain the electricity-related parameters of feeders, then to statistics the 24 hours of a day of former data to form every hour load data which fed in PSS/E power flow simulation software to preceding the study. Chao-Liao substation selected two feeders, shape or curve to load the data in accordance with the load of formed its hourly load curve and daily users of the electricity accounts than can be broadly divided into feeders of residential-oriented and agricultural-oriented. Fig1 is the daily load curve of the selected feeder that named FD07 that according to its curve shapes and statistics power consumption of customer type to be a feeder of residential-oriented. In this Fig including two curves, one for real power curve, another for reactive power curve. Focus on the real power daily load curve which obtain the real power maximum P and minimum value difference of about 1419 kW and reactive power daily load curve obtain reactive power maximum Q and minimum value difference of about 763 kVAR.

Fig2 is the daily load curve of selected feeder which name FD08 that be a feeder of agricultural-oriented. Aim at the real power daily load curve that the real power at the 21st hour as the largest 3510kW, reactive power at the 20th is the largest 1043 kVAR, which obtain the real power maximum P and minimum value difference of about 1296 kW and reactive power daily load curve obtain reactive power maximum Q and minimum value difference of about 25 kVAR.

2.2. Nodes set

Feeder distance and concentration of load will affect the distribution of feeder losses so that the node set for DG to supply the real power or reactive power share the power system power providing at the same time where will influence the losses reduction. Fig 3 is FD07 feeder of residential-oriented single line diagram with nodes setting which contain 11 nodes and Fig4 is FD08 feeder of agricultural-oriented single line diagram with nodes setting which contain 7 nodes.

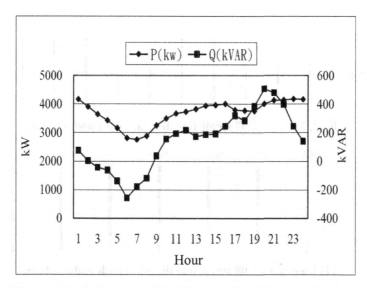

Fig1. Daily load curve of FD07 feeder of residential-oriented

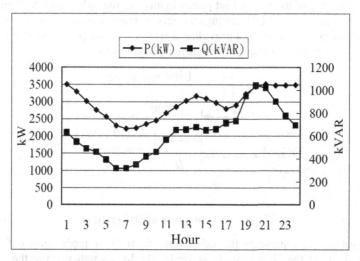

Fig2. Daily load curve of FD08 feeder of agricultural-oriented

3. Feeder losses sensitivity

Complex power is consist of real power and reactive power that quantity varying will change the line current and affect the feeder's losses. In this paper, using distribution generators installations and providing the real and reactive into the feeder node make the feeder line currents changed to reduce the loss of the feeder, which

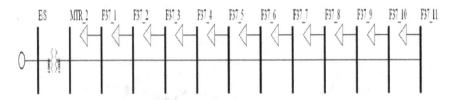

Fig3. Feeder FD07 single line diagram with nodes setting

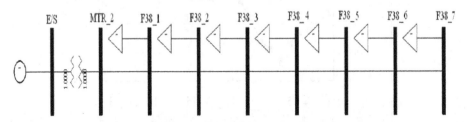

Fig4. Feeder FD08 single line diagram with nodes setting

take feeder 24 hours a day for each hour feeder maximum real power or reactive power sensitivity of the optimal set points to analyze the reduction rate of feeder loss. Fig5 is a schematic of DG settings at feeder node affected feeder loss. Use of decentralized power generators providing a real or reactive power, so that the current path of the series current changes to reduce the feeder loss.

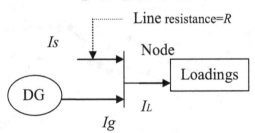

Fig5.   A schematic of DG settings

Equation (1) represents the calculation formula of feeder loss before setting DGs. And to set the DGs at node of feeder the losses will varying the calculative formula as Equation (2) which Ig is closely related to distributed power generators provide real and reactive power.

$$LineLoss = (Is)^2 R \qquad (1)$$
Where $LineLoss$: feeder loss before setting DGs
    Is : current by power provider
    R : transmission line resistance

$$SetDGLineLoss = (Is-Ig)^2 R = (I_L)^2 R \qquad (2)$$
Where $SetDGLineLoss$: feeder loss after setting DGs
    Ig : current by distribution generator
    I_L : load current

Sensitivity formula is compared the feeder losses before and after set the DG at each node of feeder and find the node of maximum value of sensitivity to be the set point. So the sensitivity divided into real power sensitivity such as Equation (3) and reactive power sensitivity as Equation (4).

$$PS\% = \frac{NLGFDP - ALGFDP}{NLGFDP} \times 100\% \tag{3}$$

Where $PS\%$: real power sensitivity of feeder
    $NLGFDP$: feeder loss without input real power of decentralized generator
    $ALGFDP$: feeder loss with input real power of decentralized generator

$$QS\% = \frac{NLGFDQ - ALGFDQ}{NLGFDQ} \times 100\% \tag{4}$$

Where $QS\%$: reactive power sensitivity of feeder
    $NLGFDQ$: feeder loss without input reactive power of decentralized generator
    $ALGFDQ$: feeder loss with input reactive power of decentralized generator

4. Solving sensitivity and feeder losses by PSS/E load flow software

4.1. Real power and reactive power sensitivity

FD07 feeder of residential-oriented single line diagram with nodes setting which contain 11 nodes and FD08 feeder of agricultural-oriented single line diagram with nodes setting which contain 7 nodes. Table1 is the feeder F07 daily real power and reactive power sensitivity which daily real power sensitivity is 12.13% and the reactive power sensitivity is 1.59%. Compare the both sensitivity, the real power sensitivity is more than the reactive power sensitivity that mean the real power injection to the feeder optimal node will more contribute to the feeder losses reduction. Table2 is the feeder F08 daily real power and reactive power sensitivity which daily real power sensitivity is 11.74% and the reactive power sensitivity is 2.47%. After statistical analysis, FD07 and FD08 real power of the distributed power of the best node injected not changed much, with respect to the reactive power optimal   injection node which are frequent changes trend in   hourly. In conclusion, feeder power sensitivity is closely with the real power and reactive power curve varying.

4.2. Loss reduction calculation

Table3 are the feeder F07 and F08 daily feeder losses reduction according to the table 1 and table2 optimal node of feeder with real power injection have obviously than reactive power injection. F07 daily loss reduction is by real power injection is 1.52% and the reactive power injection is 0.15%, and F08 daily loss reduction is by real power injection is 1.38% and the reactive power injection is 0.15%.

5. Conclusions

Apply the distributed power generation device on the feeder node injected real and reactive power to reduce the current and losses for short and middle term distances power delivery system to be a good solution. F07 daily real power and reactive power sensitivity which daily real power sensitivity is 12.13% and the reactive power sensitivity is 1.59%. F08 daily real power and reactive power sensitivity which daily real power sensitivity is 11.74% and the reactive power sensitivity is 2.47%. From view of loss reduction, F07 daily loss reduction is by real power injection is 1.52% and the reactive power injection is 0.15% and F08 daily loss reduction is by real power injection is 1.38% and the reactive power injection is 0.15%. No matter are the feeder F07 with residential-oriented or F08 with agricultural-oriented that by real power injection may have more significant influences to reactive power for feeder losses reduction. In summary, the real power and reactive power sensitivity of feeder is based on the feeder load curve more related. Further research will integrated both of real and reactive power injected to feeder node obtaining optimum set point based on the best reduction rate of feeder loss.

Table1. Feeder F07 daily real power and reactive power sensitivity

hour	Real power sensitivity		Reactive power sensitivity	
	Optimal node	Sensitivity (%)	Optimal node	Sensitivity (%)
1	F37_9	13.91	F37_9	1.96
2	F37_9	12.78	F37_1	1.03
3	F37_9	11.70		
4	F37_9	12.85		
5	F37_9	9.79	F37_11	0.37
6	F37_9	8.46	F37_5	0.93
7	F37_9	8.40	F37_5	0.90
8	F37_9	8.86	F37_9	1.23
9	F37_9	10.27	F37_2	0.78
10	F37_9	11.27	F37_7	1.24
11	F37_9	11.97	F37_6	1.21
12	F37_9	12.28	F37_9	1.34
13	F37_9	12.59	F37_6	1.34
14	F37_9	13.07	F37_5	1.28
15	F37_9	13.18	F37_9	1.53
16	F37_9	13.40	F37_10	1.48
17	F37_9	12.55	F37_2	1.55
18	F37_9	12.47	F37_9	1.54
19	F37_9	12.50	F37_9	1.90
20	F37_9	12.36	F37_7	2.50
21	F37_9	14.17	F37_5	2.48
22	F37_9	14.12	F37_9	2.20
23	F37_9	14.14	F37_7	1.98
24	F37_9	13.95	F37_7	1.39
average		12.13		1.59

Table2. Feeder F08 daily real power and reactive power sensitivity

hour	Real power sensitivity		Reactive power sensitivity	
	Optimal node	Sensitivity (%)	Optimal node	Sensitivity (%)
1	F38_7	13.97	F38_2	3.01
2	F38_7	12.84	F38_6	2.45
3	F38_7	11.50	F38_2	1.93
4	F38_7	10.34	F38_7	1.71
5	F38_7	9.40	F38_4	1.31
6	F38_7	6.95	F38_4	0.27
7	F38_7	6.72	F38_4	0.27
8	F38_7	6.74	F38_5	0.17
9	F38_7	7.08	F38_6	0.44
10	F38_6	20.74	F38_7	1.71
11	F38_7	10.03	F38_3	1.99
12	F38_7	10.89	F38_4	2.51
13	F38_7	11.71	F38_7	2.66
14	F38_7	12.40	F38_6	2.85
15	F38_7	12.00	F38_7	2.68
16	F38_7	11.46	F38_7	2.64
17	F38_7	10. 17	F38_6	2.66
18	F38_7	11.20	F38_4	2.83
19	F38_7	13.00	F38_7	4.17
20	F38_7	14.21	F38_7	4.86
21	F38_7	14.57	F38_7	4.81
22	F38_7	14.22	F38_4	4.22
23	F38_7	14.07	F38_7	3.87
24	F38_7	13.98	F38_6	3.27
average		11.74		2.47

Table3. Feeder F07 and F08 daily feeder losses reduction

hour	F07		F08	
	By real power (%)	By reactive power (%)	By real power (%)	By reactive power (%)
1	1.54	0.15	1.40	0.15
2	1.53	0.27	1.39	0.27
3	1.51	0.11	1.39	0.10
4	1.50	0.04	1.38	0.04
5	1.48	0.05	1.37	0.05
6	1.44	0.01	1.29	0.01
7	1.46	0.01	1.29	0.01
8	1.48	0.04	1.29	0.04
9	1.50	0.09	1.28	0.09
10	1.52	0.05	1.89	0.05
11	1.52	0.20	1.37	0.20

12	1.53	0.13	1.36	0.13
13	1.53	0.07	1.37	0.07
14	1.53	0.41	1.38	0.41
15	1.54	0.07	1.38	0.07
16	1.54	0.07	1.37	0.07
17	1.53	0.52	1.35	0.52
18	1.54	0.16	1.36	0.16
19	1.52	0.13	1.35	0.13
20	1.53	0.14	1.35	0.14
21	1.54	0.13	1.36	0.13
22	1.54	0.19	1.37	0.19
23	1.54	0.09	1.39	0.09
24	1.54	0.39	1.39	0.39
average	1.52	0.15	1.38	0.15

References

1. T.Y. Jyung, Y.S. Baek, Y.G. Kim, "Improvement of a power quality of microgrid system interconnected to distribution system in emergency condition,"Transmission & Distribution Conference & Exposition: Asia and Pacific, 2009, pp. 1-4.
2. D. Sharafi, "Distribution Design Using Transmission Concepts,"Power and Energy Engineering Conference, 2009, pp. 1-5.
3. D.H. Spatti,I.N. da Silva,W.F. Usida, R.A. Flauzino,"Real-Time Voltage Regulation in Power Distribution System Using Fuzzy Control,"IEEE JOURNALS, 2010, pp. 1112-1123.
4. M. Moattari, M.A. Nouri, A. Argha,"Adopting fuzzy iterative learning control method for reactive power planning in distribution systems," IEEE CONFERENCES, 2010, pp. 528-532.
5. Weixin Gao, Nan Tang, Xiangyang Mu, "A Distribution Network Reconfiguration Algorithm Based on Hopfield Neural Network, "IEEE CONFERENCES, 2008, pp. 9 - 13.
6. K. Butler-Purry, S.K. Srivastava, "Expert system-based reconfiguration of shipboard power distribution systems," IEEE CONFERENCES, 2008, pp. 1-5.
7. Yu-Lung Ke, "Rule-expert knowledge-based petri network approach for distribution system temperature adaptive feeder reconfiguration switching operation decision reasoning," IEEE CONFERENCES, 2005 ,pp. 17-31.
8. Tai-shan Yan,"An Improved Genetic Algorithm and Its Blending Application with Neural Network,"IEEE CONFERENCES, 2010, pp.1-4.
9. Dong Liu, Yunping Chen, Rongxiang Yuan, Guang Shen, "Backtracking based algorithm in hierarchical time-extended Petri net model for power system restoration,"IEEE CONFERENCES, 2005, pp. 1035-1040.

The Design of Distributed Hardware Platform for Complex Electronic System Testing

Chang'an Wei, Gang Wang, Yunlong Sheng, Shouda Jiang, and Xiangyu Tian

Department of Automatic Test and Control, Harbin Institute of Technology,
Xidazhistr. 92, 150080 Harbin, China

Abstract. Due to complex interface in the complex electronic system which contains embedded software, such as RS422, CAN, analog and digital input/output interface. In addition, embedded software in complex electronic system usually controls a large number of peripherals, there are usually no enough external test resources in one computer, according to this problem, we provide distributed structure to construct the software testing system. In hardware, a high-speed information transmission mechanism based on memory reflective technique is studied, and a multi-functional reflective memory communication cards is developed, using optical fiber to form a high speed data transmission network, realizing automatic data input and test result obtain.

Keywords: complex electronic system, embedded software testing, memory reflective technique

1 Introduction

An embedded system is a special purpose computer that is used inside of a device. Embedded systems generally use microprocessors that contain many functions of a computer on a single device. Software engineers will often program directly to the microprocessor hardware without using a host operating system[1]. As embedded systems have become more sophisticated, and often have the characteristics of real-time, interface complexity and high reliability[2], and the embedded software is often associated with a fixed hardware system, which greatly increases the complexity and difficulty of embedded software testing[3]. On the other hand, the hardware interface of the embedded software, also provides favorable conditions for automated testing[4].

2 Principle and test method of Embedded Software Testing Platform

Fig. 1 shows the principle of the Embedded Software Testing System.

This system uses test cases with high quality to incentive embedded software under test in order to get the response of the embedded software, at the same time it incentives the virtual comparison model of the software under test to

© Springer International Publishing AG 2017   269
J.-S. Pan et al. (eds.), *Advances in Intelligent Information Hiding
and Multimedia Signal Processing*, Smart Innovation, Systems and Technologies 64,
DOI 10.1007/978-3-319-50212-0_33

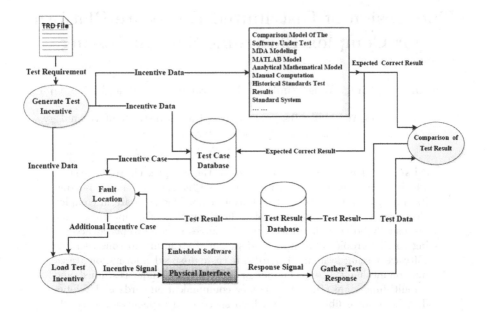

Fig. 1. Principle of the Embedded Software Testing System

get the expected correct result. Finally, we get test result by comparing them in real time. When the error is found in the running results, the system analyses the cases which are out of order and generates additional test cases, then the embedded software will be tested again, we can locate the software error after several iterations.

3   System Architecture of Test Platform

In recent years, the complexity of the embedded software also increases, the characteristic of complex embedded software reflects in: (1) Various types of interface, various kinds of dynamic parameter with great varying range, such as vibration, temperature, voltage and other physical signals, which also contains digital, analog and other forms of data. (2) Many input and output parameters, complex coupling relationship between the input parameters. With full coverage testing, the number of test cases increases exponentially with the increase of the input parameters, full test requires a lot of test cases to cover all possible parameters combination of logic and timing sequence. (3) High real-time requirement, especially for the application domain with the high requirement of real-time performance, the real-time control ability of complex embedded software testing system sets higher demands. Embedded software in electronic system usually controls a large number of peripherals, one computer usually cannot provide all the external test resources, we provides distributed structure to construct embedded software testing system. Fig. 2 shows the overall structure of Embedded Software Testing Platform.

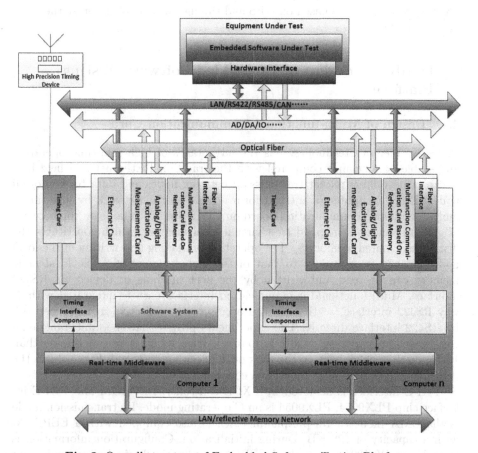

Fig. 2. Overall structure of Embedded Software Testing Platform

Embedded software requires high real-time performance, so the distributed system should meet the running real-time requirements between the real-time data transmission of embedded software. Using computer software to handle various bus data and transfering data between the nodes through the Ethernet will make it hard to guarantee the systems real-time performance. Therefore, this article adopts optical fiber communication solutions to achieve data transmission between distributed nodes , uses hardware as the main implementation method and uses the reflective memory technology to achieve the data transformation between serial interface such as RS422 and optical fiber. Every node adopts reflective memory technology with the ability to build the reflective memory network, The serial communication data can be automatically stored inside the card to the reflective memory, and then be transmitted to the various nodes of the network in real-time through the optical fiber. In this paper, we also provide multi-channel analog and digital input/output resources to achieve the analog and digital testing of the complex electronic system. Furthermore, we use a high precision timing device based on GPS and timing cards to synchronize the clock in every nodes.

4 Hardware design of Embedded Software Testing Platform

4.1 Design of Multi-functional Communication Card

Reflective memory technology has the advantages of high real-time operation and stable performance compared with the traditional Ethernet and field bus, It can achieve large packet transmission between nodes and data backup of all nodes. Every computer online can form a node on the network by inserting a multi-functional communication board on CPCI slot. Each node on the network of global memory is mapped to a virtual local memory, which constitutes the distributed shared memory[5]. The operation of reading and writing of the local node memory is equivalent to that of the global memory[6]. The local node memory can be read and written by the host machine or device via RS422 interface. Multi-functional communication interface card is integrated with 4-way RS422 interface, so that we can access to the reflective memory network by RS422 interface directly. The experiments proved that the multi-functional communication cards bandwidth is up to 30 MB/s, its bit error rate is lower than 10-15, its time delay is up to the rank of the deep microsecond. Fig. 3 shows the principle diagram of multi-functional communication card.

PXI-E interface is achieved by PXI-E controller, EEPROM and PXI-E interface chip PLX9054, PLX9054 is on C operating mode. Its transmission mode is the DMA mode[7]. Its peripheral block is also equipped with a EEPROM with a capacity of 128 KB. During initialization, Configuration information is loaded into PLX9054 from the EEPROM automatically, including equipment signs, the base address of the local bus, I/O space etc. PXI-E controller implements data parsing, framing and packaging, and other functions. Photoelectric converter and high-speed transceiver implements physical interface of reflective memory network, Optical fiber controller is responsible for controlling of optical fiber hardware interface, codec and pre-preprocessing of data. SDRAM controller achieves the management of 256M DDR2 SDRAM of the multi-functional communication card. RS232 interface makes the external device can access memory unit via the RS232 bus and reflects the data to each node of the reflective memory network. Its electrical interface is achieved by the bus controller and transformer. RS232 controller implements coding and decoding. RS422 interface makes the external device can access memory unit via the RS422 bus and reflects the data to each node of the reflective memory network. Its electrical interface is achieved by photoelectric isolation and level transformation. RS422 controller implements coding and decoding. RS485 interface makes the external device

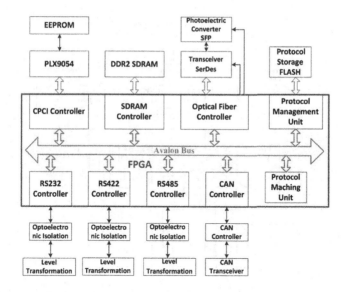

Fig. 3. Hardware principle diagram of multi-functional communication card

can access memory unit via the RS485 bus and reflects the data to each node of the reflective memory network. Its electrical interface is achieved by photoelectric isolation and level transformation. RS485 controller implements coding and decoding. CAN interface makes the external device can access memory unit via the CAN bus and reflects the data to each node of the reflective memory network. Its electrical interface is achieved by photoelectric isolation and level transformation. CAN controller implements coding and decoding. PXI-E controller, SDRAM controller, RS232 controller, RS422 controller, RS485 controller and CAN controller are all achieved by FPGA, every module mounts on the Avalon bus with the aid of the bus for data interaction[8].

4.2   Fast recognition of embedded software interface protocol

With the method of interface control protocol, three-level data structure is designed : data block, data elements and data bits for all kinds of serial interface protocols to complete description of the different interface type of tested software. Based on the idea of hierarchical management, protocol is represented as following: protocol model, protocol entry, protocol frame head/frame tail, element entry and element bit. Fig. 4 shows the hierarchical description of protocol format.

The description template of the communication protocol is designed based on the hierarchical description about above protocol format. Four-layer data structure is designed as following : protocol model, protocol entry, protocol frame head/frame tail/ element entry and element bit ,in order to describe the relevant information and the transmission characteristics of the protocol.

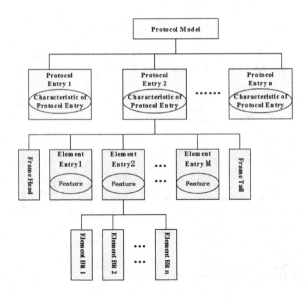

Fig. 4. Hardware principle diagram of multi-functional communication card

4.3   Identification method of embedded software interface protocol

The fundamental principles of protocol identification is setting storages to store the content and the length of each frame head, and setting a number of protocol matching units, Each protocol matching unit is responsible for a identification of a protocol. When the hardware receiving serial data packets, protocol matching unit compares the first received byte in the packets with the first byte of the frame head, if the first byte is same, the second byte will be compared. The protocol matching unit will finish matching and get the comparison results under 2 circumstance: a byte is diffirent , bytes that have matched are same and have matched all bytes of the frame head. After all the protocol matching units have completed matching, we can get the final matching result. Fig. 5 shows the functional block diagram of protocol matching unit.

4.4   Clock synchronization scheme between the distributed test nodes

As shown in Fig. 6, the clock synchronization system is built using SYN2306A, which can receive signal from GPS, the timing accuracy is better than 100 ns. The equipment has four-channel RS232 output interface, each provides accurate time information, and it has a TTL output signals whose pulse synchronization error is less than 100 ns.

The second pulse signal is received via input interface of the Analog/digital excitation/measurement card, time information is received via RS232 interface of Industrial Personal computer. Interrupt occurs through PXI - E interface after receiving the second pulse signal, The computer corrects time by using the

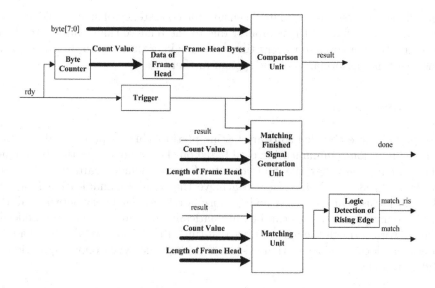

Fig. 5. Hardware principle diagram of multi-functional communication card

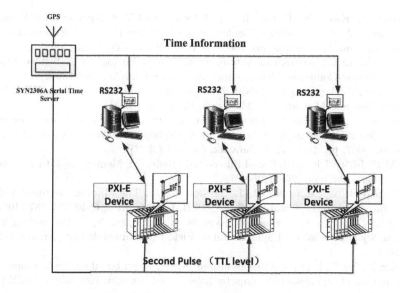

Fig. 6. Clock synchronization system

second pulse signal as standard and uses the last time received information plus 1 s as an absolute time. Correct the time of nodes every one minute. In the process of test system operation, which can make timestamp on the test data depending on the absolute time of each node computer. Start timing by using the second pulse signal received as benchmark and make the relative timestamp of the whole test systems test data to ensure the consistency of space and time. No need to correct time of the computer due to the relative timestamp or read the absolute time of computer every time when making time stamp, It can improve the real-time performance of the system further.

5 Conclusion

We provide an embedded software testing platform in this paper, and designed a multi-function communication card based on the memory reflection technique, the card not only has the function of reflective memory card, but also has RS232/422/485, CAN interface to achieve the serial communication function. The card is used at every test node to form a reflective memory network, all the serial communication data can be automatically transmitted to all the nodes in the network through the optical fiber real-time. In addition, we also build a high precision clock synchronization system to ensure the synchronous operation in different nodes.

References

1. Kang, B., Kwon, Y. J., Lee, R. Y.: A Design and Test Technique for Embedded Software. Acis International Conference on Software Engineering Research, Management and Applications pp. 160–165. IEEE Press, New York (2005)
2. Wang X., Li L.: Research of Automatic Testing Technology of Embedded Software of Satellites. Computer Measurement and Control, 20, 267–269 (2012)
3. Grenning J.: Applying test driven development to embedded software. IEEE Instrumentation and Measurement Magazine, 10, 20–25 (2008)
4. Ziyuan W., Changhai N., Baowen X.: Generating combinatorial test suite for interaction relationship. In: International Workshop on Software Quality Assurance, Soqua 2007, pp. 160–165. Dubrovnik, Croatia (2007)
5. VMIPCI-5565 Ultrahigh Speed Fiber-Optic Reflective Memory with Interrupts, GE Hardware Reference,26–57(2008)
6. Bershad B. N., Zekauskas M. J., Sawdon W. A.: The Midway distributed shared memory system. In: Compcon Spring 1993, Los Alamitos, Calif 528–537 (1993)
7. Bittner R.: Bus Mastering Pci Express In An Fpga. Fpga Proceeding of the Acm/sigda International Symposium on Field Programmable Gate Arrays. 273–276 (2009)
8. Qian Z., Yu S., Duo Z.: A design of multi-core system based on Avalon bus. In: International Conference on Computer Science and Network Technology. IEEE (2011)

PCB Level SI Simulation Based on IBIS Model for FPGA System with DDR2 Memory Devices Application

Xiao-Yu Cheng[1] and Wei-Min Zheng [2]

[1] S Shijiazhuang Flying College of the PLA Air Force,
Hebei Province 050071, China
[2] Institute of Information Engineering, Chinese Academy of Sciences,
Beijing 100093,China
zhengweimin@iie.ac.cn

Abstract. It is essential to perform SI(SI) simulation in the high speed FPGA system with DDR2 memory devices. IBIS model is chosen for simulation because IBIS model could easily be download from website of related integrity circuit(IC) or created according to datasheet of IC. The PCB-level SI simulation flow and analysis methods based on IBIS models are illustrated in this paper, especially some useful application notes. Proper SI simulation which a good guide to routing could solve problems caused by high speed transmission in PCB.

Keywords: IBIS, FPGA, SI simulation.

1   Introduction

The development of integrity circuit design and manufacturing approach have contributed to faster switch rate, steeper rising edge and falling edge of digital chip, especially in the high density, high speed FPGA system with memory devices for data buffering and transport. Besides, in pursuit of lower power makes lower power supply voltage. These factors causes a severe challenge to SI, such as reflection, monotonicity, crosstalk, etc. So the better way to solve SI problem is to perform SI simulation analysis before making PCB. At present, SPICE model, IBIS model, Verilog model are popular as the model for SI simulation. IBIS model is chosen for simulation because of its advantages mentioned in the following part.

In this paper, PCB level SI simulation flow and analysis methods based on IBIS model for FPGA system with DDR2 memory devices application are investigated. The rest of the paper is organized as follows: Section 2 describes the three models for simulation and how to choose the proper model for simulation; Section 3 describes the introduction of FPGA system with DDR2 memory devices; Section 4 introduces the simulation flow and some useful application notes; Section 5 presents experimental results and analysis methods and lastly section 6 concludes the paper.

© Springer International Publishing AG 2017   277
J.-S. Pan et al. (eds.), *Advances in Intelligent Information Hiding*
and Multimedia Signal Processing, Smart Innovation, Systems and Technologies 64,
DOI 10.1007/978-3-319-50212-0_34

2   Choice of SI simulation model

Creating PCB board level SI model, which different from other traditional design methods, plays an important role in the PCB board level simulation. The correctness of SI simulation model will determine the valid of the design, while the SI model generating will determine the feasibility of the method.

Presently, there are various SI models used for PCB board level SI simulation, especially SPICE model, IBIS model, Verilog-AMS model (VHDL-AMS model) are most commonly used in modern electronics design. The three kinds of SI models will be explained as follow and the difference among them will be concluded, as well how to choose proper SI model in our design will be discussed.

2.1   SPICE model

SPICE is an acronym for Simulation Program with Integrated Circuit Emphasis and was inspired by the need to accurately model devices used in integrated circuit design. It has now become the standard computer program for electrical and electronic simulation. The majority of commercial packages are based on SPICE2 version G6 from the University of California at Berkeley although development has now progressed to SPICE3. The increased utilization of PCs has led to the production of PSPICE, a widely available PC version distributed by the Micro-Sim Corporation whilst HSPICE from Meta-Software has been popular for workstations and is now also available for the PC.

SPICE model is composed of model equations and model parameters.   Because of model equations, it is easy to connect the SPICE model with simulator algorithm closely and get the better analysis efficiency and results. The analysis precision of SPICE model depends on the source of model parameters and the application scope of the model equations.

SPICE   simulation   software   such   as   Hspice,   Pspice,   Spectre,   Tspice, Smartspice,Ispice, and so on, is frequently used for simulation, of which Hspice from Synopsys and Pspice from Cadence are much more commonly used. In fact, Hspice as SPICE industrial standard simulation software is popular in the electronics field. Although it has characteristic of high precision and powerful function, Hspice do not provide front- end input panel so that netlist should be prepared before simulating. So Hspice does not fit for beginners and is mainly applied to integrity circuit design. While Pspice which provides graphic front end input panel and friendly user interface is best for individual user and used for PCB and system level design.

When SPICE model is used for PCB level SI analysis, it is necessary to provide correct SPICE models of integrity circuit IO cells and semiconductor manufacture parameters of   semiconductors by IC designer and manufacturer. While these models and parameters belong to their own intellectual property rights, so it is hard for them to provide SPICE model when providing chip products.

2.2   IBIS model

IBIS, or the "Input/Output Buffer Information Specification," is a device modeling standard that was developed in 1993 by a consortium of companies from within the electronic design industry. IBIS allows the development of device models that preserve the proprietary nature of IC chip designs, while at the same time providing information-rich models for SI and EMC analysis.

IBIS provides a fast and nicety IO buffer modeling method based on I/V curve and a standard file format which includes output impedance of driver, rising/falling time, input load, and so on. So IBIS model is best for high speed circuit simulation such as ringing and crosstalk. Now IBIS standard is certificated by EIA and defined as ANSI/EIA-656-A standard.

It is impossible for IBIS model which only describes behavior of driver and receiver to reveal intellectual property rights of circuit internal architecture. In other words, chip supplier could express their up to data gate level design using IBIS model and not expose excessive information of products. Compared to SPICE model, IBIS model using look-up table to calculation will save calculation quantity by ten to fifteen times.

A standard IBIS model file consists of three sections: Header Info which contains basic information about the IBIS file and what data it provides; Component, Package, and Pin Info which contains all information regarding the targeted device package, pin lists, pin operating conditions, and pin-to-buffer mapping;V-I Behavioral Model which contains all data to recreate I-V curves as well as V-t transition waveforms, which describe the switching properties of the particular buffer.

However, IBIS model is not absolutely perfect. Although IBIS file could be created by user or described from translating SPICE model, any translation tool will be of no use if it is impossible to get the minimum rising time. IBIS could not ideally deal with driver circuit of which rising time is controllable. IBIS model is short of modeling ground bounding noise.

2.3   Verilog-AMS model

Compared with SPICE model and IBIS model, behavior model language Verilog-AMS model and VHDL-AMS model are late to appear. Verilog-AMS and VHDL-AMS are superset of Verilog and VHDL respectively, while Verilog-A is a subset of Verilog-AMS.

Verilog-AMS is a true mixed signal language that uses most of the constructs available in Verilog and Verilog-A. The Verilog-AMS language can easily model mixed signal hard blocks and analog blocks that interface to the digital content of the IC. These models can be simulated with the Verilog-AMS simulator at a much faster speed than simulating the analog circuits or Verilog-A net lists with the digital net lists using mixed-mode simulators. The Verilog-AMS language can greatly simplify model creation and also be the interface between the analog and the digital circuits.

Models are created to allow for faster verification and simulation, but the models will only help if they accurately represent the circuit they are meant to model.

Verilog-AMS allows the designer to model many of the important characteristics of a circuit.

Some of the features that may be important to include in the model in order to accurately verify the circuit are:

(1) Polarity of I/O and control signals – This ensures that there are no inversions in the expected I/O and the actual signals.

(2) Dependency on control signals – This is an important feature to include in the models as it will ensure that all of the blocks are enabled when expected. It also covers polarity of the control signals.

(3) Dependency of analog signals – This can modify operation with changing supply voltages and biases that are required for correct operation.

(4) Timing – The inclusion of timing will help to verify the top level operation between the analog and the digital. Some uses of timing can include ensuring proper clock speed, and timing of the delays on the inputs and outputs that connect to the digital logic.

(5) Bandwidth/ Slew Rate – Can accurately model some timing aspects of the analog circuits and their output signals.

(6) Voltage Limits – Model the voltages at which the output or input saturates.

2.4   Choice of SI simulation model

It is impossible to carry out all PCB level SI analysis using only one kind of model up to present time. So in the high speed digital PCB design, it is necessary to mix various models above-mentioned to create transmission models of key signals and sensitive signals. Next comes the discussion about how to choose models for various ICs or transmission line.

(1) Discrete passive device. It can get the SPICE model of discrete passive device from IC manufacturer, or build the simplified model via experiments, or modeling by special tools such as three dimension electromagnetic field model extracting software.

(2) Critical digital integrity circuit. It must use the IBIS model or SPICE model from IC manufacturer for more precise. At present, most of the IC designers and manufacturers provide products and related IBIS models simultaneously by the way of web or other ways.

(3) Non-critical integrity circuit. If IBIS model could not easy to get from IC manufacturer, you can choose similar or default IBIS model according to pin function of chip. Of course, you can also build simplified IBIS models by experiments.

(4) Transmission line. Reduced SPICE model of transmission line is adopted in the SI preanalysis and solution space analysis, while whole SPICE model of transmission line which is based on actual layout design is used after routing. It is necessary to accurately modeling transmission line by three dimension model extracting software for more precise.

3   Introduction of FPGA system with DDR2 memory devices

In this paper, FPGA-DDR2 architecture of radar imaging signal processing system is taken for an example. As we can see from the block diagram, FPGA V4FX140FF1517 produced by Xilinx and DDR2 Mt47H256M8HG-3 produced by Micron are used in the design. The scheme supports two ranks of memory, each rank containing four 2Gbits (256M x 8) DDR2 memory devices. Total memory supported in the system is 2 GB. Now, the characteristics of FPGA and DDR2 technology will be simply illustrated as follow. First, FPGA   V4FX140FF1517 has the bright characteristics of high performance, high logic integration and reduced power consumption. There are some parameters to indicate the excellent performance, such as 500 MHz system clocking; 1+ Gbps SelectIO™ technology for parallel I/O; 622 Mbps~6.5 Gbps RocketIO™ transceivers; 256 GMACS (18x18) Digital Signal Processing circuitry; 450 MHz, 680 DMIPS PowerPC® processing, that is ,up to 1,360 DMIPS in a single device; up to 200,000 logic cells; triple-oxide technology to achieve performance goals while reducing power consumption as much as 50%; hard IP integration to implement key system functions with up to 80% lower power than equivalent functions constructed in logic cells; save 1 to 5 Watts per FPGA and achieve performance goals while staying within your power budget. Second, DDR2 is the second generation of double data rate (DDR) synchronous dynamic random access memory (SDRAM) capable of significantly higher data bandwidth. DDR2 improvements include lower power consumption, improved signal quality, and on-die termination schemes. Compared to the previous generation single data rate (SDR) SDRAM memories, DDR2 SDRAM memories transfer data on every edge of the clock, use the SSTL18 class II I/O standard with memories from up to 4 Gbits of data, and is widely available as modules such as dual in-line memory modules (DIMMs) or as components.

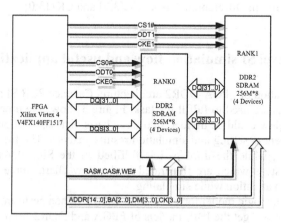

Fig. 1. FPGA and DDR2 interface diagram.

When performing SI analysis, we concentrate on the connectivity between FPGA and DDR2. As seen in the block diagram, the controller FPGA address and control signal drives 8 DDR2 receivers. The topology for these signals should be carefully planned to avoid reflections. The data and DQS signals are easier to handle, as there are only two devices on these lines. Careful placement of memory devices will help the designer to achieve an efficient topology of both address/control and data/DQS bus. So the relative placements of FPGA and DDR2 should also be concerned, as shown in figure 2.

Fig. 2. Relative placement of FPGA and DDR2.

The signals that compose a DDR2 memory bus can be broken into four unique groupings, each with their own configuration and routing rules.

(1) Data Group: Data Strobe DQS[3:0], Data Strobe Complement DQS#[3:0], Data Mask DM[3:0], Data DQ[31:0]

(2) Address and Command Group: Bank Address BA[2:0], Address A[14:0], and Command Inputs RAS#, CAS#, and WE#.

(3) Control Group: Chip Select CS# [1:0], Clock Enable CKE[1:0], and On-die Termination ODT[1:0]

(4) Clock Group: Differential Clocks CK[3:0] and CK#[3:0]

4   PCB-level SI simulation flow and useful application notes

In the SI simulation of FPGA-DDR2 architecture, Cadence PCB SI and SigXplorer as simulation tools are used to fulfill the task. Figure 3 shows the processing flow of SI analysis. The flow could be divided into three parts, that is, PCB preferences setting, analysis preferences setting and simulation results analysis. The first two parts are set in the PCB SI, while the third part is fulfilled in the SigXplorer. Each part also include some steps which are illustrated in the flow chart. Some application notes should be taken attention while simulating.

(1) Assign the logic model for FPGA and DDR2: When performing the fourth step of first part, we can get the IBIS models of FPGA and DDR2 from their own website separately. The models should be verified firstly in the Model Integrity and also edited furthermore to meet requires of simulation. The IC manufacturers always provide IBIS model of their products for general purpose, that is, the information of IBIS model files often include various IO models and part numbers of the same series

products. So we should choose proper IO logic model to fit our design. Let's take the IBIS model files of FPGA Virtex 4 and DDR2 MT47H256M8HG for examples.

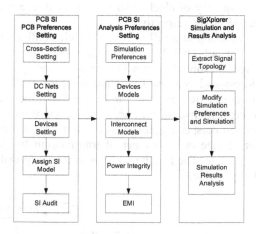

Fig. 3. The processing flow of SI analysis

First, when opening the IBIS model file of Virtex 4, we can find that it doesn't define the IO model for every pin because the pin of FPGA could be programmed. So we can edit the IBIS file to assign the desired IO model for the pins which are connected with DDR2. In the FPGA-DDR2 architecture, SSTL_18 IO cell model is used to fit for the specification of DDR2. Figure 4 shows the V/I curve of SSTL_18.

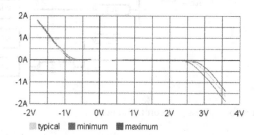

File: virtex4.ibs, Model: SSTL18_I (Pullup + POWER Clamp, Vcc re

Fig. 4. The V/I curves of SSTL18_I

Second, we can also find that the IBIS model file of DDR2 include the models of different part numbers and different data rates of every part. For example, the file includes two kinds of model name: *_533 and *_800, but actually MT47H256M8HG-3 which belongs to *_800 is used. Thus, it can only retains *_800 so that it is not necessary to modify the IO model when extracting the signal topology.

(2) Review the simulation parameters: Simulation parameters in PCB SI should be matched with that in SigXplorer.

(3) Stimulus signals: In reference to DDR2 read and write timing diagram from datasheet of MT47H256M8HG, the rate of data stream is 667Mbit/s for DDR2-667, So the simulating frequency of clock, address/command, control, data are 333MHz, 167MHz, 167MHz, 333MHz, respectively.

(4) Topology: According to a full range of module configurations that generate a standardized base for all DDR2 Small Outline Dual In-line Memory Modules (SODIMMs) defined by Joint Electron Device Engineering Council (JEDEC), the topologies for address and command nets are designed with symmetrical routing, which is similar to binary tree, as shown in Figure 5. All the address and command nets are routed to eight individual memory devices, and each net utilizes a 50 ohm stub resistor. From a SI perspective, it is easy to control the routing length and ensure robust SI. In figure 5, the order number of transmission line represents the same length of transmission line. The length of TL1 is as long as possible, while the length of TL6 is as short as possible.

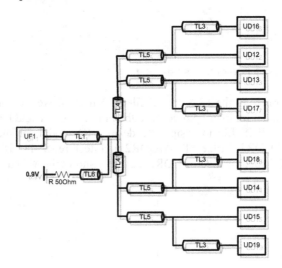

Fig. 5. The topologies for address and command nets

5   Signal topologies and simulating results analysis

5.1   Signal topologies and simulating results

According to DDR2 signal grouping mentioned in section 3, one signal topology and simulating results of each group will be illustrated as follow. The frequency of stimulus signals are in accordance with those mentioned in section 4.3, both the duty cycle of the stimulus signals are 50%. The IO logic model of FPGA data signals and differential clock signals are SSTL18_II, the others of FPGA are SSTL18_I; the IO logic models of DDR2 are DDR2_800.

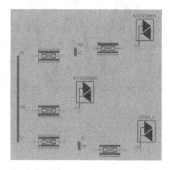

Fig. 6. The topology of data signal

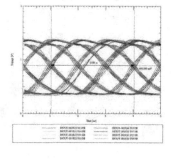

Fig. 9. The simulation results of address/command signal

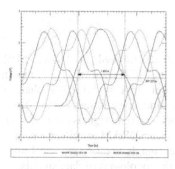

Fig. 7. Simulation results of data signal

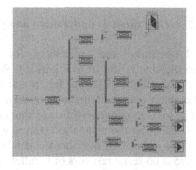

Fig. 10. The topology of control signal

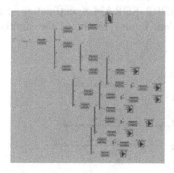

Fig. 8. The topology of address/command signal

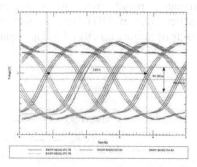

Fig. 11. The simulation results of control signal

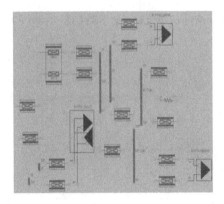

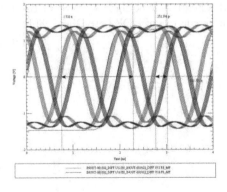

Fig. 12. The topology of clock signal   **Fig. 13.** The simulation results of clock signal

5.2   Overshoot & undershoot analysis

AC overshoot and undershoot for DDR2 signals defined in DDR2 SDRAM specification is shown in Figure 14. With respect to MT47H256M8HG-3, maximum peak amplitude allowed for overshoot/undershoot area is no more than 0.5V. Maximum overshoot/undershoot area of address and control signals are 0.80V/ns, while maximum overshoot/undershoot area of clock and data signals are 0.23V/ns.

Comparing the simulation results with the DDR2 SDRAM specification, we find that the maximum overshoot/undershoot of data signals are beyond the limitation. There is no serial resistors or parallel resistors in the topologies of data signals. So the IO logic models of data signals are changed from DQ_FULL_800 to DQ_FULL_ODT50_800 for matching impedance of termination, that is, ODT function is enabled to ensure the matching between drivers and receivers. The simulation results meet the requirements of DDR2 SDRAM specification. So in practical system, ODT function should be enabled. Figure 15 shows the simulation results of turning on ODT for data signals.

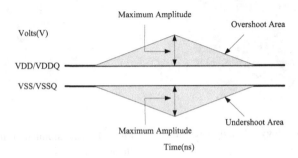

Fig. 14. AC overshoot and undershoot definition

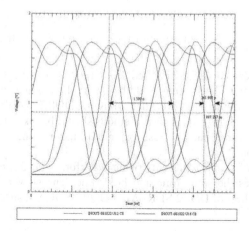

Fig. 15. The simulation results of turning on ODT for data signals.

5.3   Monotonicity analysis

To some extent, monotonic signal means that both rising side and falling side of signal waveform are monotonic between Vil_max (maximum input low threshold voltage) and Vih_min (minimum input high threshold voltage). Some causes of non-monotonic edge are large fan-out, excessive load, and short traces with impedance discontinuity. In our design, address/command signals are connected to eight receivers so that the quality of these signals could not always be monotonic. So proper topology such as binary tree should be applied to these nets.

5.4   Overshoot & undershoot analysis

The purpose of timing analysis is to get maximum available tolerance of timing between clock and data in worst case by circuit principle and datasheet of chip, and then apply the timing limitation to placement and routing.

From Figure 1, we know that the style of clock signals and data signals between FPGA and DDR2 memory is belong to source synchronous connection type, whose clock is provided by FPGA. The timing calculation is based on the following formula:

$$T_{ft\_data\_fast\_Switchdelay} > T_{hold} - T_{co\_min} + T_{ft\_clk\_fast} \tag{1}$$
$$T_{ft\_data\_slow\_Settledelay} < T_{cycle} - T_{setup} - T_{co\_max} + T_{ft\_clk\_slow}$$

$T_{ft\_data\_fast\_Switchdelay}$ — the flight time of data signal when FTS mode is fast, that is switchdelay of data signal; $T_{ft\_data\_slow\_Settledelay}$ — the flight time of data signal when FTS mode is slow, that is settledelay of data signal; $T_{ft\_clk\_fast} / T_{ft\_clk\_slow}$ —

the flight time of clock signal when FTS mode is fast or slow; T_{hold} — the hold time of receiver; T_{setup} — the setup time of receiver; and T_{co} — the valid time of clock to data output;

These parameters could be acquired from related datasheet of chip and DDR2 SDRAM specification. The timing could be justified so long as the simulation results comply with the formula. However, the timing of DDR2 signals does not always fit for the specification at first. And some methods should be taken to improve the timing, such as serial resistor, parallel resistor, modifying routing length, changing the topology, and so on.

The slew rate of signal represents rising waveform and falling waveform change rate, which ensures the setup time and hold time. The slew rate can be calculated according to datasheet of DDR2 SDRAM or DDR2 SDRAM specification.

6   Conclusion

The development of new process and new devices have contributed to more and more common application of high speed devices. SI analysis takes an important role in the high speed system design. SI analysis methods based on IBIS model, especially some useful application notes are illustrated in this paper and have been successfully applied to the SI simulation of FPGA system with DDR2 memory devices, which is part of radar imaging system.

References

1. Bob Ross，Syed Huq，Jon Powell，IBIS Models for Signal Integrity Applications，Electrical Engineering Times, A CMP Publication September 2, 1996
2. Ken Kundert. A Formal Top-Down Design Process for Mixed-Signal Circuits. www.designers-guide.org, 2001,10
3. Aaron M. Shreeve. Verilog-AMS for Mixed-Signal Integrated Circuits. http://www.cdnusers.org/
4. Xilinx, Inc. Virtex-4 User Guide. April 10, 2007
5. Xilinx, Inc. Virtex-4 FPGA Data Sheet:DC and Switching Characteristics. DS302 (v3.2) ,April 10, 2008
6. Micron Technology, Inc. the datasheet of DDR2 SDRAM MT47H256M8.2006
7. Micron Technology, Inc. DDR SDRAM Point-to-Point Simulation Process.2005.07
8. Micron Technology, Inc. DDR2 Simulation Support. 2005.07
9. JEDEC SOLID STATE TECHNOLOGY ASSOCIATION. JEDEC STANDARD: DDR2 SDRAM SPECIFICATION(JESD79-2C). MAY 2006
10. Zhou Runjing，Yuan Weiting. Cadence high speed PCB deign and simulation(Rev 2). Publishing house of electronics industry.pp483~714
11. Micron Technology, Inc. DDR2 SODIMM Optimized Address/Command Nets. 2005
12. Yu Changsheng, Liu Zhongliang. Timing calculation and Cadence simulation results application. www.cnidz.com. 2007.09.18

Part IV
Evolutionary Computing and Its Applications

Automotive air-conditioning systems performance prediction using fuzzy neural networks

Rong Hu[1], Ye Xia[2],

[1]College of Information Science and Engineering, Fujian University of
Technology,Fuzhou, China.
hurong@fjut.edu.cn
[2]College of Ecological Environment and Urban Construction, Fujian University of
Technology, Fuzhou, China.
2751808409@qq.com

Abstract. This study developed a FNN model for air-conditioning system of a
passenger car to predict the cooling capacity; compressor power input and the
coefficient of performance (COP) of the automotive air-condition (AAC) system.
An experimental rig for generating the require data is development., the
experimental rig was introduced at steady-state conditions while varying the
compressor speed, air temperature at evaporator inlet, air temperature at
condenser inlet and air velocity at evaporator inlet. A computer simulation has
been conducted. The results of predicted by FNN are compared with the values
obtained from experiments. It has been demonstrated that FNN model for
automotive air-conditioning systems performance prediction has high coefficient
in predicting the AAC system performance.

Key words: Air-condition system, Fuzzy neural network, coefficient of performance

1 Introduction

The analysis of the performance and operational strategies of Heating Ventilation and
Air Conditioning (HVAC) systems becomes very important for the effective usage of
energy [1]. HVAC systems have been discussed with the goal being performance
improvement over classical control [2-3]. The techniques include expert systems,
neural networks, fuzzy logic, and genetic algorithm. Automotive air-conditioning
(AAC) system is used to maintain comfortable condition in the compartment of
passenger cars. To achieve such condition, the panel outlet for airflow direction,
velocity, volume and temperature has to be adjustable over a large range of climatic
and driving conditions. The compressor of the AAC systems is belt driven by the
engine hence its speed is directly governed by the engine speed and causes the
cooling capacity of the system to vary with the engine speed. The AAC system must
need to have capable of lowering the air temperature in the passenger compartment
quickly and quietly. These conditions make it complicated to analyze the AAC
systems compared to that of the stationary air-condition systems, as described by
Kargilis [4]

© Springer International Publishing AG 2017   291
J.-S. Pan et al. (eds.), *Advances in Intelligent Information Hiding
and Multimedia Signal Processing*, Smart Innovation, Systems and Technologies 64,
DOI 10.1007/978-3-319-50212-0_35

Traditionally, the performance of an AAC system has been done experimentally by many researchers such as Rubas and Bullard [5] on determination of COP of refrigerant cycle for AAC system, Ratts and Brown [6] on effects of the R-134a refrigerant charge level on the performance of the AAC system, A1-Rabghi and Niyaz [7] on COP determination using two types of refrigerants, R-12 and R-134a, as a function of compressor speed and Kaynakli [8]on determination of optimum operating conditions for AAC system using refrigerant R-134a. Hosoz [9] carried out experimental work on an AAC system operating in both air-conditioning and heat pump modes, with varying compressor speed and air temperatures at the inlets of both outdoor and indoor coils.

ANN has been proven to be a useful tool in modeling in refrigeration and air-conditioning system for performance consumption [10]. Chang[11] studied on relationship of power consumption, chilled water temperature and cooling water temperatures of chillers in air-conditioning system. Ertunc [12] applied both ANN and adaptive neuron-fuzzy inference system (ANFIS) models to predict performance of evaporative condenser such as the heat rejection rate, temperature of the leaving refrigerant along with dry and wet bulb temperature of the leaving air stream. A complete review on application of ANN for energy and energy analysis of refrigeration, air conditioning and heat pump systems has been done by Mohanraj et al. [13]. Recently, Abdel-Hanmid [14] investigated fuzzy logic control of the air conditioning system of building for efficient energy operation and comfortable environment. Ning Li [15] reported an ANN-based on-line adaptive controller to control indoor air temperature and humidity simultaneously within the entire expected controllable range by varying compressor and supply fan speeds.

Although there has been many researchers of ANN applications in air-conditioning system, quite a few studies being done on AAC with fuzzy neural network. It is well known that Fuzzy neural network (FNN), which incorporates the advantages of fuzzy inference and neuron learning, has been exploited by many researchers. FNN combine the human inference style and natural language description of fuzzy systems with the learning and parallel processing of neural networks [1-2]. In this study, FNN model for an automotive air-conditioning (AAC) system was developed which employs a vapor-compression circuit working with refrigerant R-134a. The model was developed using several steady-state input data of the AAC system obtained from the actual experiments. The presented FNN model is then used to predict the AAC system performance namely cooling capacity of the evaporator, the input power to the compressor and the coefficient of performance (COP) of the AAC system. The result values to the real experimental data are analyzed.

2 Fuzzy neural networks

Generally, a wide class of MIMO nonlinear dynamic systems can be represented by the nonlinear discrete model with an input-output description form [16]:

$$y(n) = \mathbf{f}\left[y(n-1), y(n-2), ..., y(n-k+1); \mathbf{u}(n), \mathbf{u}(n-1), ..., \mathbf{u}(n-p+1)\right] \quad (1)$$

Where y is a vector containing m system outputs, $\mathbf{u}$ is a vector of system inputs; $\mathbf{f}$ is a nonlinear vector function, representing m hyper surfaces of the system, and k, p is the maximum lag of the output and input respectively.

Selecting $[y(n-1),...,y(n-k+1); \mathbf{u}(n), \mathbf{u}(n-1),...,\mathbf{u}(n-p+1)]$, $y(n)$ as the fuzzy network's input-output x_n, y_n at time t, the above equation can be put as

$$y_n = \mathbf{f}(x_n) \qquad (2)$$

The aim of the new FNN algorithm is to approximate $\mathbf{f}$ such that :

$$\hat{y}_n = \hat{\mathbf{f}}(x_n) \qquad (3)$$

Where $\hat{y}$ is the output of FNN. The objective is to minimize the error between the system output and the actual output $\|y_n - \hat{y}_n\|$. The structure of FNN is described below.

The structure of FNN illustrated by Fig.1 consists of five layers to realize the following fuzzy rule model:

Rule:

if $(x_1$ is $A_{1k})$ $AND(x_2$ is $A_{2k})$ $AND...AND(x_r$ is $A_{rk})$,

$then(y_1$ is $a_{ik})....(y_m$ is $a_{mk})$

Where $a_{jk}(j=1,2,....,m, k=1,2,...,u)$ is a constant parameter in rule k, $A_{ik}(i=1,2,...,r)$ is the membership value of the input variable x_i in rule k , r is the dimension of the input vector $\mathbf{x}(\mathbf{x} = [x_1,...x_r]$, u is the number of fuzzy rules, m is the dimension of the output vector $\hat{y}(\hat{y} = [\hat{y}_1,...,\hat{y}_m]$. In FNN, the number of fuzzy rules u changed. Initially, there is no fuzzy rule and then they are added and removed during learning.

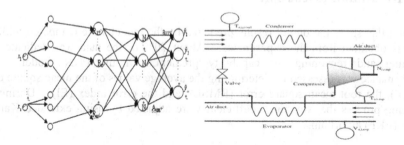

Fig.1.Structure of FNN   Fig.2. schematic diagram of the AAC system experimental rig.

Where $\hat{y} = [\hat{y}_1, \hat{y}_2,...,\hat{y}_m]^T$ is the output of the system, $a_k = [a_{1k}, a_{2k},..., a_{mk}]^T$ is its connecting weight to the output neuron. The inputs to the FNN model is(x_1, x_2, x_3, x_4), x_1, is compressor speed represented by N_{comp}, x_2 is air temperature at evaporator inlet represented by $T_{a,i,evp}$, x_3 is air temperatures at condenser inlet represented by $T_{a,i,cond}$, x_4 is air velocity at evaporator inlet $V_{a,i,evp}$. The output of the model is Y (y_1, y_2, y_3), y_1 is the cooling effect represented by Q_L , y_2 is the compressor input power represented by W, y_3 is the coefficient of performance (COP) of the system. The FNN model is developed through 2 stages: training stage and testing stage. The network is trained to predict an output based on input data during the training stage. To validate the result, the model is tested using difference sets of input. In this paper, we use MATLAT environment.

In this network, all parameters are modified by EKF algorithm [17].

3 Description of the experimental

The schematic diagram of the AAC experimental rig developed in this study is illustrated as figure 2.   It consists of three sections which are a vapor-compression refrigeration circuit, a closed air duct for evaporator section, and open air duct for condenser section respectively.   The ducts are designed according to the British Standard for rating of duct mounted air cooling coils.   Original components of a Denso air-conditioning system are used to construct the AAC system.   The Denso air-conditioning system is used in a typical compact size car working with refrigerant R-134a.   As seen in this figure, the main components of vapor compression refrigeration circuit consist of a swash-plate compressor, an evaporator, a condenser, an expansion valve, a receiver drier, a sight glass and insulated interconnecting pipes.

During the experiments, the AAC system was run until a steady-state condition was attained.   During series of experiments, four operating parameters were varied, each with their respective range as shown below.   The condenser air velocity was held constant at about 1.6 m/s at all time.

4 Result and discussion

During this study the output results predicted by the model were compared with the results obtained from the experiments.   About 70% of the data was dedicated for training and remaining for capability prediction of the FNN model.   The performance was assessed by determining the average values of the mean square error (MSE), the root mean square error (RMSE), and the error index (EI).   During the training process the weighing coefficient were adjusted by using extended Kalman filter (EKF) algorithm.

Table 1: Performance of FNN model

Layer	Node 1	Node 2	Avg. MSE	Avg. RMSE	Avg. IE
3	3	1	0.022	13.39	0.23
		2	0.004	5.60	0.10
		3	2.06×10^{-4}	1.25	0.022
		4	5.32×10^{-4}	2.05	0.037
		5	1.07×10^{-4}	0.91	0.017
		6	2.30×10^{-4}	1.50	0.027
		7	2.50×10^{-4}	1.47	0.027
		8	3.64×10^{-4}	1.74	0.032
		9	4.66×10^{-4}	2.16	0.038
		10	7.06×10^{-4}	2.56	0.046

The properties of the FNN model for the AAC system are tabulated in table 1. Figs. 3 are the plots of performance of the AAC system predicted by the FNN model vs. the corresponding values obtained from experiments.   Note that in all cased, the correlation coefficients R are very close to unity.   This indicates that the FNN model was able to predict the performance parameters of the AAC system with a very good accuracy.   As shown in Fig.3, the plot of cooling effects predicted by the FNN model vs. the values obtained from the experiments.   The FNN predictions yield a mean square error (MSE) of about 1.09×10^{-5}kw, root mean square error (RMSE) of 0.33%, error index of 0.56% and correlation coefficient, R of 0.99.   These values show that the FNN model is able to predict the heat absorbed by the refrigerant in the evaporator with a very good accuracy.

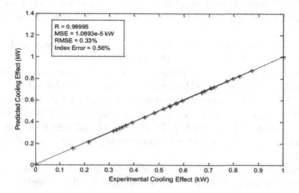

Fig.3. FNN prediction of the cooling effect   vs. the experimental results.

5   Conclusions

In this paper a Fuzzy neural network (FNN) based on the functional equivalence between a RBF and a FIS has been developed to predict the performance parameters of an experimental AAC system.   The FNN model contains five layers.   Training and testing data set for the FNN model were obtained from steady state tests conducted on the AAC experimental rig.   The cooling load, compressor power input and the coefficient of performance of the AAC system experimental rig are predicted using the trained FNN model.   The mean square error, root mean square error, error index and the correlation coefficient were used to assess the performance of the FNN model.   From the result of the experiments, the FNN model was found to be capable of accurately predicting the performance parameters of the AAC system.   All performance parameters are found to be very close to unity so that it indicates that the FNN model can predict the performance parameters of the AAC system with a very good accuracy.

Acknowledgements. The authors would like to acknowledge the supports from Fujian university of science and technology (NO. GY-Z13103) and fund provided by Fujian geography and information center (NO. 2016JX04.)

References

1. Ashrae Handbook, HVAC Systems and Equipment, ASHRAE Publications, 2008.
2. Navale R L, Nelson R M. Use of genetic algorithms and evolutionary strategies to develop and adaptive fuzzy logic controller for a cooling coil – Comparison of the AFLC with a standard PID controller[J]. Energy & Buildings, 2011, 45(1):169–180.
3. N. Etik, N. Allahverdi, I.U. Sert, I. Saritas, Fuzzy expert systemdesign for operating room air-conditioncontrol systems, ExpertSyst. Appl. 36 (2009) 9753–9758.
4. A. Kargilis, Design and Development of Automotive Air Conditioning Systems,ALKAR Engineering Company, 2003.
5. P.J. Rubas, C.W. Bullard, Factors contributing to refrigerator cycling losses,International Journal of Refrigeration 18 (3) (1995) 168e176.
6. E.B. Ratts, J.S. Brown, An experimental analysis of the effect of the refrigerantcharge level on an automotive refrigeration system, International Journal ofThermal Science 39 (2000) 592e604.
7. O.M. Al-Rabghi, A.A. Niyaz, Retrofitting R-12 car air conditioner with R-134arefrigerant, International Journal of Energy Research 24 (2000) 467e474.
8. O. Kaynakli, I. Horuz, An experimental analysis of automotive air conditioningsystem, International Communications in Heat and Mass Transfer 30 (2003)273e284.
9. M. Hosoz, M. Direk, Performance evaluation of an integrated automotive airconditioning and heat pump system, Energy Conversion and Management 47(2006) 545e559.
10. Kamar H M, Ahmad R, Kamsah N B, et al. Artificial neural networks for automotive air-conditioning systems performance prediction[J].Applied Thermal Engineering, 2013, 50(1):63-70.
11. Y.-C. Chang, Sequencing of chillers by estimating chiller power consumptionusing artificial neural networks, Building and Environment 42 (2007)180e188.
12. H.M. Ertunc, M. Hosoz, Comparative analysis of an evaporative condenserusing artificial neural network and adaptive neuro-fuzzy inference system,International Journal of Refrigeration 31 (2008) 1426e1436.
13. M. Mohanraj, S. Jayaraj, C. Muraleedharan, Applications of artificial neuralnetworks for refrigeration, air-conditioning and heat pump systems -a review, Renewable and Sustainable Energy Reviews 16 (2012) 1340e1358.
14. Attia A H, Rezeka S F, Saleh A M. Fuzzy logic control of air-conditioning system in residential buildings[J]. AEJ - Alexandria Engineering Journal, 2015, 36(3):395-403.
15. Li N, Xia L, Deng S, et al. On-line adaptive control of a direct expansion air conditioning system using artificial neural network[J]. Applied Thermal Engineering, 2013, 53(1):96–107.
16. Rong H, Ye X, Xiang X. A Self Adaptive Incremental Learning Fuzzy Neural Network Based on the Influence of a Fuzzy Rule[C]// International Conference on Intelligent Information Hiding and Multimedia Signal Processing. IEEE, 2015.
17. Hu R, Sha Y, Yang H Y. A Novel Uninorm-Based Evolving Fuzzy Neural Networks[M]// Intelligent Data Analysis and Applications. Springer International Publishing, 2015:469-478.

An Optimization Approach for Potential Power Generator Outputs Based on Parallelized Firefly Algorithm

Chia-Hung Wang[1,2,*], Trong-The Nguyen[3,#], Jeng-Shyang Pan[1,2,+], Thi-kien Dao[3]

[1]College of Information Science and Engineering, Fujian University of Technology,
Fuzhou City, Fujian Province 350118, China
[2]Fujian Provincial Key Laboratory of Big Data Mining and Applications,
Fuzhou City, Fujian Province 350118, China
[3]Department of Information Technology, Hai-Phong Private University, Vietnam

[*]jhwang728@hotmail.com, [#]vnthe@hpu.edu.vn, [+]jengshyangpan@fjut.edu.cn

Abstract. The parallel processing plays an important role in efficient and effective computations of function optimization. This paper proposes a parallel optimization algorithm for the committed generating electric powers of thermal plants based on Firefly Algorithm (FA). An economic condition for the power systems can be determined through the optimization techniques for transmission loss, power balance and generation capacity. The aim of the proposed parallel algorithm with communication strategies is to correlate agents in subgroups and to share the computation load over several machines. The expense criteria for each generation unit and the coefficient matrix are formulated as the objective function, which is to be computed in our parallel optimization for electricity flow of the transmission losses in the power systems. Four selected functions and two cases of six units and fifteen units of thermal plants are tested in our experiments for optimization. Through a comparison between different methods in related works, our experimental results show that the proposed method results in the higher effect and accuracy.

Keywords: Parallel firefly algorithm, Electric power generating plant outputs, Economic load dispatch

1   Introduction

The parallel processing is an essential requirement for optimum computations in modern equipment. The parallelized strategies simply share the computation load over several processors [1]. The sum of the computation time for all processors can be reduced compared with the single processor works on the same optimum problem. For optimal combination, the parallel computation is a more significant to determine power outputs for all generating units in the economic load dispatch problem (ELD), because of it is a large-scale non-linear constrained optimization problem [2]. In traditional methods, the cost function of each generator was represented approximately by a single quadratic

© Springer International Publishing AG 2017   297
J.-S. Pan et al. (eds.), *Advances in Intelligent Information Hiding
and Multimedia Signal Processing*, Smart Innovation, Systems and Technologies 64,
DOI 10.1007/978-3-319-50212-0_36

polynomial. These methods offered good results but when the search space was non-linear and it had discontinuities they became very complicated with a slow convergence rate, and they were not always seeking to the feasible solution. So, the obtained global optimal solution became the challenging ELD [3]. A numerical method, dynamic programming method [4] was one of the approaches to solving this drawback through the non-linear and discontinuous ELD problem, but it suffered from the problem of the curse of dimensionality or local optimality.

Recently, many swarm intelligence algorithms have been developed to solve the optimization problem, e.g., genetic algorithm (GA) [5], particle swarm optimization (PSO)[6], and firefly algorithm (FA) [7]. These have proved to be very effective in solving nonlinear problems without any restriction on the shape of the cost curves. These algorithms are a promising answer to overcoming the above-mentioned drawbacks. However, these heuristic methods do not always guarantee to discover the globally optimal solution in finite time. The parallel algorithm with communication strategies can improve the diversity solutions to avoid of dropping locally optimal, but to get globally optimal solution in the optimization. In this paper, the concept of parallel processing and a communication strategy are applied to parallel FA for solving for the committed electric power plant planning problem.

2   Problem description

ELD problem assumes quadratic cost function along with system power demand and operational limit constraints [8]. The objective of ELD problems is to minimize the fuel cost of committed generators (units) subjected to operating constraints. Practically, the economic power dispatch problem can be formulated as:

Minimize

$$F_T = \sum_{i=1}^{n} F_i(P_i) \tag{1}$$

Subject to

$$\sum_{i=1}^{n} P_i = P_D + P_L \tag{2}$$

$$P_i^{Min} \leq P_i \leq P_i^{Max} \quad i = 1,2..n, \tag{3}$$

where, F_T is the total fuel cost, n is the number of units, Fi and Pi are the cost function and the real power output of ith unit respectively, P_D is the total demand, P_L is the transmission loss, P_i^{Min} and P_i^{Max} and are the lower and upper bounds of the ith unit respectively. The equality constraint, Eqs.(2) states that the total generated power should be balanced by transmission losses and power consumption while Eqn.(3) denoting unit's operation constraints. Traditionally, the fuel cost of a generator is usually defined by a single quadratic cost function.

$$F_i(P_i) = \gamma_i P_i^2 + \beta_i P_i + \alpha_i, \tag{4}$$

where, $\alpha i, \beta i,$ and γi are cost coefficients of the ith unit. One common practice for including the effect of transmission losses is to express the total transmission loss as a quadratic function of the generator power outputs in one of the following forms [1]:

$$P_L = \sum_{i=1}^{N} \sum_{j=1}^{N} p_i B_{ij} p_j \tag{5}$$

Kron's loss formula:

$$P_L = \sum_{i=1}^{N}\sum_{j=1}^{N} P_i B_{ij} P_j + \sum_{j=1}^{N} B_{0j} P_j + B_{00}$$

(6)

where B_{ij} is called the loss coefficients. B-matrix loss formula:

$$P_L = P^T B_P + P^T B_0 + B_{00},$$

(7)

where, P denotes the real power output of the committed units in vector form and B, B_0 and B_{00} are loss coefficients in matrix, vector and scalar respectively, which are assumed to be constant, and reasonable accuracy can be achieved when the actual operating conditions are close to the base case where the B-coefficients were derived.

3   Parallelized Firefly Algorithm

Parallelized firefly algorithm (pFA) is extended from the basic version of firefly algorithm (FA). Therefore, the basic version of FA needs briefly to review and pFA will be then presented respectively.

Firefly Algorithm

FA was developed by the inspiration of behavior of fireflies [9]. Three idealized rules were simulated in FA including fireflies brightness is attractive to each other ones; less bright one will move towards the brighter one, and attractiveness is proportional to the brightness and they both decrease as their distance increases. The brightness of a firefly is affected or determined by the landscape of the objective function. A firefly's attractiveness is proportional to the light intensity seen by adjacent fireflies.

$$\beta = \beta_0 e^{-\gamma r^2},$$

(8)

where β is a variation of attractiveness with the distance r; β_0 is the attractiveness at $r = 0$. The movement of a firefly i is attracted to another more attractive (brighter) firefly j is determined by:

$$x_i^{t+1} = x_i^t + \beta_0 e^{-\gamma r_{ij}^2}(x_j^t - x_i^t) + \alpha_t \epsilon_i^t,$$

(9)

where x_i and x_j are locations of fireflies i and j. The movement of firefly i is attracted to another more attractive (brighter) firefly j is determined by the second term is due to the attraction. The third term is randomization with α_t being the randomization parameter, and ϵ_i^t is a vector of random numbers drawn from a Gaussian distribution or uniform distribution at time t. If $\beta_0 = 0$, it becomes a simple random walk.

Parallel Firefly Algorithm

In the parallel structure, several groups are created by dividing the population into subpopulations to construct the parallel processing. Each of the subpopulations evolves independently in regular iterations. They only exchange information between subpopulations when the communication strategy is triggered. It results to achieve the benefit of cooperation. The parallelized FA is designed based on original FA optimization. The fireflies in FA are divided into G subgroups. Each subgroup evolves by FA optimization independently, i.e. the subgroup has its own fireflies and finest solution. These finest fireflies among all the fireflies in one group will be traveled to another group to replace the poorer fireflies and update after running some fixed iterations.

The detail processing steps of the pFA are as follows.

1. Initialization: Generate fireflies' population and divide them into G subgroups. Each subgroup is initialized by pFA independently. Assign R-the number iterations for executing the communication strategy, N_j the number fireflies and X_{ij}^t the solutions for the j-th group, i = 0, 1,..., N_j–1; j = 0, 1,..., G–1, where G is the number of groups, Nj is the subpopulation size and t is the current iteration and set to 1.

2. Evaluation: Evaluate the value of objective function $f(X_{ij}^t)$ for fireflies in *j-th* group.

3. Update: if light intensity, move firefly *i* toward firefly *j* by using the update the Global firefly, Eq. (9).

4. Communication Strategy: Migrate *k* best fireflies among G_j^t to (*j*+1)-th group G_{j+1}^t, mutate G_{j+1}^t by replacing *k* poorer fireflies in that group and update every group in each R iterations.

5. Termination checking: Repeat step 2 until the predefined value of the function is achieved or the maximum number of iterations has been reached. Record the best value of the function $f(X_{ij}^t)$ and the best firefly among all the fireflies X_{ij}^t.

Testing results for numerical functions

Four benchmark functions are used to test the accuracy and the convergence of the proposed algorithm. All the benchmark functions for the experiments are averaged over different random seeds with 25 runs. The goal of the optimization is to minimize the outcome for all benchmarks. The initial range and the total iteration number for all test functions are listed in Table I.

Table 1. The initial range and the total iteration of four tested standard functions

Test functions		Range [x_{min}, x_{max}]	Max iteration
$f_1(x) = \sum_{i=1}^{n-1}(100(x_{i-1} - x_i^2)^2 + (1 - x_i)^2$	Rosenbrock	[−100,100]	200
$f_2(x) = \sum_{i=1}^{n}(\sum_{k=1}^{i} x_i)^2$	Quadric	[−100,100]	200
$f_3(x) = \sum_{i=1}^{N}[10 + x_i^2 - 10\cos 2\pi x_i]$	Rastrigin	[−30,30]	200
$f_4(x) = \sum_{i=1}^{N} x_i^2$	Sphere	[−100,100]	200

The final result is obtained by taking the average of the outcomes from all runs. The results are compared with the FA. Table 2 compares the quality of performance and time running for numerical problem optimization between parallel FA and original FA. Clearly, almost these cases of testing benchmark functions for parallel FA are faster convergence than original FA. The average of four benchmark functions evaluation of minimum function 25 runs is 7.75E+06 with average time consuming 7.48E+00 seconds for parallel FA and 1.03E+07 with average time consuming 7.87E+00 for original FA get better 36% in accuracy and 5% in time speed respectively.

Table 2. Comparison between the FA and pFA in terms of quality performance evaluation and speed

Test Functions	Performance evaluation		Accuracy % Comp- arison	Time consumption (seconds)		Speed % Comp- arison
	FA	Parallel FA		FA	Parallel FA	
$f_1(x)$	4.10E+07	3.10E+07	24%	6.30E+00	6.08E+00	4%
$f_2(x)$	2.47E+03	1.55E+03	37%	1.27E+01	1.22E+01	4%
$f_3(x)$	1.66E+02	8.57E+01	48%	6.54E+00	6.18E+00	6%
$f_4(x)$	4.70E+03	3.16E+03	33%	5.96E+00	5.48E+00	9%
Average	1.03E+07	7.75E+06	36%	7.87E+00	7.48E+00	5%

4 Optimization for committed electric power generating plant outputs

The search space in the ELD consists of two kinds of points: feasible and unfeasible. The feasible points satisfy all the constraints, while the unfeasible points violate at least one of them. Therefore, the solution or set of solutions obtained as the final result of an optimization method must necessarily be feasible, i.e., they must satisfy all constraints. It is common to handle constraints using concepts of penalty functions which penalize unfeasible solutions, i.e., one attempt to solve an unconstrained problem in the search space *solution* using a modified fitness function [10]. The penalty function is applied to the fitness function in Eq. (1) as follows:

$$Min \ f = \begin{cases} f(P_i), & \text{if } P_i \in F \\ f(P_i) + penalty(P_i), & \text{otherwise} \end{cases} \tag{10}$$

where penalty *(Pi)* equals zero if there is not any violated constraint; otherwise it is greater than zero. The penalty function is usually measured based on a distance of the nearest solution in the feasible region or to the effort to repair the solution.

$$Min \ f = \sum_{i=1}^{n} F_i(P_i) + q_1 \left(\sum_{i=1}^{n} P_i - P_L - P_D \right)^2 + q_2 \sum_{j=1}^{n} V_j \tag{11}$$

where q_1 and q_2 are penalty factors, positive constants associate with the power balance and prohibited zones constraints, respectively. These penalty factors were tuned empirically and their values could be q_1 set to 1000 and q_2 set to 1 in the studied cases in simulation section. The V_j is expressed as follows:

$$V_j = \begin{cases} 1, & \text{if } P_j \text{ violates the prohibited zones} \\ 0, & \text{otherwise} \end{cases} \tag{12}$$

The process of pFA for ELD is depicted as follows:

Step 1. Initialize the firefly population, the attractiveness β with the distance r is defined.

Step 2. Update the velocities to update the location of the fireflies.

Step 3. Rank the fireflies according to their fitness value of the function as Eq. (11), find the current near best solution found so far and then update the locations and the emission rate.

Step 4. Check the termination condition to decide whether go back to step 2 or end the process and output the result.

5   Experimental results

Two cases of six-unit and fifteen-unit systems are tested for the proposed method of pFA. The experimental results are compared to other methods such as GA, FA, and PSO [11][12][13]. The optimization goal for the objective function as Eq. (11) is to minimize the outcome and then to dispatch ELD for power outputs. The parameters setting for pFA is the initial attractiveness β with the distance r equals to 0.5, and 0.2, the total population size $n = 20$ and the dimension of the solution space $D = 6$ for six-unit system ($D = 15$ for fifteen-unit system). Objective function contains the full iterations of 200 is repeated by different random seeds with 10 runs. The final results are obtained by taking the average of the outcomes from all runs. The results are compared with the GA and PSO for ELD respectively.

A. Case study 1- Six units system
A system with six thermal units is used to demonstrate how the proposed approach works. The load demand is 1200MW. The characteristics of the six thermal units are given in Tables 3.

Table 3. The generating units capacity with 1200MW power demand and coefficients

Unit	γ \$/MW^2	β \$/MW	α \$	Pmin MW	Pmax MW
1	0.0070	7.0	240	100	500
2	0.0095	10.0	200	50	200
3	0.009	38.5	220	80	300
4	0.0090	11.0	200	50	150
5	0.0080	10.5	220	50	200
6	0.0075	12.0	120	50	120

In this case, each solution S contains six generator power outputs, such as P1, P2, P3, P4, P5 and P6. Initialization of solution for power generating units is generated randomly. The dimension of the population is equal to 6. In normal operation of the system, the loss coefficients for the 100-MVA base capacity are listed as follows.

$$B_{ij} = 10^{-3} \times \begin{bmatrix} 0.14 & 0.17 & 0.15 & 0.19 & 0.26 & 0.22 \\ 0.17 & 0.60 & 0.13 & 0.16 & 0.15 & 0.20 \\ 0.15 & 0.13 & 0.65 & 0.17 & 0.24 & 0.19 \\ 0.19 & 0.16 & 0.17 & 0.71 & 0.30 & 0.25 \\ 0.26 & 0.15 & 0.24 & 0.30 & 0.69 & 0.32 \\ 0.22 & 0.20 & 0.19 & 0.25 & 0.32 & 0.85 \end{bmatrix},$$

$$B_0 = 10^{-3}[-0.390 - 0.129\ 0.714\ 0.059\ 0.216 - 0.663],$$

$$B_{00} = 0.056,$$

and $P_D = 1200\text{MW}$.

Table 4 provides the statistic optimal results of the evaluation fitness function contain the full iterations of 200 is repeated by different random seeds with 10 runs. That involved the average of 10 runs for generation power outputs, generation total cost, total power loss value, and total computation times respectively.

Table 4. The best power output for six-generator systems

Unit Output	FA[12]	GA[11]	PSO[13]	pFA
P1 (MW)	459.5422	459.5422	458.0056	459.2242
P2 (MW)	166.6234	166.6234	178.5096	171.5735
P3 (MW)	258.035	253.035	257.3487	255.4936
P4 (MW)	117.4263	117.4263	120.1495	119.8301
P5 (MW)	156.2482	153.2482	143.784	154.7214
P6 (MW)	85.88567	85.88567	76.75549	73.76758
Total Power Output (MW)	1239.761	1235.761	1234.553	**1234.530**
Total Generation Cost ($/h)	14892	14862	14861	**14845**
Power Loss (MW)	37.7610	35.7610	34.5531	**34.5300**
Total CPU time (sec)	296	256	201	202

Fig. 1a shows the comparison of the proposed method of pFA for optimizing the committed electric power generating plant outputs with six-generating unit system in distribution outline of the best solution for 200 iterations is repeated 10 trials, with GA and PSO methods in the same setting of the condition.

B. Case study 2- fifteen-unit system

The system contains fifteen thermal units [4]. The load demand is 1700MW. The characteristics of the fifteen thermal units are given in Tables 5. Each solution S contains fifteen generator power outputs, listed from P1 to P15. The dimension of the population is equal to 15.

Table 5. The generating units capacity with 1700MW power demand and coefficients

Unit	γ \$/MW^2	β \$/MW	α \$	P$_{min}$MW	P$_{max}$MW
1	0.000230	10.5	670	150	455
2	0.000185	10.6	575	150	455
3	0.001125	8.5	375	20	130
4	0.001125	8.5	375	20	130
5	0.000205	10.5	460	150	470
6	0.000300	10.0	630	135	460
7	0.000362	09.7	549	135	465
8	0.000338	11.3	227	60	300

9	0.000807	11.2	173	25	162
10	0.001203	10.7	175	25	160
11	0.003587	10.3	187	20	80
12	0.005513	09.9	231	20	80
13	0.000371	13.1	225	25	85
14	0.001929	12.3	309	20	60
15	0.004547	12.4	325	15	55

In normal operation of the system, the loss coefficients with the 100-MVA base capacity are as follows.

$B_{i0} = 10^{-3} \times [-0.1 - 0.2 \ 2.8 - 0.1 \ 0.1 - 0.3 - 0.2 - 0.2 \ 0.6 \ 3.9 - 1.7 \ 0.0 - 3.2 \ 6.7 - 6.4]$; $B_{00} = 0.0055$, $P_D = 1700$ MW. Table 6 provides the statistic results that involved the generation cost, evaluation value, and average CPU time.

Table 6. The best power output for fifteen-generator systems

Unit Outputs	FA[12]	GA[11]	PSO[13]	pFA
P1 (MW)	455.21	455.01	455.01	455.01
P2 (MW)	91.98	93.98	120.03	85.00
P3 (MW)	90.06	85.06	84.85	84.83
P4 (MW)	89.97	89.97	75.56	45.29
P5 (MW)	156.00	150.00	162.94	152.00
P6 (MW)	350.76	350.76	322.48	357.49
P7 (MW)	226.36	226.36	165.70	242.22
P8 (MW)	60.00	60.00	60.34	60.56
P9 (MW)	52.37	52.37	91.84	29.60
P10 (MW)	26.10	25.10	45.10	50.40
P11(MW)	25.96	25.96	42.70	30.60
P12(MW)	74.01	74.01	77.97	80.00
P13(MW)	61.99	66.99	45.38	66.27
P14 (MW)	36.22	34.22	47.37	26.24
P15 (MW)	52.05	51.05	55.00	55.00
Total Power Output (MW)	1847.84	1837.84	1829.27	**1827.60**
Total Generation Cost ($/h)	1241.09	1235.09	1234.61	**1234.61**
Power Loss (MW)	147.84	137.84	129.27	**127.60**
Total CPU time (sec)	410	370	303	305

Fig.1b shows the comparison of the proposed method of pFA, with FA, GA, and PSO methods for optimizing the committed electric power generating plant outputs with 15 generating unit system in convergence property distribution outline of the best solution for 200 iterations is repeated 10 trials. Compared with FA, GA, and PSO methods in the same condition, the proposed method pFA outperforms the other methods.

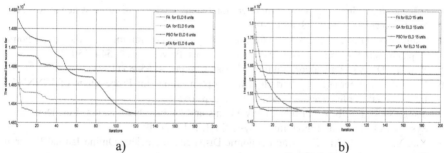

Fig. 1. Comparison of convergence characteristic between four methods (FA GA, PSO, and the proposed approach of pFA) for: a) six-generator system, b) fifteen-generator system

The observed results of quality performance in terms of convergence speed and time consumption show that the proposed method of parallel optimization outperforms the other methods.

6   Conclusion

In this paper, we presented an approach based on the parallel firefly algorithm for optimizing the committed electric power generating plant outputs. The proposed parallel algorithm with the communication strategy can share the computation load over several machines and   improve the diversity solutions to avoid of dropping locally optimal, but to get globally optimal solution in the optimization. Four selected benchmark function and two cases of six-units and fifteen units of thermal plants were implemented to evaluate the solution quality and computation efficiency of the proposed method. In generating plants, the linear equality and inequality constraints, and transmission loss were considered to formulate the objective function. Our simulation results were compared with FA, GA, and the PSO methods. It shows that the proposed pFA method obtains the better quality for optimal solutions, and less computing time than other three methods.

Acknowledgments

Dr. Chia-Hung Wang is thankful for the partial support from Fujian Provincial Department of Science and Technology, China, under Grant 2016J01330.

References

1. Dao, T.-K., Pan, T.-S., Nguyen, T.-T., Pan, J.-S.: Parallel bat algorithm for optimizing makespan in job shop scheduling problems. J. Intell. Manuf. 1–12 (2015).
2. Dao, T., Pan, T., Nguyen, T., Chu, S.: Evolved Bat Algorithm for Solving the Economic Load Dispatch Problem. In: Advances in Intelligent Systems and Computing. pp. 109–119 (2015).

3. Salgado, F., Pedrero, P.: Short-term operation planning on cogeneration systems: A survey. Electr. Power Syst. Res. 78, 835–848 (2008).
4. Liang, Z.X., Glover, J.D.: A zoom feature for a dynamic programming solution to economic dispatch including transmission losses. IEEE Trans. Power Syst. 7, 544–550 (1992).
5. Whitley, D.: A genetic algorithm tutorial. Stat. Comput. 4, 65–85 (1994).
6. Kennedy, J., Eberhart, R.: Particle swarm optimization. In: Proceedings of ICNN'95 - International Conference on Neural Networks. p. 16 (1995).
7. Yang, X.-S.: Firefly Algorithm , Levy Flights and Global Optimization. Res. Dev. Intell. Syst. XXVI Inc. Appl. Innov. Intell. Syst. XVII. 1–10 (2010).
8. Xia, X., Elaiw, a M.: Dynamic Economic Dispatch : A Review. Online Jaurnal Electron. Electr. Eng. 2, 234–245 (2009).
9. Yang, X.S.: Firefly algorithms for multimodal optimization. In: Lecture Notes in Computer Science (including subseries Lecture Notes in Artificial Intelligence and Lecture Notes in Bioinformatics). pp. 169–178 (2009).
10. Rao, S.S.: Engineering Optimization: Theory and Practice. (2009).
11. Kim, J.O., Shin, D.J., Park, J.N., Singh, C.: Atavistic genetic algorithm for economic dispatch with valve point effect. Electr. Power Syst. Res. 62, 201–207 (2002).
12. Apostolopoulos, T., Vlachos, A.: Application of the Firefly Algorithm for Solving the Economic Emissions Load Dispatch Problem, 1-23 (2011).
13. Sun, J., Palade, V., Wu, X.J., Fang, W., Wang, Z.: Solving the power economic dispatch problem with generator constraints by random drift particle swarm optimization. IEEE Trans. Ind. Informatics. 10, 222–232 (2014).

A Design of Genetic Programming Scheme with VLIW Concepts

Feng-Cheng Chang[1] and Hsiang-Cheh Huang[2]

[1] Dept. of Innovative Information and Technology, Tamkang University, TAIWAN
135170@mail.tku.edu.tw
[2] Dept. of Electrical Engineering, National University of Kaohsiung, TAIWAN
hch.nuk@gmail.com

Abstract. Genetic programming (GP) is inspired by the popular genetic algorithm (GA). The searching result of GP is a program that includes both operators and operands. The operators are the obstacle to the crossover and mutation process because invalid programs would be generated. In this paper, the concepts of VLIW is incorporated in the design of a GP scheme. A program in the proposed scheme is represented using only operands. The simulation results show that this approach is feasible and the performance could be increased by the instruction-level parallelism of the VLIW structure.

Keywords: Genetic programming; GP; Very long instruction word; VLIW

1 Introduction

Genetic algorithm (GA)[5][6] is a kind of Evolutionary computation (EC)[1]. It is an approach to find the sub-optimal solution to various kinds of problems. The basic concept is to design the data structure for representing the solution. In the optimization process, a population of the data structure instances are generated and evolved. The individuals in the population explore the solution space by evolution. The acceptable solution would be found if the evolution mechanism is properly designed.

GA is useful for the problems that the closed form solutions are not available. The process searches for a sub-optimal solution with the affordable resources by randomly explore the solution space. Genetic programming (GP)[7][10] inherits the concepts from GA for finding a program. Its computation complexity is generally higher than GA's. In many EC algorithms, including GA and GP, part of the computation can be parallelized[3]. With the advance of computer technology, concurrent computation becomes more affordable than before. This makes GP practical to solve more problems in time.

In this paper, we develop a scheme for GP that exploits the concurrency nature. In Sec. 2, we briefly introduce GP and some implementation approaches. In Sec. 3, we propose a GP scheme with the VLIW concepts. In Sec. 4, a few preliminary simulations are performed and discussed. Finally, we conclude this work and describe the future directions in Sec. 5.

© Springer International Publishing AG 2017   307
J.-S. Pan et al. (eds.), *Advances in Intelligent Information Hiding
and Multimedia Signal Processing*, Smart Innovation, Systems and Technologies 64,
DOI 10.1007/978-3-319-50212-0_37

2   Genetic Programming

In this section, the concepts of genetic programming (GP) is introduced. GP is inspired by GA and it is useful in various problem domains[11][9]. However, their properties are different in many aspects. The differences are described in the following sections.

2.1   Concepts of GA and GP

The concept of GA is to evolve the population (a set of individuals) to search for the optimal solution. An individual of GA is a bit vector representing a solution (data point) in the search space. The typical process of GA consists of the following operations: initialization, selection, reproduction, and termination.

GP specializes GA in that each individual represents a "program". A typical representation of a GP individual is an expression tree. The leaf nodes represent the input variables and constants, and the tree nodes represent operators. The complexity of a program is limited by the maximum depth of the tree. The evolution process of GP is similar to that of GA with some modifications:

- The crossover operator applies to subtrees. The bit-level operation in GA is not applicable in GP because an operator has its specific meaning.
- The mutation operator applies to a tree node or a leaf node. When mutating a tree node, the value should correspond to a valid operator. Another approach is to substitute the whole subtree with a randomly generated subtree.
- Theoretically, it is impossible to determine the *100% fit*. In GA, it is possible to find the exact solution given unlimited time and precision. In GP, we search a *program behavior* that satisfies the given training set. A training set is typically much smaller than the real data domain. Therefore, the problem of over-fit is an issue in GP. A simple approach is to terminate the evolutionary process after a given number of iterations.

2.2   Variants of GP

In addition to the tree-structured representation, several linear structures have been developed. Many of them may produce invalid expressions during evolutions. One of the exceptions is the Gene Expression Programming (GEP)[4] which uses a syntax that is safe for crossover and mutation operations.

Another popular approach is called the *Linear Genetic Programming (LGP)*[2][8]. The representation of an LGP program follows the design of CPU instructions. A program is a sequence of instructions, and each instruction consists of an operator and operands (if any). The genetic operations apply to instruction level and usually simpler and faster. However, simply reuse the instructions in a linear structure is not feasible in many cases. For example, crossover instructions inside a conditional instruction cause the shift of the branching target location(s). It is observed that LGP tends to converge to a good solution in many cases. One of the reasons is that an LGP may contain redundant instructions. After crossover or mutation, a redundant instruction may become a meaningful instruction. Thus, the probability of exploring toward the good direction is increased.

3   GP Scheme with VLIW Concepts

From the literature, we learn that GA performs well for searching sub-optimal solutions. However, GP is not effective enough to be useful in practical problems. The bit-level crossover and mutation is the key to the success of GA. There are two noteworthy assumptions of GA:

- The probability of producing an invalid solution is low after crossover or mutation.
- The more similar the bit pattern is to the optimal solution, the closer the sub-optimal solution is to it.

Although inspired by GA, the two assumptions are not easy to be satisfied in GP. A GP solution is a mixture of operators and operands. Unless a special syntax is designed, op-level (operator and operands) crossover and mutation are likely to produce invalid programs. Considering the second assumption, it is obvious that op-level similarity is not significant to program similarity. For example, $c=a+b$ is quite different from $c=a*b$ and can be treated as two different-purposed programs.

Based on the above analysis, we found that the existence of operators causes the problem. A GA solution is a sequence of data elements; the related operators are defined as part of the fitness function. A GP solution contains both the data and the related operators; the fitness function is typically used to measure the difference between the evaluated values and the expected values. In the following proposed GP scheme, we represent a program using only operands.

In modern computer architectures, there is a design called very long instruction word (VLIW). It was proposed to exploit instruction-level parallelism. A VLIW instruction encodes a few operation fields that can be executed in parallel. Each operation field specifies the operator and operands for execution. Suppose that all the supported operators are executed in each instruction, the operators can be implicit. Combining the specialized VLIW instruction and the redundant operations in LGP, the scheme for designing an operand-only program representation is developed as follows.

1. Define the available operators. For each operator, define the associated operands. For each operand, define the valid data range.
2. Define the order of the operators. The order is fixed in all the instructions. Thus the operators can be implicit.
3. An instruction is a sequence of operand groups. Each operand group is a sequence of operands for the corresponding operator.
4. A program is a sequence of instructions. It can be viewed as a table of operands (Fig. 1).
5. Define the GP crossover and mutation operations.

6. Define the normalization method to ensure an operand valid after crossover or mutation. If the operands are defined to be always valid after changes, this step can be skipped.

With this scheme, the operators are implicit in a program and cannot be changed by the GP operations. The modification to the conventional GP process is that the program normalization is performed (if necessary) whenever a program is generated or changed.

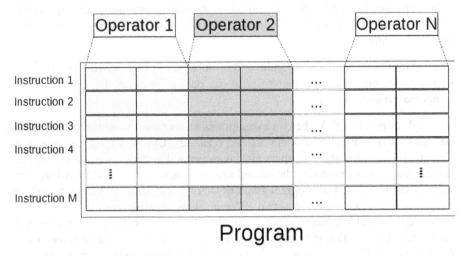

Program

Fig. 1. An example program representation of the proposed GP scheme.

4   Simulations and Discussions

In this section, a few simulations are configured and performed (Sec. 4.1). The simulation results are discussed in Sec. 4.2.

4.1   Simulation Configuration

The target algorithm is the arithmetic mean of four integers. The range of each integer is $[0, 255]$. The test dataset of 100 cases is randomly generated. Two instruction architectures are defined for the simulations. The first one (A1) provides the four primitive arithmetic operations and the load operator. The second one (A2) prepends an additional $add(\bullet)$ operator to A1. The definitions of the operators are listed below:

$$add(r_1, r_2, r_3) \qquad R[r_3] = R[r_1] + R[r_2]$$
$$subtract(r_1, r_2, r_3) \qquad R[r_3] = R[r_1] - R[r_2]$$
$$multiply(r_1, r_2, r_3) \qquad R[r_3] = R[r_1] * R[r_2]$$
$$divide(r_1, r_2, r_3) \qquad R[r_3] = R[r_1] / R[r_2]$$
$$load(r_1, v_1) \qquad R[r_1] = v_1$$

$R[\bullet]$ denotes a register of floating-pointer number. The valid index value r_k is from zero (inclusive) to the number of registers (exclusive). The value v_l is a non-negative integer, and its range $[0,9999]$ is fixed throughout the simulations. The normalization method of r_k and v_l is the modulus of the maximum value plus one. The crossover and mutation are performed at the op-level. To simplify the simulation, we choose the single-point crossover and random selection. The population size is *200*, the selection rate is *0.25*, and the mutation rate is *0.01*. The evaluation results is stored in the last register. The fitness function is the negative value of the mean square error (MSE) to the test data.

In addition to the above configuration, we extend the simulations to use another selection strategies. The original strategy randomly selects *25%* of the population for breeding the next generation. The second strategy is to keep the best one in the selected portion (the concept of elitism in GA). Combining with the instruction types A1 and A2, the elitism version is called A1e and A2e respectively.

4.2   Simulation Results

We vary the control parameters to perform the simulations: the type of instructions (A1, A2, A1e, or A2e), the number of registers (8 or 16), and the number of instructions (10 or 20). For each simulation, 5000 generations are evolved to search for the best program. We trace the MSE of the best program in each iteration and plot Fig. 2 to Fig. 5.

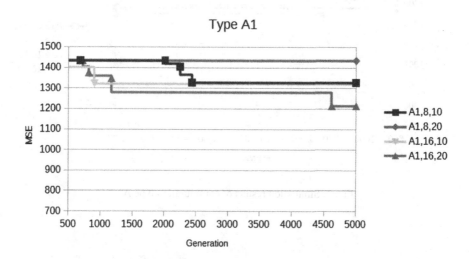

Fig. 2. Simulation results of configuration type A1.

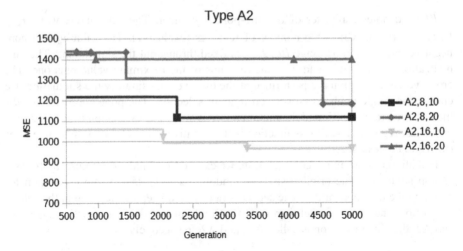

Fig. 3. Simulation results of configuration type A2.

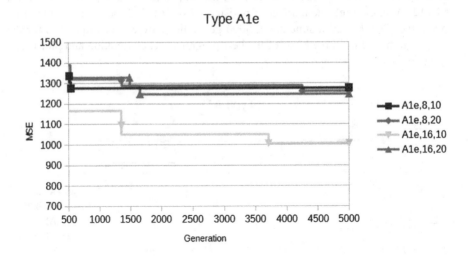

Fig. 4. Simulation results of configuration type A1e.

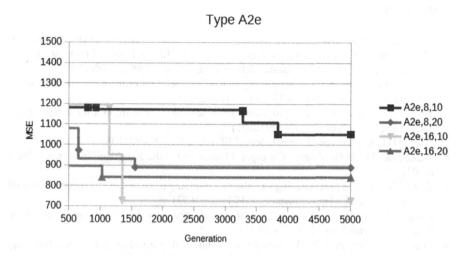

Fig. 5. Simulation results of configuration type A2e.

Comparing the results, some properties are observed:

- A1e is better than A1, and A2e is better than A2. It indicates that the elitism in GA is also applicable in our GP scheme.
- The performance of A2 is better than A1 (so for A2e and A1e) in terms of mean square error and convergence trend. It indicates that instruction-level parallelism would be a good property for GP optimization.
- Using more registers is helpful for searching a good program.
- It is not easy to determine the effects of the number of instructions. It seems that this factor is coupled with the number of registers and the types of the instructions.

5 Conclusions

In this paper, we proposed a GP scheme that incorporates the VLIW concepts. It features that a program is represented with operands. The operators are implicit.Thus, the probability of invalid programs after reproduction is decreased. The simulation results show that (1) the elitism in GA can be used in this scheme; (2) the instruction-level parallelism is helpful to the searching process; and (3) more registers is also helpful.

Due to the limited time, we only perform a small number of simulations. The preliminary simulations show that our design would be a candidate approach in GP. In the future, we would like to extend the simulation scale to verify the effectiveness in different applications.

Acknowledgments. This work was partially supported by the Ministry of Science and Technology, Taiwan, under Grant MOST 104-2221-E-032-061.

References

1. Back, T., Emmerich, M., Shir, O.: Evolutionary algorithms for real world applications [application notes]. Computational Intelligence Magazine, IEEE 3(1), 64–67 (Feb 2008)
2. Brameier, M.F., Banzhaf, W.: Linear Genetic Programming. Springer US (2007)
3. Chang, F.C., Huang, H.C.: A refactoring method for cache-efficient swarm intelligence algorithms. Information Sciences 192, 39–49 (Jun 2012)
4. Ferreira, C.: Gene expression programming: a new adaptive algorithm for solving problems. Complex Systems 13, 87–129 (2001)
5. Fogel, D.B.: Evolutionary Computation. IEEE Press, New York (1998)
6. Kantardzic, M.: Data Mining: Concepts, Models, Methods, and Algorithms. IEEE Press (2003)
7. Koza, J.R.: Genetic programming - on the programming of computers by means of natural selection. Complex adaptive systems, MIT Press (1993)
8. Oltean, M., Grosan, C.: A comparison of several linear genetic programming techniques. Complex Systems 14(4), 285–314 (2003)
9. Petrovic, N., Crnojevic, V.: Universal impulse noise filter based on genetic programming. Image Processing, IEEE Transactions on 17(7), 1109–1120 (Jul 2008)
10. Walker, M.: Introduction to genetic programming. Tech. rep., (Oct 2001)
11. Wong, H.S., Guan, L.: Application of evolutionary programming to adaptive regularization in image restoration. Evolutionary Computation, IEEE Transactions on 4(4), 309–326 (Nov 2000)

A Parallel Optimization Algorithm
Based on Communication Strategy of Pollens and Agents

Pei-Wei Tsai[1,2], Trong-The Nguyen[3], Jeng-Shyang Pan[1,2],
Thi-Kien Dao[3,*], and Wei-Min Zheng[4]

[1] College of Information Science and Engineering,
[2] Fujian Provincial Key Laboratory of Big Data Mining and Applications,
Fujian University of Technology, China
[3] Department of Information Technology, Hai-Phong Private University, Vietnam
[4] Institute of Information Engineering, Chinese Academy of Sciences,
Beijing 100093,China

vnthe@hpu.edu.vn

Abstract. Unwanted convergence to a local optimum, rather than global optimum, is possible to take place in practical multimodal optimization problems. Communication between artificial agents in the stochastic algorithms is one of the solutions to this issue. This paper proposes a novel parallel optimization algorithm, namely FDA, based on the communication of the pollen in Flower pollination algorithm (FPA) with the agents in Differential evolution algorithm (DEA) to solve the optimization problems. A communication strategy for Pollens and Agents is to take advantages of the strength points of each algorithm to explore and exploit the diversity solutions in avoiding of dropping to a local optimum. A set of benchmark functions is used to test the quality performance of the proposed algorithm. Simulation results show that the proposed algorithm increases the accuracy more than the existing algorithms.

Keywords: Parallel optimization algorithm; Differential evolution algorithm; Flower pollination algorithm.

1   Introduction

Parallel processing plays an important role in efficient and effective computations of function optimization, because of it is an essential requirement for optimum computations in modern equipment [1][2]. The parallelized strategies simply share the computation load over several processors. The sum of the computation time for all processors can be reduced compared with the single processor works on the same optimum problem. Moreover, due to physical constraints in real-world problems, there are different optimal solutions in the search space and the best results of the obtained optimum cannot always be realized [3]. Collaboration between two algorithms is to take the strength points of the two algorithms. The strength point of the algorithms is considered

© Springer International Publishing AG 2017
J.-S. Pan et al. (eds.), *Advances in Intelligent Information Hiding
and Multimedia Signal Processing*, Smart Innovation, Systems and Technologies 64,
DOI 10.1007/978-3-319-50212-0_38

motivation to merge them in parallel to overcome the issue of earlier dropping a local optimal.

The communications among agents while parallel processing would enhance the co-operating individuals, share the computation load, and increase the diversity optimizations. They can exchange information between the populations whenever a communication strategy is triggered. The parallelized structure can prove faster speed and more accuracy than the original structure, and it also extends the global search capacity than its original one [4].

Furthermore, advantages of the meta-heuristics that included the Flower pollination algorithm (FPA) [5], and Differential evolution algorithm (DEA)[6] have been considered for recently. The FPA was inspired from the pollination behaviors of flowering plants and the real-life processes of the flower pollination such as self-pollination or cross-pollination were mimicked for modeling mathematically for FPA. The cross-pollination can be considered as the global pollination because of the pollinators such as agents, bats, birds and flies can fly the long distances. However, the self-pollination is the fertilization of one flower, such as peach flowers, from the pollen of the same flower or different flowers of the same plant, thus they can be considered as the local search. Similarly, DEA is an optimization algorithm based on the stochastic direct search. This algorithm is to optimize a problem by iteratively trying to improve a candidate solution according to a fitness value. There are lots of advantages of these algorithms, and many applications have been solved successfully by them[7]. However, these algorithms also have the disadvantages such as a premature convergence in the later search period and the accuracy of the optimal value which cannot meet the requirements sometimes [8] [7].

In this paper, the communication strategy and the concepts of the parallel processing are applied to develop a diversity enhanced optimization. In the proposed method, the several weaker individuals in FPA will be replaced with the better artificial agents from DEA after fixed iterations. On the contrary, the poorer agents of DEA will be replaced with the better pollens of FPA. The benefit of this strategy is to avoid the locally converged optimal in complex constrained optimization problems.

2   Related work

Flower Pollination Algorithm.

Flower Pollination Algorithm (FPA) was emulated the characteristic of the biological flower pollination in flowering plant [5]. In this algorithm, the rules of the flowering plant were mimicked to formulate the equation of optimization in as follows.

1. The global pollination processes are biotic and cross-pollination through which the pollen transports pollinators in a way that obeys Lévy flights.

2. Local pollination is explored as abiotic and self-pollination. Reproduction probability is considered as flower constancy which is proportional to the resemblance of the two flowers in concerned.

3. The switching probability $p \in [0, 1]$ can be used to control between the local and global pollination.

4. Local pollination can have fraction p that is significant in the entire processes of the pollination because of physical proximity and the wind. To simplify the proposed algorithm development, it was assumed that each plant has a single flower and each flower emit only a single pollen gamete. This means that a flower or pollen gamete is viewed as a solution xi to a problem. FPA was designed with major stages as global and local pollination. To model the local pollination, both rule 2 and rule 3 can be represented ass.

$$x_i^{t+1} = x_i^t + u(x_j^t - x_k^t) \tag{1}$$

where x_j^t and x_k^t are pollen from different flowers of the same plant species. u is drawn from a uniform distribution in [0, 1] and it is considered as a local random walk, if x_j^t and x_k^t comes from the same species or selected from the same population. In the global pollination, the pollens of the flowers are moved by pollinators e.g. insects and pollens can move for a long distance since the insects typically fly for a long range of distances. This process guarantees pollination and reproduction of the fittest solution represented as g^*. The flower constancy can be represented mathematically as:

$$x_i^{t+1} = x_i^t + \gamma \times L(\lambda) \times (x_i^t - g^*) \tag{2}$$

where x_i is solution vector at iteration t, γ is a scaling factor to control the step size. Lévy flight can be used to mimic the characteristic transporting of insects over a long distance with various distance steps, thus, $L > 0$ from a Lévy distribution.

$$L = \frac{\lambda \Gamma(\lambda) \times \sin(\frac{\pi \lambda}{2})}{\pi \times s^{i+\lambda}}, \ (s \gg s_0) \tag{3}$$

where $\Gamma(\lambda)$ is the standard gamma function, and this distribution is valid for large steps $s > 0$.

The switching probability or the proximity probability p can be effectively used likely in the rule (4) to switch between common global pollination to intensive local pollination. The effectiveness of the PFA can be attributed to the following two reasons: In rule 1, insect pollinators can travel long distances which enable the FPA to avoid local landscape to search for a very large space (explorations). In rule 2, the FPA ensures that similar species of the flowers are consistently chosen which guarantee fast convergence to the optimal solution (exploitation). To begin with, a naive value of $p = 0.55$ can be set as an initial value. A preliminary parametric showed that $p = 0.8$ might work better for most applications.

Differential Evolution Algorithm

Differential evolution algorithm (DEA) [9] belongs to the class of genetic algorithms (GAs)[10] which use biology-inspired operations of crossover, mutation, and selection on a population in order to optimize an objective function over the course of successive generations. DEA have four main operations included initialization, mutation, crossover, and selection. Evolution is performed on a population of solutions and for a certain number of generations. The following steps show how DEA works:

Step1 Initialization: an initial population of N agents is generated randomly. Each agent is a candidate solution containing D dimension of unknown parameters. The population evolves through successive generations.

$$x_{j,i,0} = x_{j,min} + rand_j(0,1) \times (x_{j,max} - x_{j,min});$$
$$j = 1,2,..D, i = 1,2,..N; rand_j(0,1) \sim U(0,1) \tag{4}$$

where $x_{j,i,0}$ is a vector indicates an agent in a population belonging to a current generation G. $x_{i,G} = [x_{1,i,G}, x_{2,i,G},..x_{D,i,D}], i = 1,..N; G = 1,..G_{max}$. All agents in a population are generated by enforcing the constraint of boundaries in which $x_{min} \leq x_{i,G} \leq x_{max}$, where x_{min} is set to $[x_{1,min}, ..., x_{D,min}]$ and x_{max} is set to $[x_{1,max}, ..., x_{D,max}]$.

Step2 Mutation: after the initialization, DEA runs a mutation to explore the search space. There are some mutation strategies that denoted as DE/x/y/z. It specifies the DEA mutation strategies by indicating the vector /x/ to be perturbed, the number /y/ of difference vectors used to perturb /x/, and the type /z/ of crossover. In this paper, the original DEA is considered. A vector $v_{i,G}$ is computed by each vector $x_{i,G}$.

$$v_{i,G} = x_{r_{i1}G} + F \times (x_{r_{i2}G} + x_{r_{i3}G}) \tag{5}$$

where $F \in (0,2)$ is a factor of scaling variable to speed up convergence of the DEA; the indexes $r_{i,1}, r_{i,2},$ and $r_{i,3}$ are mutually exclusive integers randomly selected from the interval $[1, N]$.

Step 3 Crossover: a crossover operation recombines agents to a new solution. It can make up increasing the diversity in the population but including successful solutions from the previous generation. Usually, DEA adopts exponential or binomial crossover schemes. Here, the binomial crossover is used. It changes components that are chosen randomly from $\{1, 2, ..., D\}$ and makes the number of parameters inherited from the mutant obey a nearly binomial distribution. A new candidate solution is calculated as given.

$$u_{j,i,G} = \begin{cases} v_{j,i,G} & if \ rand_{j,i}(0,1) \leq CR \ or \ j = j_{rand} \\ x_{j,i,G} & otherwise \end{cases} \tag{6}$$

where $u_{j,i,G}$ is new a trial vector that $u_{i,G}$ assumpted as $[u_{1,i,G}, u_{2,i,G}, ..., u_{D,i,G}]$, with $u_{i,G} \neq x_{i,G}$; CR is crossover rate with $CR \in (0,1)$; $rand_{j,i}(0,1) \sim U(0,1)$; $j_{rand} \in \{1,2, ..., D\}$. The constant $0 < CR < 1$ obviously affects the amount of crossover operations. Usually, $0.6 < CR < 1$ is a good value for fast convergence.

Step 4 Selection: the population size constant is kept in consecutive generations. This operation determines if the vector $v_{i,G}$ or the vector $u_{i,G}$ survives in the next generation. The selection operation works by the following relations.

$$x_{i,G+1} = \begin{cases} u_{i,G} & if \ f(u_{i,G}) \leq f(x_{i,G}) \\ x_{i,G} & if \ f(u_{i,G}) > f(x_{i,G}) \end{cases} \tag{7}$$

where $f(.)$ is the objective function to be optimized. If the value given by $u_{i,G}$ is lower than the value of $x_{i,G}$, then $u_{i,G}$ replaces $x_{i,G}$ in the next generation, otherwise $x_{i,G}$ is kept. Therefore, the population can improve or be the same in optimization of the the $f(.)$, but it never becomes worst.

After selection, the algorithm goes back to iterate *Step 2*. Mutation, crossover, and selection are applied until a certain condition i.e. maximization of the number of generations G_{max} or minimization of the objective function stops iterations.

3   A Communication Strategy Agents and Pollens

The advantages of both FPA[5] and DEA[9] algorithms are strong robustness, fast convergence, and high flexibility. They have been applied to solve successfully many problems in engineering, financial, and management fields [11, 12]. However, the disadvantage of them also exists such as the premature convergence in the later search period. This could make the accuracy of their optimal values are not to meet the requirements in sometimes. It could be easy to converge to a local optimum if the swarm size is too small for searching solution based on its own best historical information. This issue could be overcome by applying the enhanced optimizations. Enhanced optimization can be implemented by constructing the communication between two algorithms. The exchange information among subpopulations can be figured out whenever the communication strategy triggers. The communication strategy for exchanging information between Agents and Pollen can be described as follows. The best agents in DEA could be copied to move to other subpopulations in FPA replace the poorer pollens of them, and update the positions of all subpopulations in every period of exchanging time. The flow information of communicating the agents and pollens is employed with the communication strategy. In contrast, the finest artificial pollens among all the flowers of FPA's population would migrate to the weaker agents in DEA, replace them and update all positions for each population during every period exchanging time.

A parallel structure is made up of several groups by dividing the population into subpopulations. The diversity agents for the optimal method are built based on constructing of the parallel processing. The subpopulations are evolved into regular iterations independently. The advantages of each side of algorithms are taken into account by replacing the poorer individuals of them with the finest ones, and the benefit of cooperation between them is archived. During all iterations of the proposed method FDA, the exchanging period time of communication between FPA and DEA is set to R. The population size of FDA is set to N. The numbers of the population sizes of DEA and FPA are N_1 and N_2 be set to $N/2$ respectively. The top fitness k agents of in group with N_1 will be copied to the place of worst agents in group with N_2 for replacing the same number of the agents, where t is the current iteration, during running with $\cap R \neq \emptyset$. The description of the proposed method can be summarized the basic steps as follows.

Step 1. Initialization: Population size of FDA is generated randomly by initializing the solutions of FPA and DEA. The number iteration of R is defined for executing the communication strategy. The N_1 and N_2 are the numbers of agents and pollens in solutions X_{ij}^t and S_{ij}^t for populations of DEA and FPA respectively, $i = 0, 1, .., N_{1,2} - 1, j = 0, 1, .. D$. where D is dimension of the solutions and t is current iteration with setting initializing to 1.

Step 2. Evaluation: The fitness function values of $f_2(X_{ij}^t)$, $f_1(S_{ij}^t)$ are evaluated by both DEA and FPA in each iteration according to the fitness function. The evolvement of the populations is executed by both FPA and DEA.

Step 3. Update: The global pollination and local pollination of FPA are updated by using Eqs. $(1-3)$ and the agents and food source positions of DEA are updated by using Eqs. (4) and (5). The best fitness value and their positions are memorized.

Step 4. Communication Strategy: The best pollens among all the flowers of FPA's population are copied with k the top fitness pollens in N_1, migrate to another place of group in DEA population then replace the weaker agents in N_2, and update for each population in every R iteration. In contract, do the same with agents among all the individuals of DEA's population.

Step 5. Termination checking: Go to Step 2 if the predefined value of the function is not achieved or the maximum number of iterations has not been reached, otherwise, ending with minimum of the best value of the functions: $Min(f(S^t), f(X^t))$, and the best bee position among all the agents S^t or the best pollen among all the agents X^t. are recorded.

4   Experimental results

The performance quality of the proposed algorithm of FDA is evaluated by using a set of multimodal benchmark functions [13][14] to test the accuracy and the speed of it. The outcome values of the test functions in the experiments are averaged over 30 runs with different random seeds. All the optimizations for the test functions are to minimize the outcome.

Table 1. The initial range and the total iteration of the benchmark functions

Test Functions	Ranges	Dimensions	Iterations
$F_1(x) = \sum_{i=1}^{n} [x_i^2 - 10 \times \cos(2\pi x_i) + 10]$	±5.12	30	1000
$F_2(x) = \sum_{i=1}^{n} \sin(x_i) \times (\sin(\frac{ix_i^2}{\pi}))^{2m}$, $m = 10$	$0, \pi$	30	1000
$F_3(x) = \sum_{i=1}^{n} -x_i \times \sin(\sqrt{|x_i|})$	±5.12	30	1000
$F_4(x) = [e^{-\sum_{i=1}^{n}(x_i/\beta)^{2m}} - 2e^{-\sum_{i=1}^{n}x_i^2}]\prod_{i=1}^{n} \cos^2 x_i$, $m = 5$	±20	30	1000
$F_5(x) = -\sum_{i=1}^{4} c_i \times \exp(-\sum_{j=1}^{6} a_{ij}(x_j - p_{ij})^2)$	0,10	4	1000
$F_6(x) = -\sum_{i=1}^{5} [(X - a_i) \times (X - a_i)^T + c_i]^{-1}$	0,10	4	1000

The simulation results of the proposed method are compared with those obtained results of the previous algorithms as the DEA[6], FPA[5] and Genetic algorithm (GA) [15], in terms of their performance of the accuracy and running speed. Let $S = \{s_1, s_2, \ldots, s_m\}$, $X = \{x_1, x_2, \ldots, x_m\}$, and $G = \{g_1, g_2, \ldots, g_m\}$ be the real value vectors of m-dimensional for DEA, FPA and GA respectively. The optimization goal is to minimize the outcome for all benchmarks. The outcome of the performed optimal for the test benchmark function is a minimizing problem. The population size for the methods of FDA, DEA, FPA and GA are set to 40 for all runs in the experiments. The setting parameters for DEA, FPA, and GA could be found in [3,4,10]. Table 1 lists the initial range, the dimension and total iterations for all test functions.

Table 2. The comparison quality performance evaluation of DEA, FPA, and FDA for solving the optimization problems

Test func-tions	Function values			Comparison performance	
	DEA[4]	FPA[3]	**FDA**	with DEA	with FPA
1	1.65E+02	1.73E+02	1.29E+02	27%	34%
2	1.45E+00	1.48E+00	1.05E+00	36%	41%
3	-4.87E+03	-4.10E+03	-6.09E+03	19%	33%
4	1.60E-03	1.60E-03	1.10E-03	42%	43%
5	-2.94E+00	-3.04E+00	-3.32E+00	8%	5%
6	-7.15E+00	-8.15E+00	-9.72E+00	25%	16%
Avge.	**-7.86E+02**	**-6.56E+02**	**-9.95E+02**	**26%**	**29%**

The parameters setting for FDA with DEA side is initial with setting Limit to 10. The percentage of the onlooker and employed agents are set to 50%. The total population size N_l is set to $N/2$ as equal to 20 and the dimension of solution space d is set to 30, as in ref.[4]. Corresponding to the parameters setting with FPA side is the initial probability p is set to 0.55, λ is set to 1.5, the total population size N_2 set to 20 and the dimension d is set to 30, as in ref. [9][3]. Each benchmark function is tested with 1000 iterations per a run. The performance is evaluated in the average of the results from all runs. Comparing percentage is set to abs (FDA-original algorithm) *100/(FDA).

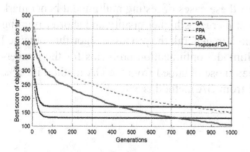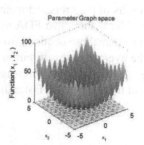

Fig. 1. The experimental results of function F1

Table 2 compares the performance quality for the multimodal optimization problems of three methods of DEA, FPA and the proposed FDA. Observed, the results of the proposed method on all of these cases of testing multimodal benchmark problems show that FDA method almost increases higher than those obtained from original methods of DEA and FPA. The maximum case obtained from FDA method increases higher than those obtained from the DEA and FPA methods are up to 42% and 44% respectively. However, the figure for the minimum cases is only the increase 07% and 06% for DEA and FPA respectively. Thus, in general of the proposed algorithm, FDA obtained the average cases of various tests multimodal optimization problems for the convergence, and accuracy increased more than those obtained from the DEA and FPA methods are 26% and 29% respectively.

Figures 1 and 2 show the experimental results for the first three multimodal benchmark functions over 30 runs output obtained from GA, DEA, FPA and proposed FDA methods with the same iteration of 1000.

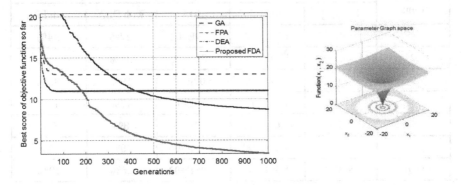

Fig. 2. The experimental results of function F2

The above figures show clearly that, all of the cases of testing functions for FDA have performance quality highest in terms of the accuracy and convergence.

Table 3 shows the performing quality and running time comparison of the proposed FDA with GA method for the multimodal optimization problems. The columns of comparison times and qualities are calculated as absolute of the obtained from FDA minus that obtained from GA then divided the obtained value of the FDA method. Clearly, the results of the proposed method on all of these cases of testing multimodal benchmark problems show that FDA method almost increases higher quality and shorter running time than those obtained from GA method. In general, the proposed algorithm obtained the average cases of various tests multimodal optimization problems for the convergence, and accuracy increased more than those obtained from the GA method is 35%, and for the speed is faster than that got from GA method is 3%.

Table 3. The comparison quality performance of GA and FDA
for solving the optimization problems

Test functions	Consumption Time		Comp. times	Performances		Comp. qualities
	GA[10]	FDA		GA[10]	FDA	
1	2.4794	2.4370	2%	1.92E+00	1.26E+00	51%
2	0.896	0.8686	3%	-5.27E+03	-6.19E+03	14%
3	0.9874	0.9501	4%	1.89E+02	1.29E+02	45%
4	0.7775	0.7931	2%	1.710E-03	1.11E-03	39%
5	0.8917	0.9071	2%	-2.23E+00	-3.28E+00	47%
6	0.9892	0.9794	1%	-8.04E+00	-9.71E+00	21%
Avg.	**1.1703**	**1.1543**	**3%**	**-8.48E+02**	**-1.01E+03**	**35%**

5   Conclusion

This paper, a novel proposed method for the optimization problems was presented with
the parallel optimization based on Agents and Pollens communication strategy, namely
FDA. The enhanced diversity agents by the parallel process to optimization could take
an important significance in the solutions for the issue of losing the global optimum in
the optimal algorithms for the multimodal or complex constrained optimization problems. The proposed communication strategy is innovative because of the introduction
of strength points of DEA and FPA in the cooperation of optimization algorithms. By
this way, the poorer pollens in FPA could be replaced with new best agents from DEA
after running the exchanging period. In contrast, the worst agents in DEA could be
replaced with fresh finest pollens from FPA in every exchanging period. Compared
with original DEA, FPA, and GA, the quality performance of the proposed FDA algorithm shows the better results of the testing set than those obtained from the DEA, FPA,
and GA methods in terms of convergence and accuracy. For the maximum cases of
testing set increase higher than those obtained from the DEA, FPA, and GA methods
are up to 42%, 43%, and 45% respectively. However, these figures for the minimum
cases are only the increase 08%, 05%, and 15%. Thus, in general of the proposed algorithm obtained the average cases of various test problems for the convergence, and accuracy increased more than those by 26%, 29%, and 35% from DEA, FPA, and GA
respectively.

References

1.   Chopard, B., Pictet, O., Tomassini, M.: Parallel and distributed evolutionary
      computation for financial applications. Parallel Algorithms Appl. 15, 15–36 (2000).
2.   Tsai, C.-F., Dao, T.-K., Yang, W.-J., Trong-The, N., Pan, T.-S.: Parallelized Bat
      Algorithm with a Communication Strategy. In: Ali, M., Pan, J.-S., Chen, S.-M., and
      Horng, M.-F. (eds.) Modern Advances in Applied Intelligence, Iea/Aie 2014, Pt I. pp.
      87 95. Springer International Publishing (2014).

3. Pan, T.-S., Dao, T.-K., Nguyen, T.-T., Chu, S.-C.: Optimal Base Station Locations in Heterogeneous Wireless Sensor Network Based on Hybrid Particle Swarm Optimization with Bat Algorithm. J. Computer, 254. 14 (2015).
4. Dao, T.-K., Pan, T.-S., Nguyen, T.-T., Pan, J.-S.: Parallel bat algorithm for optimizing makespan in job shop scheduling problems. J. Intell. Manuf. (2015).
5. Yang, X.-S.: Flower Pollination Algorithm for Global Optimization. In: Unconventional Computation and Natural Computation. pp. 240–249. Springer (2013).
6. Storn, R., Price, K.: Differential Evolution -- A Simple and Efficient Heuristic for global Optimization over Continuous Spaces. J. Glob. Optim. 11, 341–359 (1997).
7. Gandomi, A.H., Yang, X.S., Talatahari, S., Alavi, A.H.: Metaheuristic Algorithms in Modeling and Optimization. In: Metaheuristic Applications in Structures and Infrastructures. pp. 1–24 (2013).
8. Baghel, M., Agrawal, S., Silakari, S.: Survey of Metaheuristic Algorithms for Combinatorial Optimization. Int. J. Comput. Appl. 58, 975–8887 (2012).
9. Storn, R., Price, K.: Differential Evolution - A simple and efficient adaptive scheme for global optimization over continuous spaces. Science (80-.). 11, 1–15 (1995).
10. Holland, J.H.: Adaptation in Natural and Artificial Systems: An introductory analysis with applications to biology, control, and artificial intelligence. U Michigan Press (1975).
11. Das, S., Suganthan, P.N.: Differential evolution: A survey of the state-of-the-art. IEEE Trans. Evol. Comput. 15, 4–31 (2011).
12. Chiroma, H., Shuib, N.L.M., Muaz, S.A., Abubakar, A.I., Ila, L.B., Maitama, J.Z.: A Review of the Applications of Bio-inspired Flower Pollination Algorithm. Procedia Comput. Sci. 62, 435–441 (2015).
13. Jamil, M., Yang, X.-S.: A Literature Survey of Benchmark Functions For Global Optimization Problems Citation details: Momin Jamil and Xin-She Yang, A literature survey of benchmark functions for global optimization problems. Int. J. Math. Model. Numer. Optim. 4, 150–194 (2013).
14. Suganthan, P.N., Hansen, N., Liang, J.J., Deb, K., Chen, Y., Auger, A., Tiwari, S.: Problem Definitions and Evaluation Criteria for the CEC 2005 Special Session on Real-Parameter Optimization Problem Definitions and Evaluation Criteria for the CEC 2005 Special Session on Real-Parameter Optimization. Nat. Comput. 1–50 (2005).
15. Whitley, D.: A genetic algorithm tutorial. Stat. Comput. 4, 65–85 (1994).

A Comparative Analysis of Stock Portfolio Construction by IABC and GARCH with the ISCI in Taiwan Stock Market

Pei-Wei Tsai[1,2],Ko-Fang Liu[3], Xingsi Xue[1,2], Jui-Fang Chang[3,*],
Cian-Lin Tu[3], Vaci Istanda[4], and Chih-Feng Wu[1]

[1]College of Information Science and Engineering
[2]Fujian Provincial Key Laboratory of Big Data Mining and Applications
Fujian University of Technology
peri.tsai@gmail.com, xxs@fjut.edu.cn, tfengwu@hotmail.com

[3]Department of International Business
National Kaohsiung University of Applied Science
1103346116@gm.kuas.edu.tw; rose@kuas.edu.tw; cianlin.tu@gmail.com

[4]Yilan County Indigenous Peoples Affairs Office
Yilan Precinct
vaci@mail.e-land.gov.tw

Abstract. The most recent return in the investment is one of the most popular element that investors take into consideration because of its fast and direct return characteristic. With the ongoing discussion about stock portfolio design, we construct the investment model by two phases and draw an analogy in this research. The first step is using ISCI in selecting the potential stocks; and the second step is designing the stock portfolio by the GARCH model and the IABC algorithm aiming for gaining the highest return in the investment. The analysis data used in our experiments include the stock price of daily return in continuous five years in 2011 to 2015. The experimental results indicate that using IABC in constructing the stock portfolio is a stable investment strategy and the gained maximum return is also higher than the portfolio constructed by the GARCH in the Taiwan stock market.

Keywords: Portfolio, ISCI, GARCH, IABC

1   Introduction

An economic depression is recession that affects many countries around the world including Taiwan, hence it becomes the major issue to finance and invest management. Markowitz Mean-Variance Model (MV) is proposed and analyzed for a market consisting of one bank account and multiple stocks. The MV model's lower estimation risk is most striking in small samples and for investors with a low risk tolerance.

© Springer International Publishing AG 2017   325
J.-S. Pan et al. (eds.), *Advances in Intelligent Information Hiding
and Multimedia Signal Processing*, Smart Innovation, Systems and Technologies 64,
DOI 10.1007/978-3-319-50212-0_39

Investment Satisfied Capability Index (ISCI) is derived from Process Capability Index (PCI). Chang (2009) applies ISCI to select stocks at Taiwan stock market. The experimental result shows that stock portfolio construct from ISCI has the higher returns than fundamental analysis and TAIEX.

ARCH (Autoregressive Conditional Heteroskedastic) process introduced in Engle (1982) allows the conditional variance to change over time as a function of past errors leaving the unconditional variance constant. The stationarity conditions and autocorrelation structure of the Generalized Auto Regressive Conditional Heteroskedastic (GARCH) model are derived (Bollerslev, 1986). Using an empirical example of uncertainty of the inflation rate the paper demonstrates that the GARCH model provides a better fit and a more plausible learning mechanism than the ARCH model.

Artificial Bee Colony (ABC) is one of the most recently defined algorithms by Dervis Karaboga in 2005, motivated by the intelligent behavior of honey bees. Tsai et al. presented IABC in 2009. The formula, which takes the location of the employed bees in the consideration for moving the onlooker bees, is modified, and the concept of universal gravitation is introduced into the process in IABC to calculate the interactive affection between different numbers of employed bees. Compared with standard ABC algorithm, IABC algorithm simplifies the parameter setting and achieves better performance. This research applies IABC to obtain the optimal portfolio.

2   The experiment design

There are two phases in this research, and the selecting data of the stock daily return comprise five years of 2011 to 2015. In the first phase of Stock Selection that includes two stages. We set up five criteria in the investment information website "HINET MONEYDJ" to filter the stock and acquire 110 stocks in the first stage. Then applying investment satisfied capability index to execute opting for stocks in the second stage.

According to Evans and Archer (1968), investment risk will ease when the stock portfolio involves 10 to 15 stocks. Chang (2013), contrast stock portfolios with 10 and 15 stocks, the portfolio with 15 stocks has the better return on investment and less fluctuation of current events. In this research, we choose stock portfolio with 15 stocks to proceed GARCH and IABC return predictability, and comparative analysis with Taiwan stock market.

2.1   Stock selection

With investment satisfied capability index to sift the stocks, and take the lower limit return of the rate of treasury, the following formula of investment satisfied capability index:

$$C_{SL} = \frac{\mu - LRL}{3\sigma} \qquad (1)$$

where μ stands for the daily average return rate of the stock, σ indicates the standard deviation of daily rate of return of stock, and LRL is the investors anticipate the lower rate limit of return.

2.2   GARCH model construction

Time-Series Model analysis includes four steps, as follows:

STEP 1: Jarque-Bera Test. Examine the normal distribution of Time-Series data.

$$JB = \frac{T - n}{6}\left[S^2 + \frac{1}{4}(K - 3)^2\right]. \qquad (2)$$

where S stands for the skewness and K is the Kurtosis variable.

STEP 2: Ljung-Box Q test. Residual analysis.

$$Q(P) = n + (n + 2) \sum_{K=1}^{P} \frac{1}{n - k^{P_k^2}} \sim X^2(P) \qquad (3)$$

where n indicates the samples and k is the chi-squared distribution with P degrees of freedom.

STEP 3: Unit Root Test. Use ADF (Augmented Dickey-Fuller test) to examine the data of stationary state.

STEP 4: GARCH forecasting model. The last step is to predict by GARCH Model, the formula as follow:

$$y_t|\Omega_t \sim N(X_t a, \sigma^2) \qquad (4)$$

$$\varepsilon_t = y_t - X_t \qquad (5)$$

$$\sigma_t^2 = \alpha_0 + \sum_{i=1}^{q} \alpha_i w_{t-i}^2 + \sum_{i=1}^{p} \beta_i \sigma_{t-i}^2, \quad \alpha_0 > 0, \alpha_i \geq 0, \beta_i \geq 0 \qquad (6)$$

2.3   Capital allocation by IABC

In this work, we utilize IABC to decide the capital allocation into the selected stocks. To use IABC, the operation includes five steps, which are listed as follows:

STEP 1: Initialization: randomly spread n e percent of the population into the solution space, where n e indicates the ratio of employed bees to the total population.

STEP 2: Move the onlookers: move the onlookers by Eq. (8) with roulette wheel strategy based on the probabilities calculated by Eq. (7).

$$P_i = \frac{f(\theta_i)}{\sum_{k=1}^{s} f(\theta_k)} \tag{7}$$

where P_i represents the probability of selecting the ith employed bee, S denotes the total number of the employed bees, $\sum_{k=1}^{s} f(\theta_k)$ is the accumulation of fitness values of all employed bees, and $f(\theta_i)$ denotes the fitness value of the ith employed bee.

$$X_{ij}(t+1) = \theta_{ij}(t) + \sum_{k=1}^{n} \tilde{F}_{ikj} \cdot \left[\theta_{ij}(t) - \theta_{kj}(t)\right] \tag{8}$$

where X_i indicates the position of the ith onlooker bee, t stands for the number of iteration, n is the number of employed bees that taken into the consideration, θ_k represents the employed bee that have been chosen, j denotes the dimension of the solution space, and $\tilde{F}_{ikj}$ means the pulling force between the employed bees and the target onlooker.

STEP 3: Move the scouts: when the iteration matches the multiples of the predefined Limit iteration, the employed bees, whose fitness values are not improved, become the scouts.

$$\theta_{ij} = \theta_{j\,min} + \gamma\left(\theta_{j\,max} - \theta_{j\,min}\right) \tag{9}$$

where $\theta_{j\,min}$ and $\theta_{j\,max}$ denote the lower and the upper boundary of scout i in all dimensions, respectively, and γ is the random number in the range of [0, 1].

STEP 4: Update the near best solution: memorize the near best fitness value and the corresponding coordinate found so far by the bees.

STEP 5: Termination checking: if the termination condition is satisfied, exit the program; otherwise, go back to step 2.

3   Experiments and experimental results

In our experiments, we select 15 stocks from the Taiwan Stock Market based on the highest ranks of their ISCI values. The selected stocks are listed in Table 1.

Table 1   The final section screened 15 of stocks.

No.	Ticker Symbol	No.	Ticker Symbol	No.	Ticker Symbol
1	4426	6	3552	11	3004
2	1476	7	3450	12	3691
3	3131	8	2474	13	2456
4	1536	9	5490	14	9938
5	3008	10	1707	15	2395

After the stock selection process, the portfolio is generated by the GARCH model and the IABC model, respectively. The experimental results are compared year by year with the Taiwan Stock Broad Market. The experiments cover the data in 2011 to 2015. Since we present the experimental results based on the year, there are five charts of experiment results with accumulated values of the stock return demonstrate seriatim as follows, where the X-axis stands for the trading days and the Y-axis denotes the accumulated return per year.

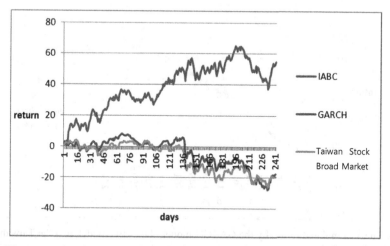

Fig. 1. The accumulated return by the GARCH model, the IABC model, and the Taiwan Stock Broad Market in 2011.

In 2011, the accumulated return obtained by the IABC model reaches 54% at the end of the year; the accumulated return obtained by the GARCH model and the Taiwan Stock Broad Market fall in -17% and -19%, respectively.

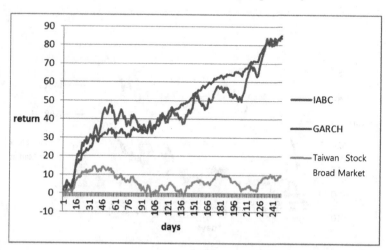

Fig. 2. The accumulated return by the GARCH model, the IABC model, and the Taiwan Stock Broad Market in 2012.

In 2012, the accumulated return obtained by the IABC model climbs up to 83%; the accumulated return obtained by the GARCH model is also with a satisfactory result at 85%, and the Taiwan Stock Broad Market reaches 9% at the end of the year.

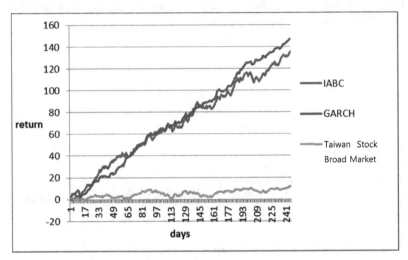

Fig. 3. The accumulated return by the GARCH model, the IABC model, and the Taiwan Stock Broad Market in 2013.

In 2013, the accumulated return obtained by the IABC model climbs up to 135%; the accumulated return obtained by the GARCH model is also with a satisfactory result at 146%, and the Taiwan Stock Broad Market reaches 11% at the end of the year.

Fig. 4. The accumulated return by the GARCH model, the IABC model, and the Taiwan Stock Broad Market in 2014.

In 2014, the accumulated return obtained by the IABC model climbs up to 55%; the accumulated return obtained by the GARCH model is 30%, and the Taiwan Stock Broad Market reaches 8% at the end of the year.

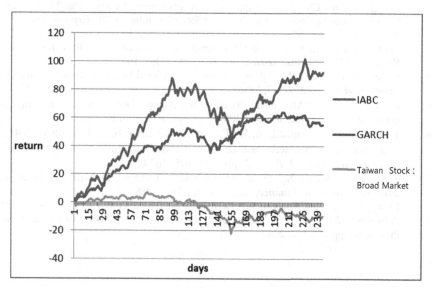

Fig. 5. The accumulated return by the GARCH model, the IABC model, and the Taiwan Stock Broad Market in 2015.

In 2015, the accumulated return obtained by the IABC model climbs up to 92%; the accumulated return obtained by the GARCH model is 55%, and the Taiwan Stock Broad Market reaches -9% at the end of the year.

As given above, it's clear to see that the portfolio constructed by IABC present higher performance in three-fifth over the test period; and the portfolio constructed by the GARCH model presents better result in two-fifth of the whole test period. Nevertheless, no matter using which method, the accumulated return is always higher than the Taiwan Stock Broad Market.

4   Conclusion and future works

With applying the ISCI in selecting the potential stocks, the stock investment portfolio are constructed by the IABC model, and the GARCH model in this research. The investors can receive higher return by selecting the optimal stock portfolio and proceeding the capital allocation. The experimental results show that the asset allocation by IABC is better than the prediction of GARCH and market. It implies that the IABC model has the investment strategy with steady and the maximization of return.

5   References

1.   Markowitz, H., "Portfolio Selection". *Journal of Finance*, vol.7, no.1, pp.77-91, 1952.
2.   Chang, J. F. and Chen, K.L., "Applying New Investment Satisfied Capability Index and Particle Swarm Optimization to Construct Stock Portfolio," *ICIC Express Letters*, vol.3, no.3, 2009, pp.349-355.
3.   Engle, R. F., "Autoregressive Conditional Heteroscedasticity with Estimates of the Variance of UK. " *Econometrica*, vol.50, pp.987-1008, 1982.
4.   Bollerslev, T., "Generalized Autoregressive Conditional Heteroskedasticity." *Journal of Econometrics* ,31 : 307-327, 1986.
5.   Karaboga, D., "An idea based on honey bee swarm for numerical optimization, "Technical Report-TR06, Erciyes Univ. Press, Erciyes,2005.
6.   Tsai, P. W., Muhammad, K. K., Pan, J. S. and Liao, B. Y. Interactive Artificial Bee Colony Supported Passive Continuous Authentication System. *IEEE Systems Journal*, vol. 8, no. 2, pp. 395-405, 2014.
7.   Evans, J. and Archer, S., "Diversification and the reduction of dispersion: an empirical analysis," *Journal of Finance,* vol.23, no.5, p.761, 1968.
8.   Chang, J. F. et al., "Applying Interactive Artificial Bee Colony to Construct the Stock Portfolio," 2013 Second International Conference on Robot, Vision and Signal Processing, pp.129-132, 2003

An Improved Learnable Evolution Model for Discrete Optimization Problem

Wenguo Wu[1] and Shih-Pang Tseng[2] *

[1] Network Information Center, Shandong Business Institute, Yantai, China
[2] Department of Computer Science and Information Engineering, Tajen University, Pingtung, Taiwan
wuwenguo@sdbi.edu.cn, tsp@mail.tajen.edu.tw

Abstract. Learnable evolution model (LEM) is an efficient evolution-ary algorithm for function optimization proposed by Michalski. However, because the design of LEM is for continuous optimization, how to ap-ply it to discrete optimization has become an important issue. In this paper, we present a discrete version of LEM for discrete optimization of traveling salesman problem (TSP), called LEMTSP. The proposed algo-rithm makes LEM capable of solving problems in the discrete domain while preserving its convergence speed. More important, the proposed algorithm can use all the crossover operators designed particularly for discrete problems.

Keywords: learnable evolution model, evolutionary computation, trav-eling salesman problem

1 Introduction

Evolutionary computation (EC) is one of the most efficient techniques for finding a near-optimal solution for the optimization problems in a reasonable time iter-atively. EC can be categorized into two types in terms of the problems they try to solve: *continuous* and *discrete*. The solutions can be encoded as floating-point numbers if the problem in question is continuous; otherwise, they are gener-ally binary or integer encoded. For example, ant colony optimization (ACO) [1] was proposed for discrete optimization, and particle swarm optimization [2] was proposed for continuous optimization. Moreover, the evolutionary compu-tation algorithms can be divided into two types: *Darwinian* and *non-Darwinian* types. The main difference is how to select the parents to generate offspring. The *stochastic* selection is used in Darwinian type of evolutionary computation. On the other hand, the *deterministic* selection is used in non-Darwinian type of evolutionary computation.

The Learnable Evolution Model (LEM) [3][4][5][6][7] is a fast non-Darwinian methodology for evolutionary computation. Unlike the standard evolutionary

---

* Corresponding author: Shih-Pang Tseng is with Department of Computer Science and Entertainment Technology, Tajen University; No.20, Weixin Rd., Yanpu Town-ship, Pingtung County 90741, Taiwan; E-mail: tsp@mail.tajen.edu.tw

© Springer International Publishing AG 2017   333
J.-S. Pan et al. (eds.), *Advances in Intelligent Information Hiding and Multimedia Signal Processing*, Smart Innovation, Systems and Technologies 64,
DOI 10.1007/978-3-319-50212-0_40

computation, such as genetic algorithms (GA) , LEM is guided by a machine learning process to speed up the convergence process. Over the past ten years, LEM has been successfully applied to theoretical and application domains, such as numerical function optimization [4], heat exchange design [8], finned-tube evaporators design [9], classification[7] and so on. But all these applications are in the continuous domain. This study is aimed at making LEM not only capable of solving discrete optimization but also using other efficient genetic algorithm operators.

The remainder of the paper is organized as follows. Section 2 will first introduce the learnable evolution model and then gives a brief introduction to the learning step of LEMTSP. Also given in Section 2 is a simple example to demonstrate how LEMTSP works. Section 3 describes in detail the instantiation of LEMTSP. Performance evaluation of LEMTSP is presented in Section 4. Conclusion is given in Section 5.

2 The Proposed Algorithm

2.1 Learnable Evolution Model

A simple outline of LEM is as shown in **Algorithm 1**. Unlike genetic algorithm, which uses crossover to generate the offspring, LEM uses the *learning* and *instantiation* steps to generate new individuals. In the deterministic selection step, the individuals with high-fitness are assigned to the high-performance group, denoted H-group, from the current population. In addition, the individuals with low-fitness are assigned to the low-performance group, denoted L-group. The learning step is used to reason *descriptions* that distinguish between H-group and L-group in the current population. The descriptions are the results of the learning step. Subsequently, the instantiation operator uses these descriptions to create new individuals. The Darwinian evolution in the LEM is to maintain the diversification by applying some form of mutation.

Although the learning method is symbolic, we try to extend the LEM to solve problems in the discrete domain, with a focus on the so-called combinatorial optimization problems. We will use the traveling salesman problem (TSP) as an example. Using LEM to solve the TSP, two issues need to be addressed are *learning* and *instantiation*. Learning is the most important step of LEM. LEM concentrates on searching a small search space bound by the learning result. This can eventually speed up the convergence speed of LEM. The original learning step of LEM is only suitable for problems in the continuous domain. Thus, the first issue is how to make learning work for problems in the discrete domain. Another issue in the discrete domain is how to perform instantiation to generate the new offspring that satisfy the descriptions and the constraints of the TSP. In other words, learning and instantiation need to be modified for the TSP.

2.2 The LEMTSP

As far as the learning step of traditional LEM is concerned, it is required that the fitness value of any individual in the high fitness group (H-group) is better

---

Algorithm 1 Learnable Evolution Model

1: Generate a population
2: **while** The terminate condition is not met **do**
3:     Deterministic selection
4:     Learning
5:     Instantiation
6:     Execute Darwinian evolution
7: **end while**

---

than any individual not in the H-group. Similarly, it is required that the fitness value of any individual in the low fitness group (L-group) is worse than any individual not in the L-group. Fig. 1 shows that the selection process of H-group and L-group is deterministic and is strictly based on if the fitness value is better or worse than a predefined threshold. Another way is to select the best *H-group ratio* individuals into H-group from the current entire population and the worst *L-group ratio* individuals into L-group. In general, LEM uses only the H-group because individuals in the L-group are not enough good but not really bad, especially after several iterations.

In this paper, we will use only the H-group too and do not consider the L-group. The learning step is used to find which properties are common in the H-group. These common properties of H-group are called *descriptions*. They are called *elementary* in LEM and are in the following form:

$$[\textbf{L rel R}] \tag{1}$$

where

L *(left side)* is a single attribute
R *(right side)* is a single value from the domain of attribute in **L**.
rel is a relational symbol from the set $\{=, \neq, >, \geq, <, \leq\}$.

Another type of description is called *composite* in LEM.

Fig. 2 shows that a smaller and more valuable region is bounded by the descriptions. In this region, all the solutions satisfy the descriptions. LEM would concentrate its search on this region. It is not sure that the region of the later generation is included in the region of the previous generation.

In this paper, we will use the *edge-set representation* (ESR) for the TSP to simplify the learning step. An ESR is a set of edges, which forms a legal tour. For instance, the example given above in Eq. (**??**) can be represented by the following:

$$G = \{S_0, S_1, s_2\} \tag{2}$$

where

$$\begin{aligned}
S_0 &= \{e_{01}, e_{03}, e_{12}, e_{24}, e_{34}\} \\
S_1 &= \{e_{01}, e_{04}, e_{13}, e_{23}, e_{24}\} \\
S_2 &= \{e_{01}, e_{03}, e_{14}, e_{23}, e_{24}\}
\end{aligned} \tag{3}$$

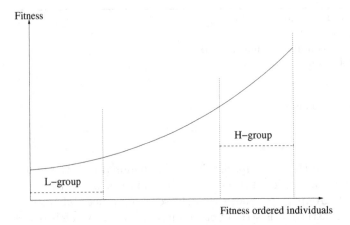

Fig. 1. H-group and L-group

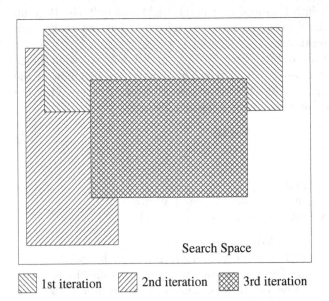

Fig. 2. Search space and learning

After all the individuals are converted to ESRs, all the learning step has to do is to compute the intersection of all the ESRs of H-group, as follows:

$$D = \bigcap_{i=0}^{|G|-1} S_i \tag{4}$$

For example, for G in Eq. (??), the intersection is given below.

$$D = S_0 \cap S_1 \cap S_2 = \{e_{01}, e_{24}\} \tag{5}$$

Then, D is interpreted into the form of descriptions, like

$$[E_{ab} = 1] \tag{6}$$

where the edge e_{ab} is in D. For example, in Eq. (??), the descriptions are $[E_{01} = 1]$ and $[E_{24} = 1]$. The two description are true and satisfied by all the individuals in the H-Group. And all the individuals in H-Group contain the two edges e_{01} and e_{24}.

3   Instantiation for TSP

Instantiation is an important step of LEM to generate new individuals according to the descriptions generated in the learning step. In the numerical optimization problems, the descriptions would restrict the range of attributes. In the instantiation step, a value in the restricted range is selected and assigned to the attribute of new individual. But this is not suitable for the TSP. Traditionally, there are two types of methods to construct a TSP tour; one is *order-based* and the other is *edge-based* [10]. The order-based methods construct a TSP tour based on the order of cities. The edge-based methods consider which edges are selected into the tour. For the viewpoint of the edge-based methods, the descriptions of LEMTSP just contain which edges must be in the tour. There is no information about how to select the other edges to complete a legal TSP tour. From the viewpoint of the order-based approaches, the descriptions are just the partial fragments of the order of cities. The descriptions contain no hints about the order of the other cities. A random method can be used to decide the order of the other cities. It is reasonably to expect that a tour generated in this way is usually bad in that TSP is very sensitive to the randomness. Additionally, such a random selection of the other edges could construct an illegal tour, or a worse tour.

An alternative is to combine the descriptions and TSP-specific heuristics, such as *nearest-neighbor*. But this alternative is too specific for the TSP, and will be not easy to extend to the other combinatorial optimization problem. In this paper, we combine the crossover and mutation operators of the evolutionary computation and the descriptions to generate the new individuals. This has three advantages:

1. It may reduce the randomness in the instantiation.
2. It is easier to extend.
3. It can inherit the results of the previous iterations.

In the instantiation step, two parents are randomly selected from H-group to crossover to generate one offspring. Next, the LEMTSP applies the mutation operator to the offspring. The offspring would be inserted into the population to replace the old individuals.

For the TSP, there are too many different crossover and mutation operators, such as edge-recombination crossover (ERX) , heuristic crossover (HX), order crossover (OX) , partial-mapping crossover (PMX), and so on. Each operator has its advantages and disadvantages. Our goal is to make LEMTSP independent of the crossover and mutation operators so that different operators can be chosen by different requirements because it is impractical to develop an LEMTSP for each operator.

4   Experimental Result

In this section, we evaluate the performance of the proposed algorithm LEMTSP by using it to solve the traveling salesman problem. The benchmarks for the TSP, from TSPLIB. Unless stated otherwise, all the simulations are carried out for 30 runs, with the population size fixed at 40, and each run contains 200 iterations.

   To improve the quality of the final results of LEMTSP, we use several useful techniques to solve the TSP. The nearest-neighbor method is used in creating the initial solution for all the algorithms involved in the simulation. The LinKernighan local search method is employed as the local search method for fine-tuning the quality of the final results. Using the nearest-neighbor method to create the initial solution is important to LEMTSP. Because the descriptions found in the learning step are regarded as an explanation of why the individuals in H-group are better, a better initial solution can be used to ensure that the descriptions have a higher probability to be correct at the early generations. If the initial solution is randomly created, the descriptions may lead the search to a wrong direction because of the randomness used in creating the initial solution at the early generations. For instance, one or more edges are randomly selected for all the individuals in the initial solution, even though the probability is low, the chance is still there.

4.1   H-group ratio

To balance the intensification and diversification is an important issue of evolutionary computation. The *H-group ratio* of LEM can influence the balance between the intensification and diversification. The LEM may fall into the premature convergence because of the too small H-group ratio. If the LEM has the bigger H-group ratio, the not enough good individuals are with more probability into the H-group. The learning process would be noised and misled by these not enough good individuals. Fig. 3 shows the experimental results of LEMTSP with varied H-group ration from 0.5 to 0.1. The vertical axis of Fig. 3 is represented the relative percentage of result to the H-group ration 0.5. In this experiment, the ERX operator is used by LEMTSP. We can find that the H-group ratio 0.25 to 0.15 is appropriate to the most benchmarks.

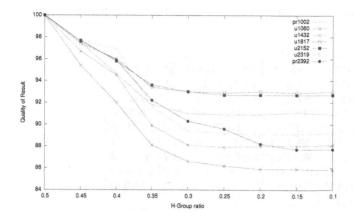

Fig. 3. The impact of H-group ratio

4.2 Crossover operators

Because LEMTSP is operator independent, we use the partial-mapping crossover (PMX), order crossover (OX), edge-recombination crossover(ERX), and heuristic crossover (HX). PMX is the most popular and simplest crossover operator for the TSP, and it is order-based. OX is another popular order-based crossover operator. On the other hand, ERX and HX are edge-based crossover operators. The Table 1 shows the comparison of LEMTSPs with different operators. The H-group ratio is the same 0.2. In the Table 1, the star mark denotes the best one in this benchmark. In most benchmarks, ERX is the most suitable operator for LEMTSP. In the 21 benchmarks, ERX is the best one in the 15 benchmarks.

5 Conclusion

LEM is a non-Darwinian methodology for evolutionary computation. The most attractive characteristic of LEM is its fast convergence speed. Originally, LEM is used to solve the numerical function optimization or similar problems. In this paper, we propose an extension of LEM to solve the traveling salesman problem. We propose an *edge-set representation* (ESR) for the TSP to simplify the learning step because for the TSP, all the learning step needs to do is to compute the intersection of all the ESRs of the H-group. In the instantiation step, the composite cities is proposed. Thus, we do not modify any operators to make them suitable for LEMTSP. The advantage is that LEMTSP is operator independent. LEMTSP preserves the fast convergence speed of the original LEM. In the future, we will extend the LEM to the other combinatorial optimization problems, like knapsack problem, vehicle routing problem, and so on.

Table 1. Experimental results of LEMTSP using PMX, OX, ERX and HX

	PMX	OX	ERX	HX
u574	38282.9	38230.6	38189.3*	38193.3
u724	43131.2	43134.2	43057.6*	43068.3
pr1002	266829.4	267122.2	266419.5*	266585
u1060	231748.8*	232441.4	231773.4	232998
vm1084	247714.8	247310.7	247059.4	246941.6*
pcb1173	59612.3	59688.1	59545*	59786.2
rl1304	266129.3*	266821.2	267033.7	266428.8
rl1323	279506.8	279099	278338.4*	278384.5
u1432	159468.1	159195.7	158960.7*	159099.7
#best	2	0	6	1

References

1. Dorigo, M., Stützle, T.: Ant Colony Optimization. Bradford Book (2004)
2. Kennedy, J., Eberhart, R.: Particle swarm optimization. In: Proceedings of IEEE International Conference on Neural Networks. Volume 4. (1995) 1942–1948
3. Michalski, R.S.: Learnable evolution: Combining symbolic and evolutionary learning. In: Proceedings of the Fourth International Workshop on Multistrategy Learning (MSL'98). (1998) 14–20
4. Wojtusiak, J., Michalski, R.S.: The lem3 implementation of learnable evolution model and its testing on complex function optimization problems. In: GECCO. (2006) 1281–1288
5. Michalski, R.S.: Learning and evolution: An introduction to non-darwinian evolutionary computation. In: ISMIS. (2000) 21–30
6. Michalski, R.S.: Learnable evolution model: Evolutionary processes guided by machine learning. Machine Learning 38(1-2) (2000) 9–40
7. Badran, M.E.S.K., Salama, G.I.: A generic feature extraction model using learnable evolution models (lem+id3). International Journal of Computer Applications 64(11)
8. Kaufman, K.A., Michalski, R.S.: Applying learnable evolution model to heat exchanger design. In: AAAI/IAAI. (2000) 1014–1019
9. Domanski, P.A., Yashar, D., Kaufman, K.A., Michalski, R.S.: An optimized design of finned-tube evaporators using the learnable evolution model. International Journal of Heating, Ventilating, Air-Conditioning and Refrigerating Research 10 (2004) 201–211
10. Whitley, L.D., Starkweather, T., Fuquay, D.: Scheduling problems and traveling salesmen: The genetic edge recombination operator. In: ICGA. (1989) 133–140

Management of Energy Saving for Wireless Sensor Network Node

Xu-Hong Huang

Fujian University of Technology, School of Information Science and Engineering,
Fuzhou, Fujian 350118, China
E-mail address: Huangxuh1@163.com

Abstract. The goal of the paper is to lengthen lifespan of Wireless Sensor Network(WSN) node with effective strategies to reduce energy consumption. On the basis of studying wireless sensor network node structure, energy consumption and transmission between nodes, put forward a kind of effectively energy saving management method for nodes. The scheme adopts the self-organized protocols of dynamic selection cluster head nodes with multiple hops routing, hierarchical topology structure, fast data fusion technology, and dynamic management of power for hardware under the condition of low power.

Keywords: Wireless sensor network (WSN); Energy conservation; Transmitted between nodes; Dynamic management of energy.

1   Introduction

Wireless sensor network consists of many cheap nodes. These nodes can achieve functions of data acquisition, data processing, data transmission. These abilities required power provided by the node's own Micro motor. Wireless sensors are often work in some bad or dangerous environment, replacing energy are difficult. Even if the energy of node can replace, the cost is also relatively big. Instead of replacing energy of the wireless sensor network node, the effective strategy of reducing energy consumption and trying to prolong the life cycle of the network are adopt. Using appropriate way of management wireless sensor network node will greatly improved the performance of network, effectively reduce the energy consumption.

2   The structure of wireless sensor network node

Sensors usually refers to the measured non-electricity and can convert them to electric quantity that is converted according to certain rule processing. It usually consists of sensitive components and conversion components, measurement circuit, power circuit. A wireless sensor node includes a wireless data transmission module, a data management module and the function of general sensors. It often also includes data

© Springer International Publishing AG 2017
J.-S. Pan et al. (eds.), *Advances in Intelligent Information Hiding
and Multimedia Signal Processing*, Smart Innovation, Systems and Technologies 64,
DOI 10.1007/978-3-319-50212-0_41

acquisition module that consists of sensitive components and conversion components, measurement circuit of a module.

Wireless sensor node are mainly divided into two categories: acquisition node and gathering node. They usually include the same hardware configuration, but functions of them are different. Acquisition nodes are responsible for collecting data and transmitting data to gathering nodes. Gathering nodes are responsible for gather data that are transmitted from acquisition node. The composition block diagram of the wireless sensor node is shown in figure 1.

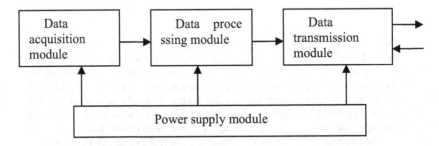

Fig. 1. composition block diagram of wireless sensor nodes.

As the common sensor, data acquisition module can collect non-electric quantity information that include temperature data, light intensity, pressure data, displacement data, flow data, liquid level data, acceleration data and so on. Then these non-electric quantity information are converted to signals that are suitable for transmission and the measuring. These signals are converted into digital signals through the A/D conversion.

Data processing module process gathering the data and usually consist of the microprocessor and memory, etc. They are responsible for the management of control node including data processing operations, data forwarding control according to the routing protocol, power management, task processing, etc.

The data transmission module is responsible for communication with other nodes, transport information collected from nodes and control exchange information for network.

Power supply circuit module provides the energy for the data processing module, data transmission module and data acquisition module. It generally composed of power supply circuit and voltage conversion circuit. The power supply provides usually use fixed batteries or solar cells at present.

3   The energy consumption and fuel consumption of wireless sensor nodes

Wireless sensor node energy consumption mainly come from the sensor data acquisition module circuit [1], micro controller and the memory of a data processing module, the data transmission module of RF circuit.

Sensor circuit consume less energy. The amount of reducing energy consumption is not large.

The power consumption of the microprocessor can be divided into two parts: the dynamic power consumption and static power consumption [2]. Reducing the dynamic power consumption is the main of reducing the energy consumption. According to the literature 3, the dynamic power is associated with the power supply voltage of the microprocessor, physical capacitance, the clock frequency. The relation between them is:

$$P_D \propto \alpha CV^2 f \tag{1}$$

P_D is as dynamic power, V as the power supply voltage, C as physical capacitance C, f as the clock frequency, α alpha factor for activities.

Therefore, reducing the clock frequency reducing the power supply voltage can reduce dynamic consumption of power. Literature 3 show that the power supply voltage and decrease the clock frequency can reduce dynamic power consumption, the working state of the processor from 200 MHz to 150 MHz and 1.5 V conversion to 1.2 V can save 52% of power consumption.

Management of dynamic power consumption can achieved by a Dynamic Voltage regulation technology (Dynamic Voltage supply, DVS) [3], in addition to by reducing Dynamic power consumption of each module itself. DVS technology can dynamically change the working voltage and frequency of the microprocessor, times changes along with the work load of nodes, thus reduce the unnecessary power output of idle period.

RF circuit energy consumption is the largest of node component. According to the requirements of wireless sensor nodes, the RF circuit generally is with low power consumption, low price, small size of mature devices. This kind of radio frequency circuit is chosen by considering the energy consumption, lowing output power and energy saving mode. For example, the Norwegian Nordic VLSI's single-chip RF transceiver nRF905, power consumption of it is low, with 10 dbm output power of the emission, 11 mA of current, 12.5 mA of current on working in the receive mode, and with closing mode and free mode that easily realize energy saving.

With controlling dynamically the operation mode of the RF module, microprocessor can change work load situation between the working model idle and free model, in order to reduce power consumption.

4   Energy saving management of wireless sensor nodes spread way

Reducing the energy consumption of wireless sensor nodes can be realized by energy saving management of spread way between nodes and reducing work load, besides it can be realized through the dynamic power consumption management way Wireless sensor network is composed of many acquisition node, number of gathering nodes ,the transfer device and the control center (PC).Among them, the acquisition node is responsible for data acquisition, data processing and communicate between nodes; Gathering node is responsible for the upload data collected by other nodes, and issued commands from transfer device; Transfer unit is responsible for uploading data collected together, transferring command from control center forward to gathering node; Control center is responsible for controlling the entire network, and the processed data to the user.

There are the following features of Wireless sensor network node distribution:

(1) Node location is random distribution in the wireless sensor network, the network protocol of WSN is self-organizing. In the actual work environment, sensor nodes are usually not accurate positioning, the relationship between nodes cannot know in advance. It would require that sensor nodes have the ability of organization, to establish and organize network itself.

(2) The wireless sensor network nodes are numerous, wide distribution. To ensure accurate information, a large number of sensor nodes are deployment in the monitoring area.

(3) The communication distance between nodes is not long. Point to point communication distance between wireless sensor network nodes is usually only a few tens of to hundreds of meters.

From the above network composition and node distribution way, reducing the work load node need to reduce the communication time and communication distance between nodes. Different ways of communication between the wireless sensor nodes have great effect on the communication time and communication distance. In the mode of transmission between nodes, a good network protocol and resource management strategy can effectively reduce the node working load, prolong the life cycle of wireless sensor network. Therefore, wireless sensor network mode of transmission should be centered on data, using self-organization, multiple hops routing, The network structure by using dynamic topology. In addition, fast data fusion technology can be use with quick information fusion and separation. it will improve the efficiency of network operation and the ability to randomly select the best path.

Power of wireless sensor network is very limited to sensor node computing power , storage capacity and carry their own energy, and is particularity by topological structure variation. Wireless sensor network protocols are quite different because of its particularity. The routing protocol and MAC protocol is different with the traditional wireless network protocols. MAC protocol of WSN decided usage mode of wireless channel. The MAC layer protocol needs to be considered in the design of energy efficiency, and design simple and efficient protocol according to the characteristics of the wireless sensor network. Routing protocol of WSN can be divided into energy awareness routing protocol, protocol of location routing based on query, reliable routing protocol [3].According to the characteristics of the wireless sensor network and application requirements, self-organizing, multiple hops routing protocol can be use appropriate.

In traditional network architecture, nodes only have transfer function for the purpose of transmission, support for the provision of application on the network transmission, without data processing. Wireless sensor network is data-centric. Its purpose is to obtain the perceived object, accurate information for a long time. The fast data fusion technology can realize to achieve sensor data of the wireless sensor network node to be combination fast and reasonable, to reduce data redundancy, to obtain more reasonable data, thus improve the network efficiency.

Dynamic topology network structure should be adopted, because wireless sensor network nodes are numerous and dense. On the premise of meet the network coverage and connectivity, it adopt dynamic topology through controlling power and selecting backbone node and remove unnecessary communication links between nodes to form a communication network structure optimization. Therefore, a good wireless sensor network topology should adopt node power control and hierarchical topology structure [4]. The node power control adjust the transmission power of each node in the network according to node communication distance and time to reduce unnecessary transmitted power of all nodes. A hierarchical topology control protocol reduces communication distance in one hop by using clustering mechanism to reduce energy consumption.

5 Wireless sensor node energy saving management solution

From above the node energy consumption and the characteristics of the mode of transmission between nodes, energy conservation and management of wireless sensor nodes could reduce the energy consumption of wireless sensor nodes through dynamic power consumption management and decrease the work load of node method to effectively achieve energy saving.

Wireless sensor node management way of energy saving measures are available from the following points to consider: dynamic power consumption management and reducing the workload of nodes are adopted to reduce the working time RF module, which reduced the traffic between nodes; Reduce RF module transmission power; Reduce the working time of the microprocessor. It is not only to consider hardware designing ,but also to solve the software management.

Wireless sensor nodes management software includes data acquisitioning and controlling of sensor network, wireless data transmission controlling, battery status monitoring, charging controlling. To reduce the power consumption of the microprocessor can be accomplished by the dynamic power consumption management of the microprocessor, and the energy consumption of the largest RF transceiver is controlled by soft of wireless data transmission.

Wireless data transmission part of the software consists of RF unit and baseband unit, and RF unit provides air interface, data communication of baseband unit which provide link of physical channel and data grouping. Microprocessor responsible for link management and control, the implementation of communication protocol of baseband and related processes, including creating links, frequency selection, link types support, media access control, power mode and security algorithm [5].Therefore, self organization, multiple hops routing, hierarchical, dynamic topology structure of network protocol are adopted in the baseband unit, in order to reduce traffic and balance the node energy, reduce energy consumption of nodes, so as to extend the life of the node.

To avoid repeated communications and information overlaps and waste of resources, hierarchical design is adopted with the organization, multiple hops routing protocol, to reduce communication time between nodes and communication distance. Due to the hierarchical cluster nodes as in the design of energy consumption is the largest, energy may be consumed ahead of time and the part of the network may be paralysis, so random selection of cluster head nodes and dynamically balance of the path are adopted. When energy consumption of cluster head node is too large, the sensor network can dynamically select nodes with less energy consumption according to the cluster head nodes energy consumption condition, balance the node energy consumption, extend the life cycle of the whole network.

Network topology can improve the efficiency of the network protocol to save energy to prolong network lifetime. Dynamic topology in the network coverage and connectivity of the premise remove unnecessary communication links between nodes through the power control and backbone node selection to achieve efficient data forwarding.

At the same time, the power mode management adopts the method of dynamic management to control the power mode in the baseband unit to reduce unnecessary power output. Different from the traditional power control, dynamic management of arousal and dormancy mechanism using heuristic node to change point state between sleep and active. This way can save the energy consumption of the free time as far as possible, strike a balance between performance and energy consumption.

Data collection control part of the software can increase the data processing part in addition to control sensors for data acquisition, based on the energy saving consideration. Data processing part adopts fast data fusion technology in the sensor nodes to rapid integrate and separate information, reduce the data redundancy in the entire network, reduce traffic, saving storage resources and network bandwidth.

6 Conclusion

A kind of wireless sensor node energy saving management scheme is put forward, after analysis of wireless sensor nodes, energy consumption and the analysis of the mode of transmission between nodes. The scheme adopts the fast data fusion technology by dynamic selecting cluster nodes with self-organization, multiple hops routing, hierarchical topology structure of routing protocol, and achieve dynamic power management under the condition of low power design of hardware.

Acknowledgments.

This research is partially supported by Natural Science Foundation of Fujian Province Of China (2008J0178).

References

1. Ju Zipei, Hua Yanqi etc., Wireless sensor network node hardware modular design [J]. Journal of Shanghai university of science and technology, 2009, (6) : 20 to 22
2. S Amit, C Anantha. Dynamic Power Managementin Wireless Sensor Networks [J]. Journal of Selected & test of computer, 2001, 19 (2) : 62-74.
3. V Vasanth, F Michael. The Power reduction techniques for microprocessor system [J]. ACM computing survey, 2005, 37 (3) : 195-237.
4. Sun Yugeng, Zhou Yin. An energy efficient clustering in wireless sensor network (WSN) network algorithm [J]. Journal of sensing technology, 2007, 20 (2) : 377-381.
5. Qiu Guoqing, Yang Zhilong, etc. Based on the ZigBee protocol of wireless sensor network node design [J]. Automation and instrumentation, 2008, (3) : 7-8 .
6. National Center for Biotechnology Information, http://www.ncbi.nlm.nih.gov

Optimum Design and Simulation of New Type Single Phase PMSG

Zhong-Shu Liu[1,2]

[1] School of Information Science and Engineering, Fujian University of
Technology, Fuzhou 350118, China
[2] The Key Laboratory for Automotive Electronics and Electric Drive of
Fujian Province, Fujian University of Technology, Fuzhou 350118, China
lzs@fjut.edu.cn

Abstract: The rare earth permanent magnet synchronous generator (PMSG) has advantages of high power density, simple structure, reliable operation etc., particularly suitable for military and civilian small generating units. According to the characteristics of the generator, the research was developed on the design theory, main performance parameters, and 2 pole single phase small power （4kVA） rare earth permanent magnet synchronous generator (PMSG) was designed based on improved genetic algorithm(IGA), by adopting a series of measures such as slot stator, big and small groove and sine winding, the voltage sinusoidal waveform distortion rate was reduced effectively. Simulation and test results show that the performance of the prototype designed reasonably.

Keywords: single phase PMSG; improved genetic algorithm; optimization design; simulation

1 Introduction

By using permanent magnet excitation instead of electrical excitation, single phase rare earth permanent magnet synchronous generator can eliminate the excitation loss without the excitation source compared with the traditional electrical excitation generator, therefore, it can improve the efficiency of the generator, and it is especially suitable for military or civilian small gasoline engine and diesel engine generator set used as power supply in the open or temporarily. The output voltage of electric excitation synchronous generators can be regulated with the load size by using automatic voltage regulator (AVR) to adjust the excitation current and compensate the armature reaction so as to keep a constant output voltage basically. However, as for the permanent magnet generator, due to its intrinsic characteristics of permanent magnet, the air-gap magnetic field cannot be adjusted by AVR[1]. Meanwhile, the optimum design for permanent magnet synchronous generator is a complex constraints, multivariable, mixed discrete, nonlinear programming problem, the objective function and constraint conditions are all nonlinear numerical function and more extreme value function of design variables, so it is difficult to solve the problem of motor optimization design fundamentally by using the traditional optimization

© Springer International Publishing AG 2017
J.-S. Pan et al. (eds.), *Advances in Intelligent Information Hiding
and Multimedia Signal Processing*, Smart Innovation, Systems and Technologies 64,
DOI 10.1007/978-3-319-50212-0_42

method. Genetic algorithm (GA) is a method based on the theory of natural selection and genetic development which is a random search and optimization algorithm. The algorithm can solve a lot of problems without many requirements and limitations, and its good robustness and parallelism make it do global search effectively, at the same time, the diversity of the algorithm for the optimization of special problem provides flexibility, which makes GA be widely used[2].However, the standard genetic algorithm (GA) is a kind of method only using basic genetic operator's group operating algorithm, derived from biological genetics and the natural law of survival of the fittest. While this kind of algorithm is simple, there exists problems of the important parameter selection difficult and slow convergence speed. In order to solve these two problems, combined with the specific requirements of the optimization design, we use the improved genetic algorithm which is a new generator structure optimization design method in this paper, meanwhile, reasonably design a single-phase permanent magnet synchronous generator stator slot and stator winding and rotor magnetic circuit structure in order to eliminate and weaken the air-gap magnetic field of tooth harmonic and higher harmonic influence, make the air-gap magnetic field is close to sine distribution, reduce the voltage waveform sine distortion rate and the voltage regulation, reach the purpose of optimization design [3].

1 characteristics of structure of a new type single-phase permanent magnet generator

The parameters of new type of single-phase synchronous permanent magnet generator (PMSG) shown as follows: power 4 kVA, frequency 50Hz, rated output voltage 230 v. and the structure is shown in figure 1. Stator winding is a single-phase concentric sine winding, the stator core adopts chute and big and small groove design, rotor magnet steel is built-in structure. Magnetic steel structure arranged as the 4 poles of radial magnetic circuit structure, but the magnets are placed as 2 poles structure in fact. This structure has the advantage of less processing cost than the magnet steel surface paste structure , for it needs only to rush in open permanent magnet rotor slot. Meanwhile, rotor radial magnetic circuit has high mechanical strength, and can increase per pole flux, so it is particularly suitable for small power single-phase permanent magnet synchronous generator.

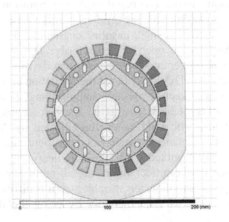

Fig. 1. Structure of single-phase PMSG

2 Main dimensions of the new generator and how to determine the size of permanent magnet

The main dimensions of the generator refers to as inside diameter D_{i1} and effective length of the stator core l_{eff}. The main dimensions determines the generator quality and material cost[4]. The main dimensions of the generator design choice of reasonable has a decisive role on the technical and economic indexes. In this paper, we use the improved genetic algorithm (IGA) for calculating the main dimensions of the single-phase permanent magnet synchronous generator and the size of permanent magnet.

2.1 Mathematical model for optimal design

Mathematical model for optimal design of generator includes three aspects: objective function, design variables and constraints,

2.1.1 Object function

Generator design optimization results reflected in whether achieving optimization goal finally. In recent years, the problem of energy crisis has become a global focus, therefore, under the premise of the generator performance, how to reduce the cost of generator has become our primary thinking[4]. Due to expensive materials such as rare earth permanent magnet used in generator, we select the corresponding cost of unit efficiency as objective function [5], namely:

$$f(x) = \min \frac{F(x)}{\eta} = \min \frac{(m_{cu}t_{cu} + m_{Fe}t_{Fe} + m_m t_m)U_N I_N}{U_N I_N - \sum p} \qquad (1)$$

In above, m_{cu} is the price for the copper materials; m_{Fe} is the price of steel materials; m_m is the price of nd-fe-b materials; t_{cu} is the total weight of copper materials used; t_{Fe} is the total weight of the steel materials used for; t_m is the total weight of nd-fe-b permanent magnet materials.

2.1.2 Design variables

We select the following parameters as optimization design of permanent magnet synchronous generator variables: the stator inner diameter D_{i1}, length of core l_{eff}, permanent magnet thickness h_m, the width of the permanent magnet b_m, rotor diameter D_2, rotor diameter D_{i2}, stator outer diameter D_1, namely:

$$X = \begin{bmatrix} x_1 \\ x_2 \\ x_3 \\ x_4 \\ x_5 \\ x_6 \\ x_7 \end{bmatrix} = \begin{bmatrix} D_{i1} \\ l_{\text{eff}} \\ h_m \\ b_m \\ D_2 \\ D_{i2} \\ D_1 \end{bmatrix} \tag{2}$$

2.1.3 Constraints

The constraint condition of rare earth permanent magnet motor mainly refers to the technical requirements and guarantee the performance of machine of some constraints, according to single-phase permanent magnet synchronous generator design engineering experience and the required performance index[6], we select constraint parameters as follows: efficiency η, voltage regulation δU, voltage waveform sine distortion rate k_γ and heat load AJ, respectively shown as follows:

$$\begin{cases} g(1) = \dfrac{\eta_0 - \eta}{\eta_0} \le 0 \\[2mm] g(2) = \dfrac{\delta U - \delta U_0}{\delta U_0} \le 0 \\[2mm] g(3) = \dfrac{k_\gamma - k_{\gamma 0}}{k_{\gamma 0}} \le 0 \\[2mm] g(4) = \dfrac{AJ - AJ_0}{AJ_0} \le 0 \end{cases} \tag{3}$$

Because the genetic algorithm is a kind of unconstrained optimization methods, we should first change the constrained optimization problem into unconstrained optimization problem. Based on this way, a penalty function is introduced, and its purpose lies in the low of the fitness of the individual risk increase its elimination, speed up the convergence speed [7]. This penalty function is shown as follows:

$$\delta(x) = \begin{cases} 0, & g(x) \le 0 \\ f_{\max} + f_{\max} \cdot \delta_0 \cdot k \cdot g(x), & g(x) > 0 \end{cases} \tag{4}$$

In above, $g(x) \le 0$ is constraint conditions; $\delta(x)$ is punish value; $f_{\max}$ is maximum generator cost; δ_0 is the importance of the constraint set of punish coefficient, among $0 \sim 1$ values; k is additional punishment for illegal restraint degree size corresponding coefficient, average value is 0.05[3] according to experience.

2.2 The optimization design of single-phase permanent magnet synchronous generator by using improved genetic algorithm (IGA)

This paper uses the improved genetic algorithm for generator structure design. Programming with existing numerical computing resources of the world's top MATLAB, use MATLAB to control, ANSYS run in batch mode, in this iterative operation to achieve the purpose of optimization. The algorithm implementation process is shown in figure 2.

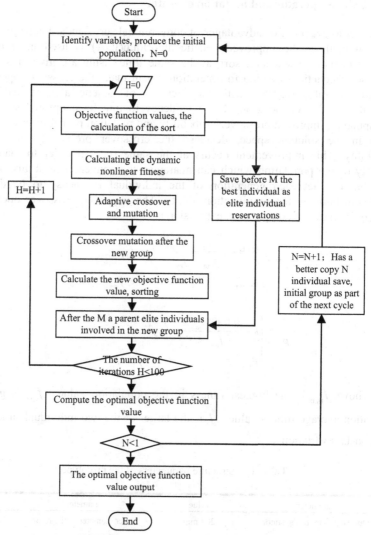

Fig. 2. IGA algorithm flow chart of PMSG

2.2.1 Fitness function

In order to reduce the cost of materials of single‾phase permanent magnet synchronous generator, we should minimize the value of objective function [4]. And improved genetic algorithm(IGA) is to find the highest fitness solution, so we should transform the objective function by changing the minimum problem into maximum problem. Therefore, we define the fitness function as follows:

$$F = \frac{1}{f(x)}$$ (5)

In above, $f(x)$ is the objective function.

2.2.2 Cross operator and mutation operator

In order to get rid of disadvantages of conventional sort standard genetic algorithm selection method, the improved genetic algorithm(IGA) is used in front of the crossing method of selection sort, at the same time, using adaptive crossover and mutation of genetic algorithm for operation, it can make the crossover operator and mutation operator regulate with the increase of genetic algebra and ongoing automatically [8]. As the group has a tendency to into local optimal solution, the corresponding improved crossover probability and mutation probability; when spread group in the solution space, decrease the crossover probability and mutation probability. This improvement occurs at the same time in order to maintain the diversity of the population which can guarantee the convergence ability of genetic algorithm that let each generation of the individual is uneasy to be destroyed, effectively improve the optimization ability. The improved crossover operator and mutation operator of computation expression is:

$$P_c = \begin{cases} 0.9 - \dfrac{0.3(f - f_{avg})}{f_{max} - f_{avg}}, & f \ge f_{avg} \\ 0.9, & f < f_{avg} \end{cases}$$ (6)

$$P_m = \begin{cases} k - \dfrac{0.099(f_{max} - f)}{f_{max} - f_{avg}}, & f \ge f_{avg} \\ k, & f < f_{avg} \end{cases}$$ (7)

In above, f_{max} is the biggest fitness in the population value; f_{avg} is generation population average fitness value; f is the larger of the two individual fitness value; k is constant variation.

Table 1.   Structure parameters of the prototype

Parameter	value	parameter	value
Outer diameter of the stator	204/mm	Outer diameter of the rotor	119/mm
Inner diameter of the stator	120/mm	Inner diameter of the rotor	30/mm
The length of the core	158/mm	The air gap length	1/mm
The thickness of the magnet	9/mm	Slot number	24
The width of the magnet	60/mm	Pole	1

In the use of improved genetic algorithm to optimize the design, if we find most of the individual fitness value of groups near a certain value, the genetic search can be considered complete. [9]

The structural parameters of single-phase permanent magnet synchronous generator for optimization design by using the improved genetic algorithm are shown in table 1.

Having determined the generator main structure parameters, we will discuss the optimum design of generator stator winding, the stator core and stator slots.

3 Stator winding and stator iron core

3.1 Sine concentric winding

Because of the air-gap magnetic field of permanent magnet generator contains rich higher harmonic, such as 3, 5, 7, 9..., therefore, to weaken the harmonic of the stator winding induction electric potential, the design of the most effective measures is using sine concentric windings [7]. Namely: under the condition of the motor stator slot number is a certain number, we can make the number of coil turns in each stator slot i.e. relative value of the total current in each slot distribute in air gap according to the sine law, the air-gap magnetic potential produced by the winding current are weaken in different extent apart from the fundamental wave. Thus we can get magnetic potential waveform sine wave[10]. The design of the prototype of the pole number is 2, i.e. $2p = 2$, the stator slots number $Z=24$, this generator per pole per phase slot number $q = Z / 2$ PM $= 24/2 \times 1 = 12$, pole pitch tau $= Z / 2p = 24/2 = 12$, coil span for 1-12, 2-11, 3-10, 4-9, 5-8, 6-7, the winding is known as sine concentric winding, and winding space distribution is shown in figure 3.

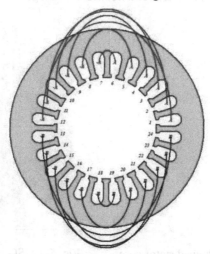

Fig. 3. PMSG stator winding spatial distribution diagram

3.2 Stator core with chute

Stator trough selects the semi-closed trapezoidal groove 24 slot uniformly distributed along the circumference of the stator, the stator iron core adopts chute design, by using Ansoft software Maxwell, we can analyze the influence of different chute width on the generator voltage waveform sine distortion rate [3]. The results are shown in table 2.

Table 2. Structure parameters of the prototype

skew width	0	0.25	0.5	0.75	1	1.25	1.5	1.75	2
Linear voltage waveform distortion rate（%）	1.547	1.382	1.065	0.773	0.619	0.301	0.217	0.334	0.481

Table 2 shows that the stator chute can effectively reduce the no-load voltage wave of tooth harmonic content so as to reduce the output voltage waveform sine distortion rate, especially when the stator chute width is 1.5times that of the stator slot pitch, the generator voltage waveform sine distortion rate of the would be the minimum, and the generator output voltage waveform sine would be best[4]. So, we chose the stator chute width as 1.5 times of the stator slot pitch.

4 Design of the stator slot

Considering the stator winding is sine concentric winding, which has per different number of turns in each slot and its distribution according to sine rule, in order to make use of stator core material reasonably, we design the prototype with the shape of each slot is directly proportional to the number of turns of coils in each slot that has shown in figure 4. This design can make every slot filled with coils rate roughly equal 73%. In addition, the stator slot type the size can reduce slot leakage, slot design leakage reactance, and it is helpful to reduce the voltage regulation [11].

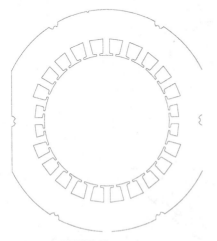

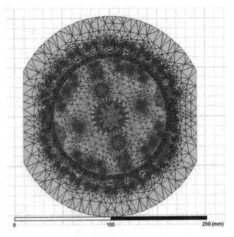

Fig. 4. PMSG Skewed stator core   **Fig. 5.** PMSG mesh generation diagram

5 Finite element analysis of magnetic field and performance simulation

5.1 Grid subdivision

Now we will analyze the prototype by using two-dimensional finite element method, which is a numerical calculation method based on discretization, it has the following two prominent advantage, which makes it particularly suitable for the calculation of the electromagnetic field distribution in the interior of the permanent magnet motor. One is that it can deal with internal medium boundary conditions very conveniently by using equivalent surface current on the side of the permanent magnet instead of permanent magnet excitation function, which can be easy to calculate the structural parameters of permanent magnet generator; The other is that the method of each section is unified which makes it easier to standardize [12]. By this way, first we use Maxwell 2D software package for grid subdivision mesh subdivision which is the most crucial step to finite element discretization. Figure 5 is the grid subdivision of prototype. It can be seen from figure 5 that the grid density of stator and rotor would be minimum while the grid density of air gap would be maximum. So through different grid density subdivision we can save computing resources and ensure the accuracy of the calculation.

5.2 Analysis of the generator internal electromagnetic field

Using Ansoft Maxwell we can analyze the flux density distribution of cloud of generator shown in figure 6 when it is in no-load. It can be seen from the diagram, the flux density value of stator yoke of generator is maximum which has reached 2.35 T, at this moment, the iron core of the generator is close to saturation. The purpose of this design is to increase the utilization rate of ferromagnetic materials so as to improve the power density of generator [9]. It can be seen from the diagram, the generator uniform magnetic field distribution, magnetic saturation area is less, only in the stator yoke department is saturated, the whole generator magnetic circuit design is reasonable. By PMSG load external circuit, apply permanent part of the perceptual load rated R = 8.81Ω, we can get the flux density distribution of cloud of generator shown in figure 7when it is in load. Compared with figure 6, we see that when the generator in load, the iron core saturation degree increase relative to the case in no-load, generator stator yoke of flux density value increase to 2.45 T, at the same time, all parts of the stator iron core flux density amplitude is increasing, the center line of the air-gap magnetic field of electric skewed compared with no-load, it is because the armature reaction cause gap magnetic field distortion [13].

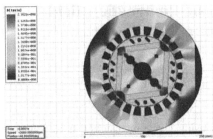

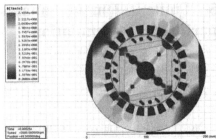

Fig. 6. No-load flux density cloud diagram of PMSG

Fig. 7. No-load flux density cloud diagram of PMSG

5.3 Performance simulation

5.3.1 Generator load voltage and load current

The output voltage and output current waveform of generator when it is in-load cases are shown in figure 8 and figure 9 respectively. As you can see by the picture, they are all sine wave, load voltage and current waveforms almost has no distortion, the peak value are 330 V and 35.4 A respectively, according to the power calculation formula, we can get the motor output power as follows:

$$P_N = U_N I_N \cos\varphi = \frac{330}{\sqrt{2}} \times \frac{35.4}{\sqrt{2}} \times 1 = 4033VA \tag{8}$$

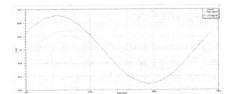

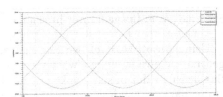

Fig. 8. PMSG output voltage waveform diagram

Fig. 9. PMSG output current waveform diagram

6 Automatic voltage regulating system

In order to remain the output voltage of prototype constant, we use automatic pressure regulating system which is designed to ensure output voltage steady at 230 V in case of load or temperature change through oil machine speed regulation. The main factors causing the change of output voltage can be classified the following three aspects: First is the generator itself within the impedance of the output voltage caused by the pressure drop of the fall; Second is the demagnetization effect of armature reaction which causes the main magnetic flux decreases with the decline of the excitation voltage output. Third is the temperature change that makes the generator impedance size change and permanent magnets working point change which cause the output voltage changes [13]. The role of the generator speed control system is to compensate the above voltage change by closed-loop feedback control.

The new type of small power 4kVA generator prototype adopt the way of coaxial connecting generator directly from a gasoline engine, and we use the generator terminal voltage closed loop feedback to control the generator speed, through the acquisition of generator terminal voltage signal instantaneous value compared with the given value, after amplification to control stepping motor positive and reverse rotation direction adjust the size of the oil motor oil door, realize automatic control of oil machine with load speed, control in high and low way. When generator voltage measured value is greater than the given value it is in low status, the relay switch action drive stepper motor reverse rotation should be turn down the gas, at same time, oil machine to slow down so as to reduce the generator voltage. On the contrary, when the generator voltage measurement value is less than the minimum value given it is in high status, the relay switching action is a stepping motor drivers come up on the gas, oil machine speed up to increase the generator voltage. When the generator voltage measurement value is between the maximum and minimum of the given value, the sealed relay switch should keep the gas constant, and the oil machine speed running within a certain range [14]. Set according to the requirement of the voltage rate w % values given the maximum $U_{g\max} = (1 + \Delta U\%)U_g$ and minimum values of a given voltage $U_{g\min} = (1 - \Delta U\%)U_g$, U_g voltage of generator terminal voltage U_N is equal to the rating of a given voltage, the voltage of the automatic control system should ensure the output voltage change between the range of $U = (1 \pm \Delta U\%)U_N$ [15].

7 Prototype test results

After the test, prototype various performance indicators are shown as in table 3, the test results show that the prototype various performance fully meet the design requirements, prototype design is successful.

Table 3.   Comparison table of the performance index from measuring and calculating of prototype

project	power	efficiency	Waveform distortion rate	Voltage regulation	Voltage	counter emf
Design value	4/kVA	92/%	5/%	18/%	230/V	220/V
The measured values	3.99/kVA	95.7/%	0.498/%	13.2/%	235.6/V	223.3/V

8 Conclusion

Improved genetic algorithm (IGA) is applied in this article to design a 4 kVA prototype of new type of small power single-phase permanent magnet synchronous

generator with minimum cost of the stator winding. By using Ansoft software
Maxwell we analyze the load magnetic field of generators and output characteristic.
The simulation results and experimental results show that, with chute of the stator
core and sine winding single-phase permanent magnet synchronous generator has a
smaller output voltage regulation, less voltage waveform harmonic component, low
waveform distortion rate, which fully meet the design requirements.

References:

1. Li Zheng, Sun Tiantian, Gao Peifeng. Design and Simulation of Disc Permanent Magnet
   Synchronous Generator for Distributed Systems [J].Electric Machines & Control
   Application, 2016,43 (1): 29-34.
2. He Biyi, Zhou Guoping, Zhang Ya. A Study on the Control of the Permanent Magnet
   Synchronous Generator Based on the Fast Terminal Sliding Mode [J].Electrical Automation,
   2015, 37(3):14-17.
3. Li Shenghu, An Rui, Xu Zhifeng, Dong Wangchao. Coordinated LVRT of IG and PMSG in
   hybrid wind farm [J].Electric Power Automation, 2015, 35(2):21-27.
4. Gong Jianfang. The Influence of Skew Width Air-Gap on PMSG's Performance [J].Large
   Electric Machine and Hydraulic Turbine, 2008, 15(4):17-20.
5. Chen Guiyu, Tang Ningping, Li Dongxu. Optimization Design of PMSG Based on Improved
   Genetic Algorithm [J].Explosion-Proof Electric Machine, 2011, 46(1):5-9.
6. Wen Huihui. Steady State Performance of Three-Phase PMSN with Single-Phase Supply
   System [J].Control Engineering of China, 2014, 21(1):142-147.
7. Liu Jian. Research on Control Technologies of Six-Phase Permanent Magnet Synchronous
   Generator [M].Harbin: Harbin Institute of Technology Ph.D. 2014:39-67.
8. Shen Xijun, Zhao Qiaoe, Xu Pengtao. Control Research on Maximum Power of the Direct-
   drive Permanent Magnet Synchronous Generator [J].Journal of Electric Power, 2014,
   29(1):28-31.
9. Yang Kun, Yang Xiu. Modeling Optimization and Simulation of Grid-connected Direct-
   Driven Permanent Magnet Wind Power System Based on Small Signal Stability Analysis
   [J].East China Electric Power, 2014, 42(10):2028-2033.
10. Zhai Yunzhuo. Research on Vector Control of PMSG in Low Temperature Waste Heat
    Power Generation System [M].Shenyang: Shenyang University of Technology
    Ph.D.2015:17-30.
11. Feng Hao, Huo Zhihong, Hu Xuebin. Constant Power Control for Wind Turbine Based on
    Finite-time Method [J].Renewable Energy Resources, 2015, 33 (12): 1831-1834.
12. Shen Jianxin, Miao Dongmin .Variable Speed Permanent Magnet Synchronous Generator
    Systems and Control Strategies [J].Transactions of China Electrotechnical Society, 2013,
    28(3):2-8.
13. An Quntao, Sun Li, Sun Lizhi. Research on Novel Open-End Wind Permanent Magnet
    Synchronous Motor Vector Control Systems [J].Proceedings of the CSEE, 2015, 35 (22):
    5891-5898.
14. Gao Jian. Research on Design Technology and Application of Direct-drive Permanent
    Magnet Generator with Wind Turbine [M]. Changsha: Hunan University Ph.D. 2013:36-55.
15. Huang Qinghua. Design and Performance of Analysis on Single-Phase Small-Power Work-
    Frequency Rare Earth Permanent magnetic Generator Set[J].Journal of Minjiang University,
    2012, 33(5):47-51.

An Unmanned Aerial Vehicle Optimal Route Planning Based on Compact Artificial Bee Colony

Tien-Szu Pan[1], Thi-Kien Dao[1], Jeng-Shyang Pan[2], Trong-The Nguyen[3],

[1]Department of Electronics Engineering,
National Kaohsiung University of Applied Sciences, Taiwan
[2]Provincial Key Lab of Big Data Mining and Applications,
Fujian University of Technology, China
[3]Department of Information Technology, Hai-Phong Private University, Vietnam
tpan@cc.kuas.edu.tw, vnthe@hpu.edu.vn, jvnkien@gmail.com

Abstract. In the inevitable trends of the modern aerial equipment, unmanned aerial vehicle (UAV) is one of the most concerned components to research. This paper presents a saving memory optimization algorithm of Compact artificial bee colony (cABC) for UAVs route planning problem. In the proposed method, route length and danger exposure are modeled mathematically as the objective function, and the compact algorithm concept is implemented to accommodate the route planning situation. In the compact algorithm, actual design variable of solutions search space of artificial bee colony algorithm is replaced with a probabilistic representation of the population. A probabilistic representation random of the collection behavior of bees is inspired to employ for this proposed algorithm. The real population is replaced with the probability vector updated based on single competition. The computational results compared with other algorithms in the literature shows that the proposed method can provide the effective way of using a modest memory for UAV route planning problem.

Keywords: Unmanned aerial vehicle, Route planning, Compact artificial bee colony

1 Instruction

An Unmanned Aerial Vehicle (UAV), commonly known as a drone is an aircraft without a human pilot on board. Its flight is controlled either autonomously by computers in the vehicle, or via remote control by a pilot on the ground or in another vehicle [1]. UAVs provide a safe alternative where a UAV with an onboard vision or thermal camera go through and scan this area. The images captured by the UAV can be sent back to a ground control station whether it can be processed either manually or autonomously to locate the object of interest [2]. The UAV technology has opened a sea of opportunities in its potential use in civilian domain. Civilian agencies or departments, as well as civilian commercial companies, can utilize this technology in emergency response, search and rescue, transportation, and surveillance, monitoring or inspection of infrastructure. As there are no pilots, the mission of UAVs should be gracefully predesigned

© Springer International Publishing AG 2017

J.-S. Pan et al. (eds.), *Advances in Intelligent Information Hiding and Multimedia Signal Processing*, Smart Innovation, Systems and Technologies 64,
DOI 10.1007/978-3-319-50212-0_43

to make sure the UAVs can complete it with less power consumption, less exposure to the enemies and many other constraints [3]. This is achieved mainly by preplanning the flying routes of the UAVs. The flight route planning in a large mission area is a typical large-scale optimization problem. The metaheuristic algorithm is one of the preferred ways to solve the large-scale optimization problem. However, in the field of route planning for UAV, no application of compact algorithm exists yet. Compact algorithm simulates the behavior of population-based algorithms by employing, instead of a population of solutions with its probabilistic representation. An effective compromise used in the compact algorithm is to present some advantages of population-based algorithms but the memory is not required for storing an actual population of solutions. In this way, the space for storing the number of parameters in the memory is smaller.

In this paper, we use an original artificial bee colony (ABC) [4] and an improved ABC as compact algorithm [5] to solve UAV route planning problem. A fitness function according to the satisfied points and avoid threats is mathematically modeled as an objective function of the optimal route planning.

2   Mathematical Model in UAV Route Planning

The goal for route planning is to calculate the optimal or suboptimal flight route for UAV within the appropriate time, which enables the UAV to pass through the dangerous threat environments, and self-survive with the perfect completion of the mission [3]. The design of optimal flight route for UAV needs to meet certain performance requirements according to the special mission objective, and subject to the constraints of such as the terrain, data, threat information, power energy, and time. UAV route planning can be modeled mathematically as follows. A UAV workplace is described as a route in the global coordinates $O-XY$, with the start position and the target position. The route planning problem is transformed into a D-dimensional function optimization problem. The original coordinate system is transformed into new coordinate $O-X'Y'$ whose horizontal axis is the connection line from starting point to target point. The corresponding transformation formula is as follows:

$$\begin{bmatrix} x' \\ y' \end{bmatrix} = \begin{bmatrix} \cos\varphi & -\sin\varphi \\ \sin\varphi & \cos\varphi \end{bmatrix} \times \begin{bmatrix} x \\ y \end{bmatrix} + \begin{bmatrix} x_{Start} \\ y_{Targ} \end{bmatrix} \tag{1}$$

where the point (x, y) is coordinate in the original ground coordinate system; the point (x', y') is coordinate in the new rotating coordinate system, where φ is the rotation angle of the coordinate system from the X-axis to the line; Start–Target, (x_{Start}, y_{Targ}) is the point Start in the coordinates $O_{x,y}$. The horizontal axis X' is divided into $n+1$ equal partitions with n points then optimize vertical coordinate Y' on the vertical line for each node to get a group of points composed by vertical coordinate of n points. After drawing n vertical lines through these points, a set of lines denoted $\{l_1, l_2, ..l_n\}$. A complete route from start point to end point through connecting these points together $\{p_1, p_2, ..p_n\}$ can be constructed by sampling at random on vertical lines of $l_1, l_2, ..l_n$. Thus, the route planning problem is transformed into optimizing the following set of points $\{Start, p_1, p_2, ..p_n, \text{and } Target\}$.

The search agents are generated in the beginning with respect to the UAV's initial position and regarding its sensing range. The objective function of performance indicators for UAV route can be formulated by as following.

$$minimize \ F = \omega \times T + (1 - \omega) \times L \tag{2}$$

where T and L are the function of threat intensity criteria. and the distance criteria of the route; $\omega \in [0, 1]$ is balanced coefficient between safety performance and distances, whose value is determined the special task for UAV. If flight safety is of highly vital importance to the task, then k will be chosen a larger, otherwise k will be set a small. The threat intensity criteria of the route can be calculated as:

$$min \ T = w_t \times \sum_{i=0}^{n} t_i \ (p_i, p_{i+1}) \tag{3}$$

where t_{ij} is the threat intensity between p_i and p_{i+1}; w_t is the threat cost for each point on the route. The length of path, supposing that the start state and the target state are p_0 and p_{n+1}, the length of a path can be approximated by

$$min \ L = w_l \times \sum_{i=0}^{n} l_i(p_i, p_{i+1}) \tag{4}$$

where L is the length of path and $l_i(p_i, p_{i+1})$ represents the distance between p_i and p_{i+1}. It is calculated as $l_i(p_i, p_{i+1}) = \sqrt{(x'_{p_i} - x'_{p_{i+1}})^2 + (y'_{p_i} + y'_{p_{i+1}})^2}$. The sample m points on the path segment between p_i and p_{i+1} can be taken evenly. $m = [l_i / interval]$; $interval$ is a predefined sampling step length. Threat point of the path segment can be calculated as:

$$t_i = \frac{1}{d_{i+1}^4} + \sum_{k=1}^{m} \frac{1}{d_{ik}} \tag{5}$$

where d_{i+1} and d_{ik} are distance of p_{i+1}, p_i, and the nearest threat point from them respectively.

3   UAV Optimal Route Planning with Compact Algorithm

This section presents the compact algorithm based on the frame of ABC algorithm for solving UAV route planning problem. We first review briefly ABC and then present the compact method processing for the constrained optimization problem of route planning for UAV.

3.1   Artificial Bee Colony Algorithm

The Artificial Bee Colony (ABC) algorithm was inspired from finding nectar and sharing the information behavior of bees [4]. The artificial agents were known as employed bee, onlooker, and scout which play different roles in the optimization process. The employed bee stays on a food source, which represents a spot in the solution space, and provides the coordinate for the onlookers in the hive for reference. The onlooker bee receives the locations of the food sources and selects one of the food sources to gather the nectar. The scout bee moves in the solution space to discover new food sources. The process of ABC optimization is listed as follows:

1. A portion of the population called the employed bees is generated into the solution space randomly. The probability of selecting a food source is calculated as.

$$P_i = \frac{F(\theta_i)}{\sum_{k=1}^{S} F(\theta_k)} \tag{6}$$

where θ_i denotes the position of the ith employed bee and $F(\theta_i)$ denotes the fitness function; S represents the number of employed bees, and Pi is the probability of selecting the ith employed bee. The roulette wheel selection method is used to select a food source to move for every onlooker bees.

$$x_{ij}(t+1) = \theta_{ij}(t) + \emptyset(\theta_{ij}(t) - \theta_{kj}(t)) \tag{7}$$

where x_i denotes the position of the *i-th* onlooker bee, t denotes the iteration number, θ is the randomly chosen employed bee, j represents the dimension of the solution and $\emptyset$ (.) produces a series of a random variable in the range from -1 to 1.
2. Update the best food source found so far.

$$\theta_{ij} = \theta_{jmin} + r \times (\theta_{jmax} - \theta_{jmin}) \tag{8}$$

where r is a random number and r ∈ range from 0 to 1. The best fitness value and the position, which are found by the bees. If the fitness values of the employed bees do not be improved by a continuous predetermined number of iterations, which is called "*Limit*", those food sources are abandoned, and these employed bees become the scouts
3. Termination Checking: Check if the amount of the iterations satisfies the termination condition. If the termination condition is satisfied, terminate the program and output the results; otherwise go back to selection step.

3.2 Compact ABC for UAV Route Planning

The methods of compact algorithm try to simulate very similarly to searching operators of the population-based methods. A probability distribution is used generating new candidate solutions that being iteratively biased toward an optimal solution. Two operations of searching and selecting could be combined considerably in the optimization algorithms. The compact method, the actual population base of ABC will be described as a virtual population by encoding within a data structure, namely Perturbation Vector (PV). PV is the probabilistic model of a population of solutions, which suggested use in [6][7]. Candidate solutions are probabilistically generated from the vector, and the competing components toward to the better solutions are used to change the probabilities in the vector. The virtual population can be configured by considering probability density functions (PDFs)[8] through the framework of facilitated EDA. A probabilistic model of the actual population is the Gaussian distribution that is adapted by truncating PDF. The distribution of the individual in the hypothetical swarms must be based on a PDF. Gaussian PDF with mean μ and standard deviation σ is used to assume a distribution each particle of swarms. The generated trial solutions are allocated in boundary constraints. If the variables of the algorithm are normalized as probability generated trials in interval arrange of -1 to +1, for $j = 1, 2, .. d$, where d is the dimension of the problem, the memory of storing particles will be reduced significantly in the boundaries

$[lb_j, ub_j]$. The design parameters of the PDF for each variable are normalized in each search interval arrange of (-1, +1). Therefore, PV is a vector of m×2 of matrix the for specifying the two parameters of the PDF of each design variable as defined.

$$PV^t = [\mu^t, \sigma^t] \tag{9}$$

where μ and σ are mean and standard deviation values of a truncated Gaussian (PDF) within the interval of (-1, +1) respectively. The amplitude of the PDF is normalized by keeping its area equal to 1. The apex t is time steps. The initialization of the virtual population is generated for each design variable i, e.g. μ_i^1 is set to 0 and σ_i^1 is set to k, where k is set as a large positive constant (e.g. referencing k set to 10). A generated candidate solution x_i is produced from $PV(\mu_i, \sigma_i)$.

$$P_i(x) = \frac{\sqrt{\frac{2}{\pi}} \times \exp\left(-\frac{(x-\mu_i)^2}{2\sigma_i^2}\right)}{\sigma_i\left(\text{erf}\left(\frac{\mu_i+1}{\sqrt{2}\sigma_i}\right) - \text{erf}\left(\frac{\mu_i-1}{\sqrt{2}\sigma_i}\right)\right)} \tag{10}$$

where $P_i(x)$ is the value of the PDF corresponding to variable xi, and erf is the error function[9]. The slot memory only needs to store vectors μ_i and σ_i on memory whenever the probability density function is trigged. The PDF in Eq. (10) is then used to compute the corresponding Cumulative Distribution Function (CDF) [10]. The codomain of CDF is arranged from 0 to 1. The CDF describes the probability that a real-valued random variable X with a given probability distribution to be found at a value less than or equal to x_i. The newly calculated value of x_i is inversed to from CDF. Two designed variables are used to compete for finding out who is winner or loser. The winner or loser vectors are according to the evaluation of fitness function value that is better or worst. The new solution is then evaluated in Eq.(2) and compared against $x_{pbest_i}^t$ and x_{gbest}^t to determine who is the winning and losing individual. The competing algorithm for winner and loser is based on the elements μ_i^{t+1} and σ^{t+1} of the PV are updated to the new solution based on the differential iterative.

$$\mu_i^{t+1} = \mu_i^t + \frac{1}{N_p}(winner_i - loser_i) \tag{11}$$

where Np is virtual population size, which is only parameter typical of the compact algorithm and it does not strictly correspond to the population size as in a population-based algorithm, for further detail about a reported Np see in[11]. Regarding σ values, the update rule of each element is given as.

$$\sigma_i^{t+1} = \sqrt{(\sigma_i^t)^2 + (\mu_i^t)^2 - (\mu_i^{t+1})^2 + \frac{1}{N_p}(winner_i^2 - loser_i^2)} \tag{12}$$

The virtual population size, in real-valued compact optimization, is a parameter which biases the convergence speed of the algorithm. A new candidate solution is obtained based on a comparison between it and the elite.

3.3   UAV Optimal Route Planning Based on cABC

The processing steps of the compact method are simulated as the behavior of population-based algorithms by sampling probabilistic models. The fitness value of the position x^t is calculated and compared with best to determine a winner and a loser. Eqs. (11) and (12) are then applied to update the probability vector PV. If rand is smaller than Prob (probability Eq. (6) is calculated from employment bee phrase), x^t will be calculated by Eq. (8). Update local and update global are implemented in Onlooker bee phrase Eqs.(7) and (8). If f(sol) < fbest the value of function is memorized the value of the global best is then updated. The route planning optimization for UAV is summarized as the pseudo-code in Fig. 1.

---

1) *Initialization probability vector (PV(μ, δ))*:
    for *i=1:n* **do** μ_i^t =0; δ_i^t=k= *10;*
2) *Initialization parameters:* trial=0; limit=10;
    while *termination is not satisfied* **do**
3) Employed Bee Phase
    generate x^t from PV; Calculate f(x^t); // Obj. function Eq.(2)
    if (f(x^t)<f(sol) then   sol=x^t; f(sol)=f(x^t); else   trial=trial+1; end if
    //Update PV
    [*winner, loser*]=**compete**(x^{t+1}, *best*) // According to Eq. (11), (12)
4) Onlooker bee phase
    x^t=sol;
    if (rand<Prob)
        for i=1:n do
        x^t (i)= x^t (i)+rand*(x^t(i)-x(k)); with random *k=1,..,n*
    end if
     Calculate f(x^t); //Update local
    if (f(x^t)<f(sol) then   sol=x^t; f(sol)=f(x^t);
    else trial=trial+1; end if   //Update global
    if (f(sol)<fbest then best=sol; fbest=f(sol);
5) Scout Bee Phase
    if (trial==limit) then
        x^t=generated from PV;   end if
    end while

Fig. 1. The pseudo code of compact (cABC) for UAV Optimal route planning

The basic steps of the optimization process are described as follows:
Step 1: Modeling UAV workspace including the threats, and the UAV's start and target positions.
Step 2: Parsing solution is as mapping search agents to a model of route planning during optimization.
Step 3: Implementing the cABC to find optimal routes of the above model.
Step 4: Guide the UAV to the target position by the optimal routes selected

4   Simulation results

In this section, a simulation of UAV route planning is conducted based on compact artificial bee colony (cABC).

4.1   Setting environment of UAV working

An optimization of route planning is carried out in order to verify the proposed method of cABC. Initialization parameters of the algorithm are as follows. The environment map of the simulation is set to grid network areas. The number of threats, grid network set or reset as GUI scheme shown in Figure 2. The parameters are set as the initial e.g. *'limit'* is set to 10 of food source, the dimension of the solution space *dim* is set to 10. The maximum number of iterations is *MexIter* and it is repeated by different random seeds with 10 runs. The final result is obtained by taking the average of the outcomes from all runs. The results are compared with the several methods in literature such as Ant colony optimization (ACO) [12], Bat algorithm (BA) [13], Biogeography-based optimization (BBO) [14], Genetic algorithm (GA)[15], Particle swarm optimization (PSO)[16], and original ABC.

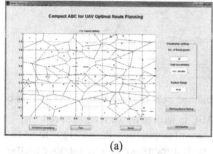 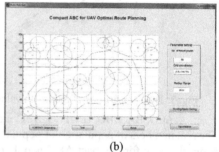

(a)   (b)

Fig. 2. Workspace initialization setting for UAV route planning simulation a) generating randomly VORONOI diagram, b) optimization for UAV route planning

4.2   Experimental results.

Figure 2 shows GUI scheme for the proposed method of UAV route planning. The initial setting for environmental simulation e.g. threats, space, UAV positions and the coordinate system is shown in Figure 2 a, and the obtained sequence UAV route planning is shown in Figure 2 b. Results of the average obtained optimal route planning of the proposed method are compared with the obtained from ACO, BA, BBO, GA, PSO, and the base version of ABC methods.

Table 1. The minimum objective function values found by algorithms averaged over 10 runs

Max Iteration	Algorithms						
	ACO	BA	BBO	GA	PSO	ABC	**cABC**
50	12.4819	13.5782	14.1072	7.3797	6.076	4.5491	4.8541
100	12.3884	11.6048	10.5705	6.0887	4.1725	3.4353	**3.3523**
150	11.1408	11.4874	9.7978	3.7267	4.5459	3.1636	3.4269
200	11.3976	10.4323	10.8224	3.6358	2.9917	2.9434	3.0080
250	11.1958	10.4213	10.0553	2.9715	4.8005	3.1409	**3.0506**
300	11.1058	11.0113	10.1553	2.9715	4.8005	3.1409	**3.0360**

Tables 1 shows the average obtained optimal values of the proposed method for fitness function of the route planning in Eq. (2) with different max iteration numbers. Clearly,

the proposed method of cABC and the base ABC perform other methods. The obtained
results by cABC are as good as those obtained by base ABC.

Figure 3 indicates the comparison of a mean of best so far of the proposed method,
ACO, BA, BBO, GA, PSO methods for the objective function in Eq. (2). Fortunately,
the proposed method also provides the better results than those obtained by other methods in terms of convergence speed.

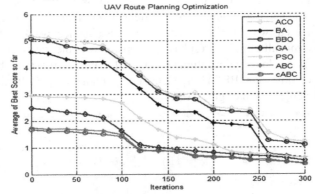

Fig. 3. The comparisons of minimum value obtained from fitness function over 10 runs of

ACO, BA, BBO, GA, PSO, ABC and cABC in term of the converged speed

The real numbers of population or population size of population-based algorithms need
N but that size for compact ABC is only one with dimension D. The number memory
variables of cABC is smaller than that of ABC in the same condition of computation
such as iterations.

5   Conclusion

This paper presented a saving memory optimization method for Unmanned aerial vehicle (UAV) route planning in complicated field environments. The implementation of
the compact for optimization algorithms could have important significance for the development of embedded devices small size, low price, and light weight. A novel type
of improved version of the artificial bee colony algorithm (ABC) as a compact ABC
has been described for a UAC route planning problem while avoiding the threat areas.
In the compact algorithm, the actual design variable of solutions search space of ABC
algorithm is replaced with a probabilistic representation of the population. The simulation results compared with other optimization methods in the literature, show that the
proposed method is a feasible and saving memory way in UAC route planning, and can
provide the performance of optimization significantly as within the category of
memory-saving algorithms. In UAC route planning, there are many issues worthy of
further study, and efficient route planning method should be developed depending on
the analysis of specific complex field environments.

References

1. Chao, H., Cao, Y., Chen, Y.: Autopilots for small unmanned aerial vehicles: A survey. Int. J. Control. Autom. Syst. 8, 36–44 (2010).
2. Xie, T., Zheng, J.: A Joint Attitude Control Method for Small Unmanned Aerial Vehicles Based on Prediction Incremental Backstepping. J. Inf. Hiding Multimed. Signal Process. 7, 277–285 (2016).
3. Chee, K.Y., Zhong, Z.W.: Control, navigation and collision avoidance for an unmanned aerial vehicle. Sensors Actuators, A Phys. 190, 66–76 (2013).
4. Karaboga, D., Basturk, B.: Artificial Bee Colony (ABC) Optimization Algorithm for Solving Constrained Optimization. Lnai 4529. 789–798 (2007).
5. Dao, T.-K., Chu, S.-C., Nguyen, T.-T., Shieh, C.-S., Horng, M.-F.: Compact Artificial Bee Colony. In: Modern Advances in Applied Intelligence. pp. 96–105. Springer (2014).
6. Harik, G.R., Lobo, F.G., Goldberg, D.E.: The compact genetic algorithm. IEEE Trans. Evol. Comput. 3, 287–297 (1999).
7. Dao, T.-K., Pan, T.-S., Nguyen, T.-T., Chu, S.-C.: A compact Articial bee colony optimization for topology control scheme in wireless sensor networks. J. Inf. Hiding Multimed. Signal Process. 6, (2015).
8. Billingsley, P.: Probability and Measure - Third Edition. (1995).
9. Abramowitz, M., Stegun, I.A.: Handbook of mathematical functions: with formulas, graphs, and mathematical tables. Courier Corporation (1964).
10. Cody, W.J.: Rational Chebyshev approximations for the error function. Math. Comput. 23, 631–631 (1969).
11. Iacca, G., Mallipeddi, R., Mininno, E., Neri, F., Suganthan, P.N.: Super-fit and population size reduction in compact differential evolution. In: IEEE SSCI 2011 - Symposium Series on Computational Intelligence - MC 2011: 2011 IEEE Workshop on Memetic Computing. pp. 21–28. IEEE (2011).
12. Zhangqi, W., Xiaoguang, Z., Qingyao, H.: Mobile Robot Path Planning based on Parameter Optimization Ant Colony Algorithm. Procedia Eng. 15, 2738–2741 (2011).
13. Yang, X.: Bat Algorithm : Literature Review and Applications. Int. J. Bio-Inspired Comput. 5, 1–10 (2013).
14. Zhu, W., Duan, H.: Chaotic predator-prey biogeography-based optimization approach for UCAV path planning. Aerosp. Sci. Technol. 32, 153–161 (2014).
15. Cheng, Z., Sun, Y., Liu, Y.: Path Planning Based on Immune Genetic Algorithm for UAV. Electr. Inf. Control Eng. (ICEICE), 2011 Int. Conf. 0–3 (2011).
16. Yong Bao, Xiaowei Fu, Xiaoguang Gao: Path planning for reconnaissance UAV based on Particle Swarm Optimization. 2010 Second Int. Conf. Comput. Intell. Nat. Comput. 2, 28–32 (2010).

A Multi-Objective Optimal Vehicle Fuel Consumption Based on Whale Optimization Algorithm

Mong-Fong Horng[#], Thi-Kien Dao[#], Chin-Shiuh Shieh[#], Trong-The Nguyen[*]

[#]Department of Electronic Engineering,
National Kaohsiung University of Applied Sciences, Taiwan
[*]Department of Information Technology, Haiphong Private University, Vietnam
jvnkien@gmail.com, csshieh@cc.kuas.edu.tw

Abstract. Due to the complex environmental constraints in the field of transport and logistics, optimization for the vehicle fuel consumption problem satisfies not only one criterion but also several criteria. In this paper, a novel multi-objective method for an optimal vehicle traveling based on Whale optimization algorithm (WOA) is proposed. In the proposed method, two criteria of distance and path-travel gasoline of the vehicle traveling issue are transformed into a minimization one. The vehicle routing and traffic status in the environmental transportation are considered to model the fitness function for the solution in WOA. The path of the globally best whale in each iteration is selected, and reached by the vehicles in sequence. In addition, the information of traffic status updates to vehicle periodically times from traffic navigation system during the traveling. Series scenarios of simulations are implemented in different traffic environments for the optimal paths when the vehicles reached the destination. The results show that the proposed method provides the confirm the practicality of the model of transport and logistics, and this proposed method may be the alternative method of optimization for the vehicle traveling in the logistics.

Keywords: Whale optimization algorithm, Vehicle fuel consumption, Multi-objective optimization.

1 Introduction

Fuel consumption, emissions, long travel time and accidents can be both direct and indirect consequences of vehicle traffic congestion and rough driving pattern [1]. Solution to alleviate the vehicle congestion problem is to build new high-capacity streets and highways. Nevertheless, this solution is very costly, time-consuming and in most cases, it is not possible because of the space limitations. Optimal usage of the existent roads and streets capacity can lessen the congestion problem in large cities at the lower cost. However, this solution needs accurate information about current status of roads and streets which are a challenging task due to the complicated and changing

© Springer International Publishing AG 2017
J.-S. Pan et al. (eds.), *Advances in Intelligent Information Hiding and Multimedia Signal Processing*, Smart Innovation, Systems and Technologies 64,
DOI 10.1007/978-3-319-50212-0_44

environments[2]. Fortunately, the advancement in electronics, communications and information technologies and the Internet have led to the rapid proliferation of wireless sensor networks (WSN) [3]. It is envisioned WSN as a promising answer to these problems. Sensor-enabled products and their networks are becoming a commonplace, and central to the everyday life, e.g., traffic control and navigation, object tracking and monitoring, and so on. A promise of providing alternative paths with accurate information about current status of roads, streets, and shortest path distances can be helpful to reduce fuel consumption and lessen traffic congestion based on WNS equipped. A large number of inexpensive sensors, small size and integrated with a sensing unit and wireless communication capabilities in WSNs can support to the traffic control and navigation system. In those systems, the sensor nodes are being deployed in a wide terrain to perform their intended tasks efficiently. Typically, the heterogeneous sensor networks that are more practical, having better network performance, and lifetime, scalability, efficient load-balancing, and are cost-efficient [4].

In the transportation, a cost of a vehicle traveling along a route depends on several factors. Among of them, two main types of the factors include an indirect and direct relationship with the traveling schedule [5]. The indirect to traveling schedule one includes depreciation of the tires and the vehicle, maintenance, driver wages, tax, etc. The direct relationship with the traveling schedule includes distance, load, speed, road conditions, fuel consumption. Fuel consumption is per unit of distance, fuel price, etc. By comparison, the factors in the type of directly related to fuel consumption can be regarded as variable cost or fuel cost. In addition, if other factors are kept constant, the fuel consumption then mainly depends on distance, time traveling and congestion. Furthermore, metaheuristic methods have been applied successfully to solve many problems in many areas including the traffic control related vehicle routing schedule [6]. Metaheuristic methods have been inspired by the natural phenomena or the evolution genetic, or swarm intelligent [7]. Whale optimization algorithm (WOA) [8] is a recent novel swarm intelligent optimization method. As a result of NP-completeness of the dynamic vehicle routing problem, metaheuristic methods have been developed increasingly to cope with the high computational costs and complexities of classic methods, especially for high degrees of freedom. The vehicle fuel consumption problem with respect to a single objective function are generally non-smooth and questionable practicality time when turning their course of moving abruptly.

In real life, there are many situations and problems that are recognized as multi-objective problems. This type of problems containing multiple criteria to be met or must be taken into account. Often these criteria are in conflict with each other and there is no single solution that simultaneously satisfies everyone. Optimization vehicle traveling along a route is affected by several constraints parameters. Therefore, in this paper, we have worked out a fuel consumed vehicle traveling to apply the two objectives simultaneously, which are the distance and edge-travel gasoline criteria. This can be implemented based on a multiobjective approach in a WOA context, namely MOWOA.

2   Vehicle Transportation in Grid Networks

Vehicle routing and scheduling models are very useful for the dynamic vehicle transportation in the urban area. Real time information of street length, lanes, vehicle density, direction, velocity restrictions of a certain road are all recorded and collected by WSNs with the aid of city traffic surveillance system and global positioning system for vehicle identification and navigation [1]. The actual need of routing is to search the optimal way from a source to a destination that satisfies the driver's needs. Used WSN in traffic navigation is an effective way in the applications of the city traffic surveillance system. For little infrastructure, a number of sensor nodes in WSNs are used to monitoring a local area and getting the traffic information. There are two different types of the equipped WSN for a traffic observed system includes the structured and unstructured ones [9]. In the unstructured style, it contains a dense collection of sensor nodes and the nodes may be deployed in an ad-hoc manner. However, the structured style, the number of sensor nodes are deployed in pre-planned manners. This structured style can be deployed with fewer sensor nodes for traffic information gathering where these nodes are placed at specific locations to provide coverage optimization. Figure 1 shows the installed sensors road in the urban traffic network, and the WSN communicates with a local area network or wide area network through a gateway.

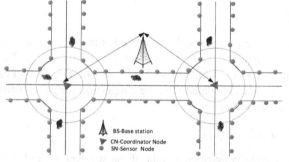

Fig. 1. An effective way in the applications of WSN for city traffic surveillance system

The roadside sensor node acts as the end device to collect the traffic information and the intersection node sensor acts as a coordinator and transfers the collected traffic information to the gateway or the base station. The gateway acts as a bridge connecting the WSN and other networks, and it enables the data to be processed or stored by other resources.

The traffic model used in this work is simulated on the grid networks. Congestion analysis and least fuel navigation are based on the traffic information collected from the WSNs of the urban area. The routing of vehicles can be modeled as a directed graph $G = (V, E)$ which V is a set of the intersections nodes and E is a set of the roads as edge in the grid network. Each navigation path candidate is composed of node-edge sequence. All the intersection nodes of the local area are labeled for the navigation and different permutation sequence of these nodes that contain a candidate path of a navigation is a potential solution. The objectives of a least gasoline consumption and the shortest paths are considered to formulate for fitness function. The first objective function is based on length of the path. Supposing that the start state and the target state are $p_0 0$ and p_{n+1}, the length of a path can be approximated by

$$L(p) = \sum_{i=0}^{n} d(p_i, p_{i+1}) \tag{1}$$

where $L(p)$ is the length of path and $d(p_i, p_{i+1})$ represents the distance between p_i and p_{i+1}. In the coordinates $Start_{x_0, y_0,}$, since the path Start –Target is divided into $n + 1$ equal segments, the value of $d(p_i, p_{i+1})$ can be calculated as follow:

$$d(p_i, p_{i+1}) = \sqrt{\left(x'_{p_i} - x'_{p_{i+1}}\right)^2 + \left(y'_{p_i} + y'_{p_{i+1}}\right)^2}$$

$$= \sqrt{\left(\frac{d(p_0, p_{n+1})}{n+1}\right)^2 + \left(y'_{p_i} + y'_{p_{i+1}}\right)^2} \tag{2}$$

The initial population is generated such that along each sensing direction, an agent is created at a certain distance from the vehicle, determined by the range of the used sensor. The objective function is shortness path that can be defined as the Euclidean distance between the agent and the goal point in each iteration:

$$F_1(p) = \sum_{i=0}^{n-1} d_i \tag{3}$$

The second objective function is based on edge-travel gasoline cost and congestion weight. The congestion weight can be determined by the traffic condition at the regular time obtained from traffic surveillance system. Edge-travel gasoline cost in the navigation path can be calculated with the congestion weight constructed according to the current traffic condition.

$$F_2(p) = \sum_{i=0}^{n-1} g(d_i) \times E(d_i) \tag{4}$$

Where $g(d_i)$ is the congestion weight of the segment d_i, and $E(d_i)$ is the gasoline consumption of the vehicle in the segment d_i. A congestion weight is set to 1 or 0 that depends on the current traffic condition.

3   Multiobjective WOA for Vehicle Fuel Consumption

The original version of the whale optimization algorithm (WOA) [8] is only for single objective optimization. In order to solve multiobjective functions of the vehicle fuel consumption, WOA needs to be extended to multiobjective whale optimization algorithm, namely MOWOA. The basic version of WOA and Pareto optimal front are first briefly reviewed, and the fuel consumption problem will be dealt with then it based on MOWOA.

3.1 The Whale Optimization Algorithm

WOA was imitated by the bubble-net hunting strategy of humpback whales [8]. It was worth mentioning here that bubble-net feeding is a unique behavior that can only be observed in humpback whales. The spiral bubble-net feeding maneuver is mathematically modeled in order to perform optimization. The humpback whales search randomly according to the position of each other.

$$\vec{X}_{t+1} = \vec{X}_{rand} - \vec{A} \times (\vec{C} \times \vec{X}_{rand} - \vec{X}), \tag{5}$$

where $\vec{X}_{rand}$ is a random position vector (a random whale) chosen from the current population. $\vec{A}$ and $\vec{C}$ are coefficient vectors which can be calculated as $\vec{A} = 2 \times \vec{a} \times \vec{r} - \vec{a}$ and $\vec{C} = 2 \times \vec{r}$ respectively. In which $\vec{a}$ is linearly decreased from 2 to 0 over the course of iterations and $\vec{r}$ is a random vector in [0,1]. The location of prey can be recognized and encircled by whales. Let X^* be the position vector of the best solution obtained so far. The humpback whales swim around the prey within a shrinking circle and along a spiral-shaped path simultaneously. To model this simultaneous behavior, there is a probability of 50% to choose between either the shrinking encircling mechanism or the spiral model to update the position of whales during optimization. The mathematical model is as follows:

$$\vec{X}_{t+1} = \begin{cases} \vec{X}_t^* - \vec{A} \times (\vec{C} \times \vec{X}_t^* - \vec{X}_t), & if\ p < 0.5 \\ \vec{X}_t^* + (\vec{X}_t^* - \vec{X}_t) \times e^{bl} \times \cos(2\pi l), & if\ p \geq 0.5 \end{cases} \tag{6}$$

where $\vec{X}$ is the position vector with t indicates the current iteration. p is a constant for defining the shape of the logarithmic spiral. $\vec{A}$ is coefficient vector with the random values greater than 1 or less than -1 to force search agent to move far away from a reference whale. In contrast to the exploitation phase, the position of a search agent is updated in the exploration phase according to a randomly chosen search agent instead of the best search agent found so far. This mechanism and $|A| > 1$ emphasize exploration and allow the WOA algorithm to perform a global search. So that A is used to smoothly transit between exploration and exploitation: by decreasing A, some iterations are devoted to exploration ($|A| \geqslant 1$) and the rest is dedicated to exploitation ($|A| < 1$).

3.2 Pareto Optimal Front

The domination of a solution vector $x = (x_1, x_2, .., x_n)^T$ on a vector $y = (y_1, y_2, .., xy_n)^T$ for a minimization problem if and only if $x_i \leq y_i$ for $\forall_i \in \{1, ..., n\}$ and $\exists_i \in \{1, ..., n\}: x_i < y_i$. It means that is no component of x is larger than the corresponding component of y, and at least one component is smaller. Similarly, the dominance relationship could be defined by

$$x \preccurlyeq y \Leftrightarrow x \prec y \lor x = y. \tag{7}$$

For maximization problems, the dominance can be defined by replacing symbol of $\prec$ with the symbol of $\succ$. Therefore, a point x_* is called a non-dominated solution if no solution can be found that dominates on it.The Pareto front PF of a multiobjective can be defined as the set of non-dominated solutions as following.

$$PF = \{s \in S | \not\exists s' \in S : s' \prec s\} \tag{8}$$

where S is the solution set. A good approximation could be obtained from the Pareto front, if a diverse range of solutions should be generated using efficient techniques [10].

3.3   Optimal Mobile Robot Planning based on MOWOA

The optimal solution of multiobjective optimization can be obtained from the Pareto optimal solution. Multiobjective optimization issue for a minimization problem with d-dimensional decision vectors and h objectives is given by

$$Minimize \ F(x) = (f_1(x), f_2(x), \dots f_h(x))$$

Subject to $\qquad\qquad\qquad x \in [x_L, x_U],$ $\qquad\qquad\qquad$ (9)

where $\mathbf{x}$ is a decision vector as a set of $(x_1, x_2, \dots, x_u) \in X \in R^d$ and $F(x)$ is the objective function with the objective vector as a set of $(f_1, f_2, \dots, f_u) \in Y \in R^h$. The decision vector x is belonging to the d-dimensional decision space X, which is corresponding to the space d dimensional of search agents in WOA. The objective function $F(x)$ belongs to the h −dimensional objective space Y, in which it is mapping functions from the decision space to the objective space. x_L, and x_U are lower and upper bound constraints of the agent range, respectively. The set of all the search agents meeting the constraints forms the decision space feasible set $\Omega = \{x \in R^d | x \in [x_L, x_U]\}$. The purpose of optimization is to find the Pareto-optimal solution. The decision space includes the dimension d and the objective space h. A population of N_p search agents is generated randomly so that these search agents should distribute among the search space as uniformly as possible. This can be achieved by using sampling techniques via uniform distributions. The model of the path planning problem with the two objective functions are defined by Eqs. (3) and (4) consist of the objective function $F(p)$. Therefore, from Eqs. (3), (4) and (9) can be formulated in the optimum mathematical form in MOWOA as.

$$Minimize \ \boldsymbol{F(p)} = (f_1(\hat{x}_\iota, \hat{y}_\iota), f_2(\hat{x}_\iota, \hat{y}_\iota))$$
$$Subject \ to \quad (\hat{x}_\iota, \hat{y}_\iota) \in (\widehat{x_L}, \widehat{y_L}), (\widehat{x_U}, \widehat{y_U}) \tag{10}$$
$$i = m + 1, \dots n,$$

where decision vectors $p = (\hat{x}_\iota, \hat{y}_\iota)$ are the estimated coordinates corresponding to solutions in WOA. $(\widehat{x_L}, \widehat{y_L}), (\widehat{x_U}, \widehat{y_U})$ are the lower and upper bound constraint values, f_1 is the objective function of the length path constraint, and f_2 is the objective function of the fuel consumption constraint. Obtaining the multiobjective Pareto optimal solution is the ultimate goal of building a multiobjective optimal model for gasoline consumption issues, which meets both the shortest path constraint and the fuel consumption path constraint. Therefore the main essence of MOWOA can be described as determining the dominant relationship according to the decision space feasible set Ω and the Pareto front $F(p^*)$ saving Pareto optimal solution set S in an archive by Eq. (10) and updating the best solution of multiobjective.

4   Simulation results

In this section, the proposed multiobjective whale optimization algorithm (MOWOA) for optimal vehicle fuel consumption is investigated. Two objective functions of f_1 and f_2, and Pareto archived evolution strategies have been done by applying MOWOA. A criterion for performance evaluation of multi-objective optimization algorithm is the error rate (ER). The error rate measures the probability whether the obtained non-domination solution is the actual Pareto frontier or not. The calculation method is given in Eq(12).

$$ER = \frac{\sum_{i=1}^{n'} x_i}{n'} \tag{11}$$

where n' is number of the obtained optimal points in Pareto frontier. If the obtained solution is an actual Pareto frontier elements, then x_i is set to 0, otherwise, x_i is set to 1.

Setting environment network for traffic navigation system.

An optimization of least fuel consumption for urban area vehicle navigation is carried out in order to verify the proposed method. Initialization parameters of the algorithm are as follows. The population size is 100. The maximum number of iterations is 300. The environment map of the simulation is set to grid network of 200*300-unit areas. The number of coordinator nodes, grid network set or reset as GUI scheme shown in Figure 2.

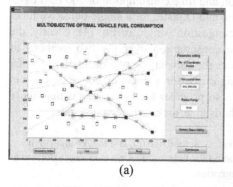
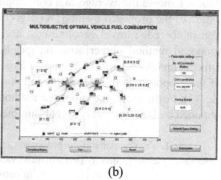

(a)   (b)

Fig. 2. Setting environments of an area vehicle navigation: a) generating randomly network space, b) Sparing solution as mapping to optimization process

Experiments have been conducted some tests with different grid network setting based on two objective functions which are the path length and path-travel fuel consumption.

The basic steps of the optimization:
Step 1: Modeling traffic navigation workspace including grid network with edge travel and coordinator nodes, and the vehicle locations.
Step 2: Sparing solution is as mapping search agents to a model of traffic navigation during optimization.
Step 3: Implementing the proposed MOWOA to find optimal fuel consumption path of the above model.
Step 3.1: WHILE   the maximum number of iterations has not been reached, i.e., $t <= T_{max}$, DO
If $(A<1)$ Update the global best location Eq. (5);
Else   Update the locations of agent Eq. (6);
          Take the best position of agent;
          Calculate the objective values and the constraint- violated degree of each
          agent by Eq. (10)
Step 3.2: Store all non-dominated feasible agents into the feasible archive, and non-dominated infeasible agents into the infeasible archive; Update the feasible archive and the infeasible archive;
Increase the loop counter, $t = t + 1$;
Step 3.4: Output optimal results.
Step 4: Guide the vehicles to the destination by the optimal path selected.

Experimental Results
Results of the average obtained least fuel consumption paths of the proposed method are compared with the obtained from Dijkstra [11] and A* algorithm [12] methods. Tables 1 shows the average fuel consumption and time consumption of 300 runs navigation of the proposed method, A* algorithm and Dijkstra for a single objective function of the edge-travel gasoline cost and congestion weight in Eq. (4). Clearly, the proposed method outperforms on gasoline consumption over the two methods of Dijkstra and A* algorithm. Figure 3a indicates the comparison of a mean of best so far of the proposed method, Dijkstra and A* algorithm for a single objective function in Eq. (3). Fortunately, the proposed method also provides the better results than those obtained by Dijkstra and A* algorithm methods in terms of convergence speed.

Table 1. The comparison the proposed method, with the Dijkstra, and the A+ algorithm methods in terms of quality performance evaluation for time and fuel costs in single objective of consumption

Methods	Ave time cost(h)	Ave fuel cost
Dijkstra [11]	1.31	1.2×10^{-2}L
A* algorithm [12]	1.25	1.3×10^{-2} L
Proposed method	1.12	1.1×10^{-2} L

The experiment is to verify the effectiveness of the paths generated by grid setting scheme based on the proposed method and the performance is analyzed by comparison with another method as multi-object genetic algorithm MOGA [13]. The probability of the obtained non-domination solution Pareto frontier is calculated according to Eq.(12), ER = 0.1 for MOWOA, however, this figure for MOGA for vehicle traveling problem is bigger as RE=0.2 that shown as in Figure 3b. Obviously, the proposed method shows the solution closer to Pareto front, that means it got results are better than the obtained of MOGA.

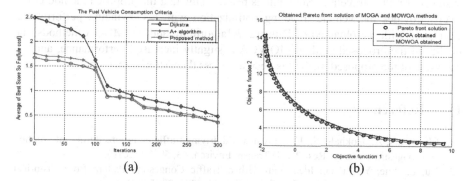

(a)   (b)

Fig. 3. Several graphical results of running the proposed method of multiobjective whale optimization algorithm (MOWOA) for the optimal vehicle traveling: a) average convergence of the proposed method, Dijkstra and A* algorithm method for a single objective function Eq. (3), b) The comparison of Pareto front solution for the path traveling between MOGA and MOWOA methods

5   Conclusion

This paper proposed a novel multi-objective whale optimization algorithm (MOWOA) for optimization of the vehicle fuel consumption problem. The environment of current traffic condition, the path distances, and the vehicle locations was modeled to the objective functions and search agents are mapped to a parsing solution in each iteration of a vehicle traveling during optimization. As the technology for traffic navigation advances and moves from simply improved traffic signs and message boards to automated online traffic information systems and vehicle specific route guidance advisories. A traffic navigation system equipped with Wireless sensor networks (WSN) and the vehicle equipped with the Global positioning system (GPS) can make up a grid network of vehicle transportation.

In the proposed method, a grid network of vehicle transportation was used to represent solution constraint, and MOWOA handles two objectives simultaneously: the shortness distance and path-travel gasoline of the vehicle traveling. The optimal solution of these two objectives can be obtained from the Pareto optimal solution by calculating the probability of the obtained non-domination solution Pareto frontier. The vehicle routing and traffic status in the environmental transportation are considered to model the fitness function for minimizing vehicle fuel consumption.

The path of the globally best whale is selected in each iteration, and reached by the vehicles in sequence permutation. In solving combinatorial optimization problems, first an initial solution must be generated randomly to be chosen as the current solutions, then a number of neighbor solutions of the current solutions can be searched and the best of them will be chosen as the new current solution. In addition, the information of traffic status updates to vehicle periodically times from traffic navigation system during the traveling. Simulations results show that the proposed method effectively provides the vehicle traveling optimization with a convincing performance. Compared with the obtained of Dijkstra, A*algorithm, and the MOGA methods, the quality of the proposed method MOWOA is slightly increased performance, and the error rate of the proposed method is decreased less than MOGA method.

6   References

1. Choudhary, A., Gokhale, S.: Urban real-world driving traffic emissions during interruption and congestion. Transp. Res. Part D Transp. Environ. 43, 59–70 (2016).
2. Xu, L., Yue, Y., Li, Q.: Identifying Urban Traffic Congestion Pattern from Historical Floating Car Data. Procedia - Soc. Behav. Sci. 96, 2084–2095 (2013).
3. García-hernández, C.F., Ibargüengoytia-gonzález, P.H., García-hernández, J., Pérez-díaz, J. a: Wireless Sensor Networks and Applications : a Survey. J. Comput. Sci. 7, 264–273 (2007).
4. Pan, T.-S., Dao, T.-K., Nguyen, T.-T., Chu, S.-C.: Optimal Base Station Locations in Heterogeneous Wireless Sensor Network Based on Hybrid Particle Swarm Optimization with Bat Algorithm. J. Comput VO - 254. 14 (2015).
5. Juran, I., Prashker, J.N., Bekhor, S., Ishai, I.: A dynamic traffic assignment model for the assessment of moving bottlenecks. Transp. Res. Part C Emerg. Technol. 17, 240–258 (2009).
6. Baskar, L.D., De Schutter, B., Hellendoorn, J., Papp, Z.: Traffic control and intelligent vehicle highway systems: a survey. IET Intell. Transp. Syst. 5, 38 (2011).
7. Mucherino, A., Seref, O.: Monkey search: A novel metaheuristic search for global optimization. In: AIP Conference Proceedings. pp. 162–173 (2007).
8. Mirjalili, S., Lewis, A.: The Whale Optimization Algorithm. Adv. Eng. Softw. 95, 51–67 (2016).
9. Guo, L., Fang, W., Wang, G., Zheng, L.: Intelligent traffic management system base on WSN and RFID. In: CCTAE 2010 - 2010 International Conference on Computer and Communication Technologies in Agriculture Engineering. pp. 227–230 (2010).
10. Zavala, G.R., Nebro, A.J., Luna, F., Coello Coello, C.A.: A survey of multi-objective metaheuristics applied to structural optimization, (2014).
11. Chen, Y.Z., Shen, S.F., Chen, T., Yang, R.: Path optimization study for vehicles evacuation based on Dijkstra algorithm. In: Procedia Engineering. pp. 159–165 (2014).
12. Takei, R., Tsai, R., Shen, H., Landa, Y.: A Practical Path-planning Algorithm for a Vehicle with a Constrained Turning Radius: a Hamilton-Jacobi Approach. Proc. Am. Control Conf. (2010).
13. Stewart, T.J., Janssen, R., van Herwijnen, M.: A genetic algorithm approach to multiobjective land use planning, (2004).

Printed in the United States
By Bookmasters